Riding Lawn Mower
Service Manual

(2nd Edition)

CONTENTS

Cover photo credit: Snapper Power Equipment

TECHNICAL PUBLICATIONS DIV.

INTERTEC PUBLISHING CORPORATION

P.O. Box 12901, Overland Park, Kansas 66212

MOWER SERVICE

FUNDAMENTALS

INTRODUCTION

Correct-fast-easy service may not be possible if the servicing mechanic does not have an understanding of how the machine is supposed to operate. Understanding these fundamentals will not, however, repair worn or damaged equipment. Knowledge of the machine design fundamentals, familiarity with standard accepted service procedures, specific repair information, proper tools and necessary new parts can be combined by service personnel to repair damaged equipment. The mechanic should also be aware that his safety and the safety of others is most important. **Do not take chances.** Be prepared to handle crises (such as spills, fires or run-away equipment) quickly and safely, even before an emergency occurs.

SAFETY INTERLOCK SYSTEMS

Riding lawn mowers may be equipped with two or more interlock safety switches which prevent the engine from starting if blade clutch or drive clutch is engaged or if transmission is in gear. Various types of interlock safety starting systems are used and safety switches may be located on blade clutch, main drive clutch, transmission, under mower seat or on recoil starter rope handle lock.

Always inspect the interlock safety starting system and make sure it is in good operating condition before checking the basic ignition system.

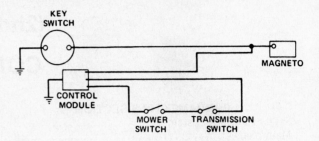

Fig. 2 — Typical wiring diagram of interlock safety starting system used on some models equipped with recoil starter. Refer to text.

NOTE: Never return a mower to service with any safety interlock system switches by-passed. The switches are installed to encourage safe operation. Defeating any of the system will permit unsafe operation, which may result in injury.

The interlock safety starting system shown in Fig. 2 is equipped with an electronic control module. Before engine can be started, mower switch and transmission switch must be closed (mower blade clutch disengaged and transmission in neutral). When engine is operating, current feedback from engine magneto switches the control module, isolating the safety switch circuit. At this time, mower blade clutch can be engaged and transmission shifted from neutral without affecting engine operation. If engine will not start with blade clutch disengaged and transmission in neutral or if engine will start with blade clutch engaged or transmission in gear, check and repair safety starting system.

CAUTION: Use extreme care when checking the system as the interlock switches may be by-passed at times during tests.

If engine will not start with blade clutch disengaged and transmission in neutral, disconnect wires to safety switches at control module. Install a by-pass wire on control module where switch wires were removed. If engine will start now, use a continuity light or ohmmeter and test for faulty safety switches or switch circuit wires. If engine will not start, disconnect control module wire from engine magneto terminal. If engine will start now, control module is faulty and must be renewed. If engine will still not start, check basic ignition system as trouble is not in safety starting system.

If engine will start with blade clutch engaged or transmission in gear, disconnect wires to safety switches at control module. If engine will not start, inspect safety switches and switch circuit wiring and renew as necessary. If engine will start with safety switch wires disconnected, but will not start with magneto terminal grounded to frame, control module is faulty and must be renewed.

NOTE: Always make certain that interlock safety starting system is in good condition and that all safety switches are connected before returning mower to service.

Another type of safety starting system is shown in Fig. 3. Safety switches

Fig. 1 — Operate and service all power equipment carefully and safely.

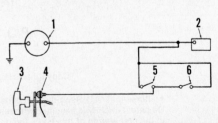

Fig. 3 — Typical wiring diagram of interlock safety starting system used on some recoil starter models. Refer to text.

1. Key switch
2. Engine magneto
3. Recoil starter handle
4. Safety switch
5. Drive clutch pedal safety switch
6. Blade disengagement safety switch

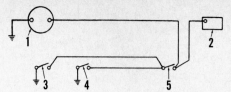

Fig. 4 — Typical wiring diagram of interlock safety starting system used on some recoil starter models. Refer to text.

1. Key switch
2. Engine magneto
3. Drive clutch safety switch
4. Blade clutch safety switch
5. Seat safety switch

(5 and 6) must be open (drive clutch pedal locked in fully disengaged position and mower lift and blade disengagement lever in fully disengaged position) before engine can be started. When recoil starter handle (3) is unlocked, spring leaf type safety switch (4) moves to ground against frame. After engine is started, recoil starter handle must be locked in position in frame pushing safety switch (4) open, before blade or drive clutch is engaged. If engine will not start with drive clutch and blade fully disengaged, check for faulty switches (5 and 6) and grounded wires (safety switches to magneto) and renew as necessary.

If engine will start with drive clutch or blade engaged, check to see that safety switch (4) makes good ground contact when recoil starter handle is unlocked. Then, with handle unlocked, check safety switches (5 and 6) and wiring using a continuity light or ohmmeter. Renew faulty components as required.

NOTE: Always make certain that interlock safety starting system is in good condition and that all safety switches are connected before returning mower to service.

On models equipped with safety starting system shown in Fig. 4, safety switch (5) is located under the lawn mower seat. When operator is off the seat, switch (5) is closed. Before engine can be started, safety switches (3 and 4) must be open (mower blade clutch disengaged and drive clutch pedal locked in fully depressed position). After engine is started and operator is seated on the riding mower, the weight of operator on the seat (normally 100 pounds is required) pushes switch (5) to open position. At this time, blade clutch and drive

clutch can be engaged without affecting engine operation. However, if operator raises off the seat while blade clutch or drive clutch is engaged, engine will stop. If engine will not start when blade clutch and drive clutch are in disengaged position, use a continuity light or an ohmmeter and check for faulty safety switches (3 and 4) or grounded switch wires. Renew faulty switches and wiring as necessary.

If engine will start with mower blade clutch or drive clutch engaged, first check to see that seat safety switch (5) closes when operator is off the seat. Then, check to see that safety switches (3 and 4) are closed when clutches are engaged. Renew faulty switches as required.

NOTE: Always make certain that interlock safety starting system is in good condition and that all safety starting switches are connected before returning mower to service.

Many electric start models are equipped with the safety starting system shown in Fig. 5. On this system, safety switches are in the electric starter solenoid switch circuit instead of the magneto ignition circuit as on recoil start models. Transmission must be in neutral and mower blade clutch in disengaged position (both switches closed) before the key switch will operate the electric starter solenoid (magnetic) switch.

If electric starter will not operate when transmission is in neutral and blade clutch is disengaged or if starter will operate when transmission is in gear or blade is engaged, use a continuity light or an ohmmeter and check for faulty safety switches. Renew faulty switches as necessary.

NOTE: Always make certain that safety starting system is in good condition and that all safety starting switches are connected before returning mower to service.

MAINTENANCE

Normal maintenance should include a complete check of the equipment, making sure that equipment is adequately lubricated and adjusting components/controls as required. These normal maintenance procedures can effectively

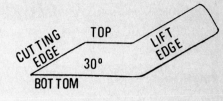

Fig. 6 — Drawing of typical blade cross-section identifying parts of blade. The cutting edge should form approximately a 30 degree angle with bottom.

reduce the amount of wear and damage that will require repair.

Encourage operators to check all systems as described in the operating instructions, each time the equipment is to be used. Additional regular inspections by service personnel should be encouraged, usually at the beginning and end of normal operating seasons.

Follow the manufacturer's lubrication recommendations. The manufacturer recommends type of lubricant and frequency of lubrication service to reduce damage to the equipment. Altering recommended lubrication intervals or changing type of lubricant may result in extensive damage.

Adjust drive belt (or chain) tension, tire pressure, controls, etc., as required. Checks may indicate the need for adjustments, repair or installation of new parts. Delaying needed adjustments or repairs may result in increased wear rate and/or extensive damage to otherwise usable parts.

CUTTING BLADES

INSPECTION. Mower blades should be inspected frequently for sharpness, balance and straightness. Dull blades will cause ragged grass cutting and can cause excessive load on engine. Bent or out-of-balance blades cause excessive vibration.

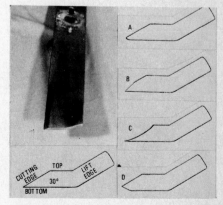

Fig. 7 — Views of typical blade. View "A" shows typical cross-section of dull blade. View "B" is incorrect because of sharpening at too much angle. View "C" is sharpened at less than 30 degree angle. View "D" shows incorrect method of sharpening caused by grinding lower edge of blade.

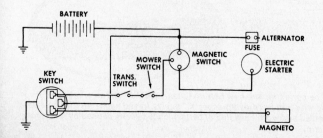

Fig. 5 — Typical wiring diagram of interlock safety starting system used on some models equipped with electric starter. Refer to text.

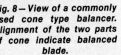

Fig. 8 — View of a commonly used cone type balancer. Alignment of the two parts of cone indicate balanced blade.

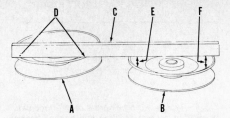

Fig. 10 — Be sure that pulleys are correctly aligned so that belt will move smoothly from one pulley to the other. Sometimes belt is used to drive pulleys that are not aligned, but special guides are employed to change direction of belt. Refer to text.

CAUTION: Always remove wire from engine spark plug before performing any inspection or service of blades.

Dull blades can normally be sharpened, but bent or cracked blades must be renewed. Refer to the following paragraphs for blade servicing.

SHARPENING. Slightly dulled blades can normally be restored with a few strokes of a file. Badly dulled or nicked blades should be removed and sharpened on a grinder. Ideal cutting edge is sharpened at a 30 degree angle as shown in Fig. 6. Blades wear on the underside as shown at (A – Fig. 7). If blade is not worn excessively, the 30 degree angle cutting edge and flat underside can be obtained without excessive grinding. Do not sharpen bottom edge as shown at (D). Sharpen both ends of blade evenly so that blade remains balanced. Always check blade for balance after sharpening.

BALANCING. An unbalanced blade can cause severe vibration resulting in damaged blade spindle bearings and cracked mower housing. Various types of blade balancers are available. A popular blade balancer is the cone type shown in Fig. 8. Before balancing blade, make sure blade is clean and properly sharpened. Place blade center mounting hole over cone balancer and check balance. Mark heavy end of blade. Resharpen blade to remove metal from heavy end. DO NOT grind away the lift edge to balance blade.

BLADE TRACKING. Mower blades should cut on a plane parallel to the level mower housing. With riding mower on a smooth floor, measure distance from end of blade to floor. Rotate blade 180 degrees and measure distance from opposite end of blade to floor. Measurements should be the same within 1/16-inch. If not, blade or blade spindle (sometimes crankshaft) is bent and should be renewed.

DRIVE BELTS, BELT GUIDES AND PULLEYS

Drive belts should be checked periodically and adjusted if necessary to prevent slippage under normal operating conditions. Belts should be kept clean and dry. Wipe belts with a clean rag to remove any oil or dirt and carefully check condition of belts. Install new belt of correct size if old belt is stretched, worn or otherwise damaged. Refer to Fig. 9. Worn or damaged belt may be normal after much use, but quick wear or damage may be caused by other failure.

Pulleys should be aligned. Straight edge (C – Fig. 10) should indicate that all are in line. If side of pulley (A) is thicker

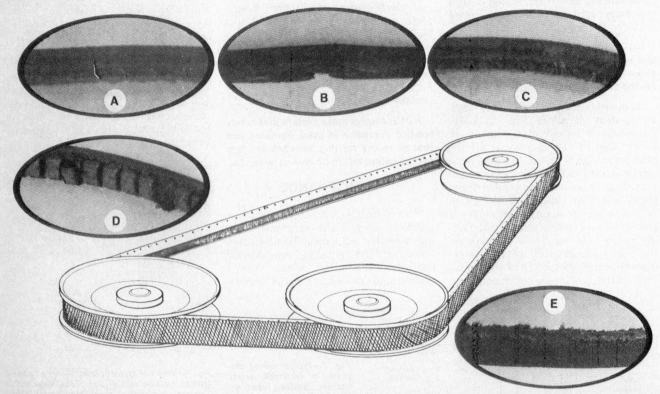

Fig. 9 — Visually check condition of belt. Problems illustrated in the insets should be corrected and new belt installed. Inset "A" shows cracks or cuts in belt. Inset "B" shows localized burned section of belt caused by drive pulley turning with belt not moving. Inset "C" shows frayed and worn friction sides. Inset "D" shows notched belt with sections broken loose or out. Inset "E" shows frayed and worn backside of belt which is usually caused by incorrectly adjusted belt guard.

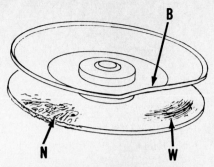

Fig. 11—Damage to pulley such as rough sur-face (N), cupped surface (W) or bent sides (B) will quickly wear belt out.

than side of pulley (B), clearance (E & F) should be the same and equal to the difference in thickness of the sides when straight edge is touching at points (D). If the sides of pulleys are the same, clearance at (E & F) indicates that pulley (B) is lower than pulley (A). Method indicated in text for moving pulley should be employed to align pulleys. If clearance (E & F) are different, pulley may be bent or one of the shafts is not in line with the other shaft. Correct alignment to improve belt operation and increase operating life. Check pulleys and install new pulley and belt if nicked (N – Fig. 11), worn (W) or bent (B). Condition and alignment of pulleys is critical to life of belt.

Belt guides and belt stops must be positioned so they will be free of belt when clutch is engaged but will hold belt free of drive pulley when clutch is disengaged. If belt guide specified clearance is not available, position belt guides for 1/8-inch clearance (clutch engaged) and check operation. Be sure to reinstall all guards to prevent injury or belt damage.

DRIVE CHAIN

On models equipped with a drive chain, make certain that chain sprockets are aligned to prevent excessive wear on chain and sprockets. Drive chain should be adjusted so that chain has about 1/4-inch slack. if chain is too tight, excessive wear and stretch will result. If chain is too loose, it may jump off the sprocket or ride into the sprocket teeth resulting in excessive wear on sprockets and chain. Under normal operating conditions, open drive chain should be lubricated with engine oil each 10 hours of operation.

Sprocket tooth profile (B – Fig. 12) is precisely ground to fit the roller diameter and chain pitch (A). When chain and sprocket are new, the chain moves around the sprocket smoothly with a minimum of friction, and the load is evenly distributed over several sprocket teeth. Wear on pins and bushings of a roller chain results in a lengthening or "stretch" of each individual chain pitch as well as lengthening of the complete chain. The worn chain, therefore, no longer perfectly fits the sprocket. Each roller contacts the sprocket tooth higher up on the bearing area (C) and that tooth bears the total load until the next tooth and roller make contact. Chain wear will therefore quickly result in increased sprocket wear.

As a rule of thumb, a new chain should be installed whenever chain stretch exceeds 2% (or 1/4-inch per foot.) Renew #40 chain when a 24 pitch length of chain measures 12 1/4 inches. Check sprockets carefully for wear if chain wear is substantially greater than 2%, and renew sprockets if in doubt. Sprocket wear usually shows up as a hooked tooth profile. A good test is to fit the sprocket to a new chain. Wear on sides of sprocket indicates misalignment. If sprockets must be renewed because of wear, always renew the chain. Early failure can be expected if a new chain is mated with worn sprockets or new sprockets with a worn chain.

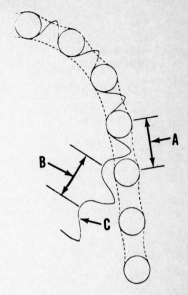

Fig. 12—Chain pitch (A) should be matched to the pitch and contour of sprocket tooth (B).

REPAIR

Adjustment and repair information for the individual mower units is included within the **MOWER SERVICE SECTION.** Engine application is listed in the Mower Service Section; however, specific engine repair information is not included in this book. Refer to applicable engine service manual such as: Small Air Cooled Engines Service Manual, Large Air Cooled Engines Service Manual or Small Diesel Engines Service Manual. Transmission, transaxle and differential application is listed in the Mower Service Section, but procedure for servicing the more common units is described in the **TRANSMISSION RE-PAIR** or **DIFFERENTIAL REPAIR** Sections.

DRIVE BELTS

Make & Model	Mfg Part No.	Dayco No.	Gates No.	Size*
ALLIS-CHALMERS				
Scamp 5				
Transmission	2028241	L431	6831	½x31
Pto	2028217	L434	6834	½x34
Mower (24 in.)	2025893	L448	6848	½x48
26 in.	2029223	L444	6844	½x44
Scamp 8				
Main Drive	2025879	L463	6863	½x63
or	2029660	L462	6862	½x62
Directional Drive	2029451	L4293	....	½x29.3
Mower	2029582	....	A133	½x135
405				
Transmission	2028241	L431	6831	½x31
Pto	2028217	L434	6834	½x34
Mower	2929223	L444	6844	½x44
508				
Main Drive	2029660	L462	6862	½x62
Directional Drive	2029451	L4293	....	½x29.3
Mower	2029582	....	A133	½x135
526				
Transmission	1665706	L439	6839	½x39
Mower	1664946	L457	6857	½x57
830				
Transmission	1665706	L439	6839	½x39
Mower	1665024	L454	6854	½x54
1036				
Transmission	1665706	L439	6839	½x39
Mower	2087732	L478	6878	½x78
ARIENS				
Emperor (925000)				
Drive	72026	L5415	6941	⅝x41.5 or 21/32x41
or	72065	L336	6736	⅜x36
Mower, 26 in.	72049	L547	6947	⅝ or 21/32x47
30 in.	72050	L544	6944	⅝ or 21/32x44
or	72052	L569	6969	⅝ or 21/32x69
38 in. (to deck)	72038	L453	6853	½x53
Blades	72052	L569	6969	⅝ or 21/32x69
Emperor (927000)				
Drive	72112	L335	6735	⅜x35
Mower, 26 in.	72115	L467	6867	½x67
28 & 30 in.	72113	L465	6865	½x65
32 in.	72124	....	....	....
BOLENS				
728, 828, 829				
Transmission	1724261	L452	6852	½x52
Mower	1724035	L468	6868	½x68
1011 GM				
Transmission	1820101	....	....	17/32x100¾
Mower	1820209	....	....	....
1011HM				
Primary	1820283	....	....	....
Secondary	1820284	L4313	....	½x31.3
Mower	1820209	L488	....	17/32x88 11/16
COLUMBIA				
Refer to corresponding MTD model.				
JOHN DEERE				
55, 56	M44498	L460	6860	½x60

JOHN DEERE (Cont.)

57
Transmission	M45863	L456	6856	½x56
Mower...................	M45862	L448	6848	½x48

66
Transmission	M82167	L451	6851	½x51
or	M49155	L452	6852	½x52
or	M80652	L451	6851	½x51
or	M81505	L451	6851	½x51
Mower...................	M80386	L464	6864	½x64
or	M49920	L499	6899	21/32x99

68
Transmission	M82167	L451	6851	½x51
or	M80652	L451	6851	½x51
or	M49155	L452	6852	½x52
or	M81505	L451	6851	½x51
Mower, 34 in.	M49920	L499	6899	½ or 21/32x99
30 in.	M80386	L464	6864	½x64

DIXON
ZTR308
Transmission	30-2426	....	3440	21/32x44
	....	....	or 6944	21/32x44
Mower...................	....	....	3740	21/32x74

ZTR 424, 425, 426
Transmission	1539	L541	6941	⅝ or 21/32x41
Mower drive	6111	....	B75	21/32x78
Blade	6109	....	A80	½x82

DYNAMARK
826, 1260, 1269
Transmission	43979	L456	6856	½x56
Mower...................	45187	L444	6844	½x44

832, 1261
Transmission	43979	L456	6856	½x56
Mower...................	45186	L463	6863	½x63
or	47390	L471	6871	½x71

1281
Transmission	39454	L488	6888	½x88
Mower...................	39456	L569	6969	⅝x69
or	42111	L570	6970	⅝x70
or	43066	L571	6970	⅝x70-71

1284
Transmission	39454	L488	6888	½x88
Mower...................	46466	L483	6883	½x83
or	48044	L582	6982	⅝x82

1288
Transmission	39454	L488	6888	½x88
Pto	44313	L452	6852	½x52
or	45509	L453	6853	½x53
Mower...................	44264	L498	6898	½x98

3180, 3181
Transmission	49405	L488	6888	½x88

3188
Transmission	39454	L488	6888	½x88
	50737	L484	6884	½x84
Mower...................	43066	L571	6970	⅝x70-71

3189
Transmission	43979	L456	6856	½x56
Mower...................	45186	L463	6863	½x63

5180, 5181
Transmission	49405	L488	6888	½x88

5184
Transmission	43979	L456	6856	½x56

5188
Transmission	39454	L488	6888	½x88
Mower...................	43066	L571	6970	⅝x70-71

DYNAMARK (Cont.)

5267
Transmission	43979	L456	6856	½x56
Mower	44039	L444	6844	½x44
	or 45187	L444	6844	½x44
	or 45186	L463	6863	½x63

5282, 5285, 5288
Transmission	39454	L488	6888	½x88
Mower	39456	L569	6969	⅝x69
	42111	L570	6970	⅝x70
	43066	L571	6970	⅝x70-71
	48044	L582	6982	⅝x82

5289
Transmission	43979	L456	6856	½x56
Mower	44148	L463	6863	½x63
	45186	L463	6863	½x63

5294
Transmission	44150	L441	6841	½x41
Mower	45186	L463	6863	½x63
	or 47611	L488	6888	½x88

5296
Transmission	39454	L488	6888	½x88
Mower	43066	L571	6970	⅝x70-71
	or 46466	L483	6883	½x83
	or 48044	L582	6982	⅝x82

5297
Transmission	39454	L488	6888	½x88
Pto	44313	L452	6852	½x52
	or 45509	L453	6853	½x53
Mower	44264	L498	6898	½x98

FORD

51
Transmission	308506	L331	6731	⅜x31
Mower	308494	L466	6866	½x66

60 (early), 65 (early)
Transmission	308475	L431	6831	½x31
Mower	308476	L470	6870	½x70

60 (late), 61, 65 (late), 66
Transmission	308491	L331	6731	⅜x31
	or 308506	L331	6731	⅜x31
Mower	308493	L470	6869	½x69 or 70

526
Transmission	308511	L438	6838	½x38
Mower	308510	L456	6856	½x56

Early Belt Drive Models -
830, 830E
Transmission	308491	L331	6731	⅜x31
Mower	308493	L470	6869	½x69 or 70

Late Belt Drive Models -
830, 830E, 1130E
Transmission	308491	L331	6731	⅜x31
Mower	308522	L468	6867	½x67 or 68

Friction Drive Models -
830, 830E, 1130E
Mower	392545	. . . .	. . . .	. . . .

GILSON*

52013, 52031, 52038
Mower	14365	

52014, 52015
Mower	17518	

52032, 52033, 52039, 52040
Mower	23472	

52051C
Transmission	36750	
Mower	11793	
	or 36751	

GILSON (Cont.)

52060
Transmission	36750		
Mower....................	210692		

52061
Transmission
Primary	210752		
Secondary	210753		
Mower....................	36751		

52064, 52065, 52072, 52073
Transmission	212931		
Mower....................	213676		

52066, 52074
Transmission	212931		
Mower....................	210692		

*Gilson does not recommend belts other than belts supplied by Gilson.

GRAVELY
830, 830-E, 1130-E
Transmission	26137	L331	6731	³⁄₈x31
Mower....................	26149	L468	6867	½x67 or 68

HECKENDORN
73901, 73904, 73905
Swinging to Main Spindle	100100	2/3VX355	2/3V355	³⁄₈x35½
Engine to transmission	100102	2/3VX475	2/3V475	³⁄₈x47½
Engine to Swinging Spindle ..	100029	2/3VX530	2/3V530	³⁄₈x53

75902, 77902
Swinging to Main Spindle	100100	2/3VX355	2/3V355	³⁄₈x35½
Engine to Transmission	100103	2/3VX500	2/3V500	³⁄₈x50
Engine to Swinging Spindle ..	100044	2/3VX560	2/3V560	³⁄₈x56
Eng. to Rear Drive Shaft	100101	2/3VX450	2/3V450	³⁄₈x45
Side Mower Spindle	100031	2/3VX630	2/3V630	³⁄₈x63

76901, 76903, 78901, 78903
Swinging to Main Spindle	100105	3/3VX355	3/3V355	³⁄₈x35½
Engine to Transmission	100103	2/3VX500	2/3V500	³⁄₈x50
Engine to Swinging Spindle ..	100040	3/3VX670	3/3V670	³⁄₈x67
Engine to Rear Drive Shaft...	100106	3/3VX475	3/3V475	³⁄₈x47½
Side Mower Spindle	100032	2/3VX750	2/3V750	³⁄₈x75

76902, 78902
Swinging to Main Spindle	100105	3/3VX355	3/3V355	³⁄₈x35½
Engine to Transmission	100104	2/3VX530	2/3V530	³⁄₈x53
Eng. to Swinging Spindle	100040	3/3VX670	3/3V670	³⁄₈x67
Eng. to Rear Drive Shaft	100106	3/3VX475	3/3V475	³⁄₈x47½
Side Mower Spindle	100032	2/3VX750	2/3V750	³⁄₈x75

HOMELITE
RE-5
Transmission	158573	L431	6831	½x31
Mower....................	174911	L444	6844	½x44
Pto	121398	L434	6834	½x34

RE-8E
Transmission	177452	L462	6862	½x62
Pto	176451	....	A133	½x135

RE-30
Transmission	1651201	....	A71	½x73
Pto	176451	....	A133	½x135

INTERNATIONAL HARVESTER
Cadet 55
Forward	549108-R1	L438	6837	½x37-38
To Deck..................	549609-R1	L450	6850	½x50
or	62817-C1	L449	6849	½x49
Mower....................	488077-R1	L448	6848	½x48

Cadet 60
Trans.-Fwd.	487994-R1	L437	6837	½x37
Trans.-Rev.	487996-R1	L435	6835	½x35
Mower....................	488082-R1, R2	L452	6852	½x52
Blade Spindle	488077-R1	L448	A46	½x48

INTERNATIONAL HARVESTER (Cont.)

Cadet 75				
Mower	488077-R1	L448	6848	½x48
Cadet 85				
Transmission	549107-R1	L439	6839	½x39
Mower	106991-C1	....	A53	½x55
Blade	106990-C1	L445	A43	½x45
Cadet 85 Spl.				
Transmission	549107-R1	L439	6839	½x39
Mower	106465-C1	L451	A49	½x51
Cadet 95 Electric				
Transmission	57842-C1	L433	6833	½x33

JACOBSEN

42635				
Transmission	546341	L525	6925	⅝ or 21/32x25
Variable	546342	L531	6931	⅝ or 21/32x31
Mower	546343	L552	6852	½x52
Other 426 Models				
Mower	330463	L476	6876	½x76
430 Models with 30 inch Mower				
Mower	330463	L476	6876	½x76
RMX				
Mower	JA392545	....	....	....

MASSEY-FERGUSON

MF5, MF6, MF626				
Transmission	523733M1	L436	6836	½x36
Mower	523734M1	L458	6858	½x58
MF8, MF832				
Transmission	530669M1	L437	6837	½x37
Mower	530756M1	L498	6898	½x98

MTD

360 (early)				
Transmission	754-101	L435	6835	½x35
Blade	754-138	L550	6950	⅝x50
360 (late)				
Transmission	754-936	L447	6847	½x47
Blade	754-107	L430	6830	½x30
362 (early)				
Transmission	754-936	L447	6847	½x47
Blade	754-935	L431	6831	½x31
362 (late)				
Transmission	754-936	L447	6847	½x47
Blade	754-107	L430	6830	½x30
380				
Transmission	754-198	L462	6862	½x62
Blade	754-138	L550	6950	⅝x50
385 (early)				
Transmission	754-164	L463	6863	½x63
Blade	754-138	L550	6950	⅝x50
385 (late)				
Transmission	754-198	L462	6862	½x62
Blade	754-138	L550	6950	⅝x50
390 (early), 395 (early)				
Transmission	754-164	L463	6863	½x63
Blade	754-167	L564	6964	⅝x64
390 (late), 395 (late)				
Transmission	754-198	L462	6862	½x62
Blade	754-167	L564	6964	⅝x64
400				
Transmission	754-101	L435	6835	½x35
Blade	754-188	L551	6951	⅝x51
402, 405, 406, 407				
Transmission	754-136	L531	6931	⅝x31
Blade	754-188	L551	6951	⅝x51

MTD (Cont.)

410, 412				
Transmission	754-101	L435	6835	½x35
Blade	754-138	L550	6950	⅝x50
420 (early), 425 (early)				
Transmission	754-136	L531	6931	⅝x31
Variable Speed	754-135	L525	6925	⅝x25
Blade	754-127	L566	6966	⅝x66
420 (late), 425 (late)				
Transmission	754-135	L525	6925	⅝x25
or	754-136	L531	6931	⅝x31
Blade	754-147	L552	6952	⅝x52
430, 435				
Transmission	754-135	L525	6925	⅝x25
or	754-136	L531	6931	⅝x31
Blade	754-127	L566	6966	⅝x66
440 (early), 445 (early)				
Transmission	754-185	L449	6849	½x49
Blade	754-178	L582	6982	⅝x82
440 (late), 445 (late)				
Transmission	754-200	L448	6848	½x48
Blade	754-178	L582	6982	⅝x82
460 (early), 465 (early), 470 (early), 475 (early)				
Transmission	754-937	L559	6959	⅝x59
Blade	754-118	L563	6963	⅝x63
460 (late), 465 (late)				
Transmission	754-136	L531	6931	⅝x31
Mower	754-138	L550	6950	⅝x50
Blade	754-151	L567	6967	⅝x67
470 (late), 475 (late)				
Transmission	754-191	L465	6865	½x65
Blade	754-151	L567	6967	⅝x67
480 (early), 485 (early)				
Transmission	754-136	L531	6931	⅝x31
or	754-147	L552	6952	⅝x52
Blade	754-145	L569	6969	⅝x69
480 (late), 485 (late)				
Transmission	754-191	L465	6865	½x65
Blade	754-151	L567	6967	⅝x67
495, 497				
Transmission	754-173	L537	6937	⅝x37
or	754-191	L465	6865	½x65
Blade	754-145	L569	6969	⅝x69
498				
Transmission	754-226	L482	6882	½x82
Blade	754-145	L569	6969	⅝x69
520, 525				
Transmission	754-198	L462	6862	½x62
Blade	754-195	L454	6854	½x54
630, 632, 638, 698				
Transmission	754-266	....	....	....
Blade	754-151	L567	6967	⅝x67
or	754-145	L569	6969	⅝x69
760, 780, 784, 786				
Primary Drive	754-245	L459	6859	½x59
Secondary Drive	754-255	L437	6837	½x37
Mower	754-230	L456	6856	½x56
Blade	754-246	....	....	....
796, 797				
Transmission	754-207	L442	6842	½x42
Blade	754-198	L462	6862	½x62
820				
Primary Drive	754-245	L459	6859	½x59
Secondary Drive	754-244	L440	6840	½x40
Mower	754-230	L456	6856	½x56
Blade	754-246	....	....	....

MURRAY

2503, 2513

Drive	20716	L437	6837	½x37
Mower	20717	L456	6856	½x56

3013, 3033, 3043, 3063
(With cross drive)

Drive	20716	L437	6837	½x37
Mower	20556	L448	6848	½x48
Cross Drive	20555	L441	6841	½x41

3013, 3033 (Without cross drive)

Drive	20716	L437	6837	½x37
Mower	23496	L474	6874	½x74
or	21649	L455	6855	½x55

3233

Drive	20716	L437	6837	½x37
Mower	21058	L463	6863	½x63
Cross Drive	21059	L444	6844	½x44

3633

Drive	20716	L437	6837	½x37
Mower	20558	L464	6864	½x64
Cross Drive	20557	L446	6846	½x46

25501, 25502

Drive	37X38	....	....	....
Mower	37X36	....	....	....

30501, 30502

Drive	37X38	....	....	....
Mower	37X37	....	....	....

31501

Drive	37X27	....	....	....
Mower	37X34	....	....	....

36503

Drive	37X35	....	....	....
Mower	37X39	....	....	....

39001

Drive	37X26	....	....	....
Mower	37X39	....	....	....

J.C. PENNEY

1820

Transmission	754-198	L463	6863	½x63
Blade	754-195	L454	6854	½x54

1824

Transmission	21615	L495	6895	½x95
Mower	21649	L455	6855	½x55
Blade	20557	L446	6846	½x46

1831

Transmission	754-198	L463	6863	½x63
Blade	754-195	L454	6854	½x54

1832

Transmission	754-936	L447	6847	½x47
Blade	754-107	L430	6830	½x30

1834

Transmission	21615	L495	6895	½x95
Mower	23748	L462	6862	½x62
Blade	23749	L499	6899	½x99

1835

Transmission	754-226	L482	6882	½x82
Blade	754-225	AP105	....	½Ax107

1839

Transmission	754-0198	L463	6863	½x63
Blade	754-0167	L564	6964	⅝x64

1840

Transmission	754-0191	L465	6865	½x65
Blade	754-0151	L567	6967	⅝x67

1841, 1842

Transmission	23347	L484	6884	½x84
Mower	23882	L459	6859	½x59
Blade	20557	L446	6846	½x46

J.C. PENNEY (Cont.)

1844

Primary Drive	754-0245	....	....	½x59
Secondary Drive	754-0244	....	....	½x40
Blade	754-0246	....	....	....

1845, 1846

Transmission	754-0226	L482	6882	½x82
Blade	754-0145	L569	6969	⅝x69

1847

Transmission	754-0248	....	....	½x89
Blade	754-0246	....	....	....

1848

Transmission	754-0226	....	....	½x82
Mower	754-0145	....	....	21/32x69

1905, 1907, 1907A, 1908, 1909

Transmission	20716	L437	6837	½x37
Blade	20717	L456	6856	½x56

1910

Transmission	20716	L437	6837	½x37
Blade	21649	L455	6855	½x55

ROPER

K511, K512, K522

Transmission	67346	L439	6839	½x39
Mower	67438	L447	6847	½x47

K831, K832

Transmission	67346	L439	6839	½x39
Mower	67398	L453	6853	½x53

K852-Belt information unavailable.

L711, L721, L722

Transmission	58406	L465	6865	½x65
Mower	70637	L440	6840	½x40

L821, L861, L863

Transaxle	71240	L492	2920	½ or 21/32x92

SEARS-CRAFTSMAN

502.256011, 502.256091

Transmission	20716	L437	6837	½x37
Mower	20717	L456	6856	½x56

502.256020, 502.256030, 502.256040

Transmission	20715	L437	6837	½x37
Mower	20717	L456	6856	½x56

502.256071

Transmission	20715	L437	6837	½x37
Mower	23496	L474	6874	½x74

502.256080

Transmission	20715	L437	6837	½x37
Mower	23536	L477	6877	½x77

502.256111, 502.256121
502.256130, 502.256141

Transmission	20715	L432	6837	½x37
Mower	21649	L455	6855	½x55

SIMPLICITY

305, 315

Transmission	158573	L431	6831	½x31
Mower	174911	L444	6844	½x44
or	108248	L448	6848	½x48
Pto	121398	L4335	6834	½x33.5 or 34

355

Transmission	158573	L431	6831	½x31
Pto	164146	L434	6834	½x34

808

Transmission	177452	L462	6862	½x62
Pto	176451	....	A133	½x135

3005

Transmission	158573	L431	6831	½x31
Mower	174911	L444	6844	½x44
Pto	121398	L4335	6834	½x33.5 or 34

SIMPLICITY (Cont.)
3008-2, 3008-3

Transmission	1651201	L473	A71	½x73
Pto	176451	....	A133	½x135

SNAPPER

All Models................	1-0749	17T700	6234	**

**Double ½ inch, 36 degree drive belt 71⅜ inches long.

WARDS
33857A, 33857B

Transmission	36750	8X1030	....	17/32x104
Mower...................	36751	L488	....	17/32x88 11/16

33867A

Primary	210752	....	....	....
Secondary................	210753	L4313	....	½x31.3
Mower...................	36751	L488	....	17/32x88 11/16

33877A

Transmission	36750	8X1030	....	17/32x104
Mower...................	210692	....	....	17/32x96½

33887A, 33889A

Transmission	212931	....	....	17/32x100¾
Mower...................	210692	....	....	17/32x96½

WHEEL HORSE
A-50 (3-0114)

Transmission	103345	L325	6725	⅜x25
Mower...................	8430	L459	6859	½x59

A-51 (05BP01, 02)

Transmission	9430	L436	6836	½x36.31

A-60 (05BF01)

Transmission	225282	....	6729	⅜x29.3

A-70 (08BP01)

Transmission	225187	....	....	....

A-81 (08BP01)

Transmission	9430	L436	6836	½x36.31

A-111 (11BP01, 02)

Transmission	9430	L436	6836	½x36.31

R-26

Transmission	1597	L325	6725	⅜x25
Reverse..................	1597	L325	6725	⅜x25
Mower...................	8430	L459	6859	½x59

RR-532, RR-832

Transmission	9430	L436	6836	½x36
Mower...................	108491	....	....	....
26MS01 (Mower)	225360	L463	6863	½x63
32MS01 (Mower)	107298	L472	6872	½x72
	108491	....	....	....
32XS01 (Mower)	107647	....	....	....
	108503	....	....	....

36MR01, 36MR02, 36MS00, 36MS01 (Mower)

Drive....................	102741	L497	6897	½ or 21/32x97
	105477	L440	6840	½x40
Spindle	8411	L477	6877	½x77
	105476	L480	A78	½x80

36XR00, 36XR01, 36XS00, 36XS01, 36XS02 (Mower)

Drive....................	102741	L497	6897	½ or 21/32x97
	105477	L440	6840	½x40
	106533	L440	A38	½x40
Spindle	8411	L477	6877	½x77
	105476	L480	A78	½x80
	106751	L481	6880	½x80

36YR01 36YR02 (Mower)

Drive....................	107230	L460	6860	½x60
	108492	....	....	....
Spindle	8411	L477	6877	½x77

WHITE

R50

Engine to Variator..........	32-0021415	L524	6924	21/32x24
Variator to Trans.	32-0021407	L531	6931	21/32x31
Mower....................	32-0021253	L551	6951	21/32x51

R80

Transmission	32-0018635	L550	6950	21/32x50
Mower....................	32-0018910	L582	6982	21/32x82

R82

Transmission	32-0055301	L442	6842	½x42
Mower....................	32-0024368	L462	6862	½x62

WIZARD

7110, 7115

Transmission	212931	....	....	17/32x100¾
Mower....................	212676	....	....	....

7380

Transmission	36750	8X1030	6990	17/32x104
Mower....................	36751	L488	....	17/32x88 11/16

*All belt dimensions are in inches.

ALLIS-CHALMERS

ALLIS-CHALMERS
P.O. Box 512
Milwaukee, WI 53201

Model	Make	Engine Model	Horsepower	Cutting Width, In.
Scamp 5	B&S	130000	5	26
Scamp 8	B&S	190000	8	30
405	B&S	130000	5	26
508	B&S	190000	8	30
526	B&S	130000	5	26
830	B&S	190000	8	30
1036	B&S	220000	10	36

STEERING SYSTEM

Models Scamp 5 and 405

REMOVE AND REINSTALL. The axle main member is also the front frame assembly (10–Fig. AC1). Pivot point is at joint of front and rear frame sections. To disassemble the steering system, support front of unit and remove front wheel assemblies. Unbolt and remove tie rods (14). Remove cotter pins from top of spindles (16 and 19), drive out roll pins (12) and remove both spindles with washers (13) and torsion springs (11). Unbolt and remove steering wheel (8) and cover (7). Remove washer (2), bushing (3), cup (4) and retaining ring (5) from upper end of steering shaft (6). Working through rear opening in front frame, remove cotter pin from steering shaft. Withdraw steering shaft from bottom of frame and remove cup (9), bushing (17) and washer (18).

Clean and inspect all parts and renew any showing excessive wear or other damage. Reassemble by reversing the disassembly procedure. Lubricate spindles and steering shaft bushings with SAE 30 oil.

Models Scamp 8 and 508

REMOVE AND REINSTALL. The axle main member is also the front frame assembly (19–Fig. AC2). Pivot point is at joint of front and rear frame sections. To disassemble the steering system, support front of unit and remove front wheel assemblies. Unbolt and remove the tie rod (17) and drag link (11). Remove retaining rings (12) and remove spindles (13 and 18). Remove steering wheel (1), washer (2) and bushing (3). Unbolt and remove steering support and cover (20 and 21). Remove cotter pin and washer (10) from lower end of steering shaft (4), then withdraw steering shaft and pinion. Unbolt and remove quadrant gear (8), special

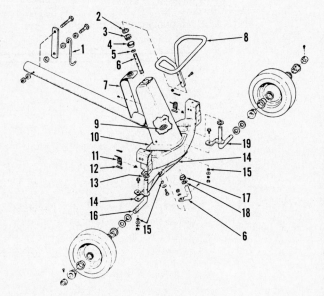

Fig. AC1 – Exploded view of typical front axle and steering system used on Models Scamp 5 and 405.

1. Mower hanger
2. Washer
3. Bushing
4. Bushing cup
5. Retaining ring
6. Steering shaft
7. Cover
8. Steering wheel
9. Bushing cup
10. Front frame & axle assy.
11. Torsion spring
12. Roll pin
13. Washer
14. Tie rods
15. Spacers
16. Spindle R.H.
17. Bushing
18. Washer
19. Spindle L.H.

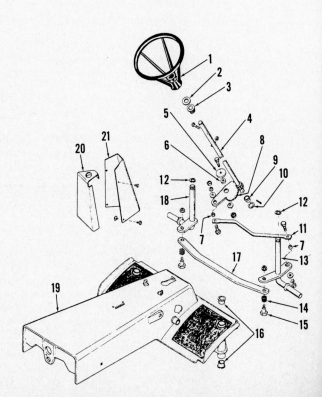

Fig. AC2 – Exploded view of typical front axle and steering system used on Models Scamp 8 and 508.

1. Steering wheel
2. Washer
3. Bushing
4. Steering shaft & pinion
5. Washer
6. Bushing
7. Spacers
8. Quadrant gear
9. Bushing
10. Washer
11. Drag link
12. Retaining ring
13. Spindle R.H.
14. Spring
15. Shoulder bolt
16. Spindle bushings
17. Tie rod
18. Spindle L.H.
19. Front frame & axle assy.
20. Steering support
21. Support cover

washer (5) and bushing (6). Remove steering shaft bushing (9) and spindle bushings (16), if need for renewal is indicated.

Clean and inspect all parts and renew any showing excessive wear or other damage. Using Fig. AC2 as a guide, reassemble by reversing the disassembly procedure. Tighten flange nut on quadrant gear center bolt to a torque of 60 ft.-lbs. Torque flange nuts on tie rod shoulder bolts (15) to 30 ft.-lbs. Apply a light coat of lithium grease to pinion and quadrant gear teeth and lubricate tie rod, drag link and steering shaft bushings with SAE 30 oil. Spindle bushings (16) are nylon and require no lubrication.

Models 526, 830 and 1036

REMOVE AND REINSTALL. Remove mower deck as needed, refer to MOVER BLADE CLUTCH AND BELT section for removal. To remove front axle and steering parts, raise and securely block front portion of chassis.

Remove locking caps securing front wheels, then slide wheel assembly off spindles (31 and 32 – Fig. AC3). Remove cotter key (12) securing drag link (28) in spindle arm (31), then lift drag link out of bushing (29) and swing clear of axle assembly. Remove nut (24) and washer (23) from pivot pin (22). While supporting axle assembly (25) withdraw pivot pin (22), then lower complete axle assembly and place to the side for inspection and repair.

With reference to Fig. AC3 disassemble and inspect spindles, axle tubes and control linkage for excessive wear, bending, cracks and any other damage. Inspect bushings 21 and 29 for excessive wear or any other damage and renew all parts as needed. Reassemble in reverse order of disassembly.

Inspect steering gear (17), pinion gear (10) and bushings (8) for excessive wear or any other damage and renew all parts as needed.

Reinstall and lubricate front wheel bushings, shouldered spindle bushings, steering shaft and wear points in linkage with SAE 30 engine oil. Wipe off excess to prevent dirt accumulation.

ENGINE

All Models

For overhaul and repair procedures on engines listed in Specification Table, refer to Small or Large Air Cooled Engines Service Manual. Refer to following paragraphs for removal and installation procedures.

Models Scamp 5 and 405

To remove the engine assembly, un-
bolt and remove engine hood on all models so equipped. On electric start models, disconnect battery cables and starter wires. On all models, disconnect ignition wire and throttle control cable. Unbolt and remove pulley and belt guard from left side and remove the belt guide. Place the blade clutch lever in disengaged position, depress clutch-brake pedal and remove belts from engine pulley. Unbolt and remove engine assembly. Reinstall engine by reversing the removal procedure.

Models Scamp 8 and 508

To remove the engine assembly, open the rear frame cover and on electric start models, disconnect battery cables and starter wires. On all models, disconnect throttle cable and ignition wire. Push transmission primary drive belt idler forward and remove belt from engine pulley. Place blade clutch control lever in disengaged position and remove mower drive belt from engine pulley. Unbolt and remove engine assembly. Reinstall engine by reversing the removal procedure.

Models 526, 830 and 1036

To remove engine assembly, open rear frame cover and on electric start models, disconnect battery cables and starter wires. On all models, disconnect throttle cable and ignition wire. For ease in repair, remove mower deck and drive belt assembly as outlined in MOWER BLADE CLUTCH AND BELT section. Loosen cap screws securing engine drive pulley belt guard, then slide clear of pulley. Loosen tension on traction drive belt idler as needed, then slip drive belt off engine drive pulley. Remove any parts that will obstruct in removal of engine, then unbolt and remove engine assembly. Reinstall engine by reversing the removal procedure. Clearance between engine drive pulley belt guard and lower pulley should be ⅛-inch.

TRACTION DRIVE CLUTCH AND BRAKE

All Models

The traction drive clutch on all models is of belt idler type. Band and drum type brake is used on all models.

Models Scamp 5 and 405

To adjust the clutch and brake linkage, remove rear hitch plate, refer to Figs. AC4 and AC5 and proceed as follows: With clutch-brake pedal in fully up (clutch engaged) position, adjust the brake set collar to a distance of 1 to 1½ inches (Scamp 5) or 1⅛ to 1¼ inches

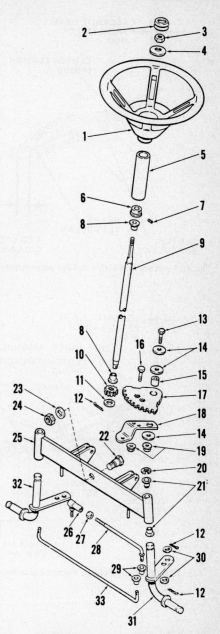

Fig. AC3—Exploded view of typical front axle and steering system used on Models 526, 830 and 1036.

1. Steering wheel	
2. Cap	18. Steering arm
3. Nut	19. Nut
4. Washer	20. "E" clip
5. Tube	21. Bushing
6. Collar	22. Pivot pin
7. Set screw	23. Washer
8. Bushing	24. Nut
9. Shaft	25. Axle assy.
10. Pinion gear	26. Ball joint
11. Washer	27. Jam nut
12. Cotter key	28. Drag link
13. Cap screw	29. Bushing
14. Washer	30. Washer
15. Spacer	31. Spindle (L.H.)
16. Cap screw	32. Spindle (R.H.)
17. Steering gear	33. Tie rod

(405) from bracket as shown. On all models, adjust locknut to ⅛-inch from end of clutch rod. On Model Scamp 5 with clutch-brake pedal in up position, adjust the clutch set collar so that clutch tension spring is preloaded 1/16-inch. On

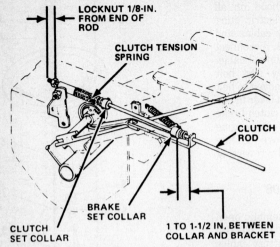

Fig. AC4 — Brake and clutch linkage adjustment on Model Scamp 5.

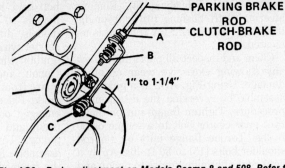

Fig. AC6 — Brake adjustment on Models Scamp 8 and 508. Refer to text.

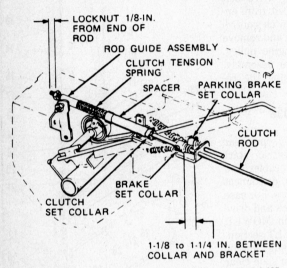

Fig. AC5 — Brake and clutch linkage adjustment on Model 405.

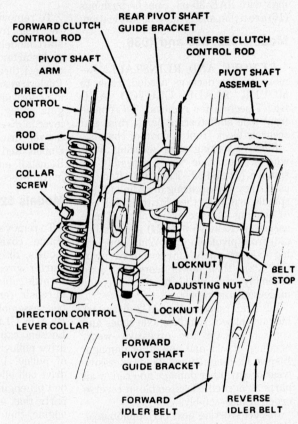

Fig. AC7 — View of clutch and forward-reverse drive linkage on Models Scamp 8 and 508. Refer to text for adjustment procedure.

Model 405 depress the clutch-brake pedal until brake is fully applied. Adjust clutch set collar so that spacer holds the clutch tension spring against the clutch rod guide. Spring should not be compressed when pedal is fully depressed.

Models Scamp 8 and 508

To adjust the clutch and brake linkage, refer to Fig. AC6 and turn nut (A) on parking brake rod to end of threads. Engage parking brake and tighten nut (B) until the spring against it is fully compressed. Adjust nut (C) to obtain a distance of 1 to 1¼ inches between flat washer and brake band bracket as shown. Disengage parking brake and check to see that spring by nut (A) pushes brake band free of brake drum. If not, adjust nut (A) as required.

Clutching occurs when the pivot shaft assembly is moved to neutral position

and belt tension is released from both the forward and reverse idler belts. Depressing the clutch-brake pedal or moving the direction control lever to NEUTRAL position will place the pivot shaft in neutral. Refer to Fig. AC7 and loosen the set screw in directional control lever collar. Place directional control lever in NEUTRAL position. Rotate pivot shaft assembly to tighten forward idler belt, applying about five pounds pressure. Place a mark on directional control rod at front edge of rod guide. Rotate pivot shaft in opposite direction to tighten reverse idler belt, once again applying about five pounds pressure. Place a second mark on directional con-

trol rod at front edge of rod guide. Distance between the two marks should be 11/16-inch. If this distance is incorrect, loosen the four cap screws securing pivot shaft in place. Move pivot shaft assembly forward or rearward as necessary to obtain the correct distance. Moving pivot shaft rearward will increase the distance. When the 11/16-inch distance is obtained, tighten the four cap screws to a torque of 15 ft.-lbs. Place a center mark on rod halfway between front and rear marks. Align front edge of rod guide with the center mark and tighten the collar set screw. Place the directional control lever in full forward position and pull pivot shaft downward.

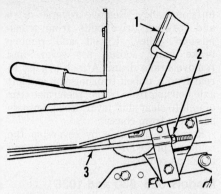

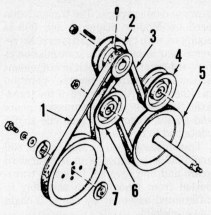

Fig. AC8 – Adjust brake-clutch pedal (1) height by turning adjustment nut (2) on linkage rod (3) on Models 526, 830 and 1036. Clearance between brake pedal arm and front edge of slot should be ¼-½ inch.

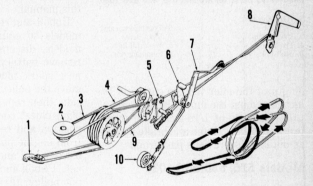

Fig. AC10 – View showing traction drive belt, pto drive belt and pulleys used on Models Scamp 5 and 405.

1. Traction drive belt
2. Engine pulley
3. Pto drive belt
4. Pto (blade) clutch idler pulley
5. Pulley and jackshaft
6. Traction drive clutch idler pulley
7. Transmission input pulley

TRACTION DRIVE BELTS

Models Scamp 5 and 405

REMOVE AND RENEW. To remove the traction drive belt (1 – Fig. AC10), unbolt and remove engine hood on all models so equipped. On electric start models, disconnect battery cables. On all models, disconnect spark plug wire, then unbolt and remove pulley and belt guard from left side. Remove the belt guide and place blade clutch lever in disengaged position. Remove the pto belt (3) from engine pulley. Depress the clutch-brake pedal and remove the traction drive belt.

Install new belt by reversing the removal procedure. Adjust clutch and brake linkage as required.

Models Scamp 8 and 508

REMOVE AND RENEW. To remove the traction drive belts (3, 4 and 9 – Fig. AC11), disconnect spark plug wire, open the rear frame cover and on electric start models, remove the battery. On all models, push the primary drive belt idler (1) forward until primary belt (3) can be removed from engine pulley (2). Remove primary belt from left side of transmission pulley. Note position of belt stops on the forward and reverse control idler pulleys and loosen the mounting bolts. Remove reverse drive belt (4) and forward drive belt (9), then complete the removal of the primary belt.

Install new belts by reversing the removal procedure. Refer to Fig. AC12 and adjust outer adjusting nut until the

Loosen the locknut and turn adjusting nut until a clearance of ¼-inch exists between the adjusting nut and the forward pivot shaft guide bracket. Tighten the locknut. Move the directional control lever to full reverse position and push pivot shaft upward. Loosen the locknut and set the adjusting nut to a clearance of ¼-inch from the reverse pivot shaft guide bracket. Tighten the locknut.

Models 526, 830 and 1036

To adjust engagement and disengagement of traction brake and clutch proceed as follows. Fully depress brake-clutch pedal (1 – Fig. AC8) using normal pressure, then measure clearance between pedal arm and front edge of frame slot. Clearance should be ¼-½ inch. To adjust, turn adjustment nut (2) on linkage rod (3) until correct clearance is attained.

With pedal in released position measure length of spring (2 – Fig. AC9). Spring should be 1-1¼ inches long. To adjust, loosen set screw in set collar (3), then slide collar on rod (1) until correct length is attained. Retighten set screw.

Fig. AC11 – View showing clutch and brake linkage and drive belt arrangement on Models Scamp 8 and 508.

1. Primary drive belt idler
2. Engine pulley
3. Primary drive belt
4. Reverse drive belt
5. Pivot shaft assy.
6. Parking brake lever
7. Directional control lever
8. Clutch-brake pedal
9. Forward drive belt
10. Brake assy.

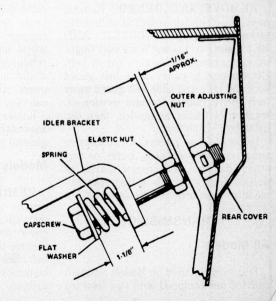

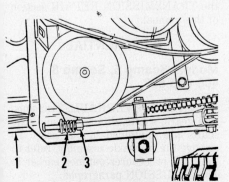

Fig. AC9 – View showing brake-clutch rod (1), spring (2) and set collar (3) on Models 526, 830 and 1036. Spring released measured length should be 1-1¼ inches long.

Fig. AC12 – View showing primary drive belt idler adjustment on Models Scamp 8 and 508.

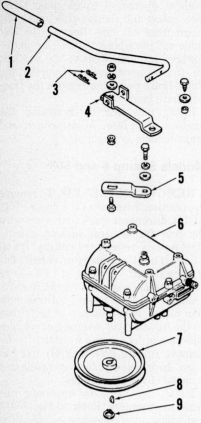

Fig. AC13—View showing transmission and associated parts on Models 526, 830 and 1036.

1. Handle grip
2. Shift rod
3. Hairpin clip
4. Pivot lever
5. Shift link
6. Transmission assy.
7. Input pulley
8. Woodruff key
9. Retaining ring

length of the idler tension spring is 1⅛ inches. Adjust the inner elastic stop nut to a clearance of 1/16-inch from frame bracket. Adjust clutch and brake linkage as outlined in previous paragraphs.

Models 526, 830 and 1036

REMOVE AND RENEW. Remove mower deck and drive belt as outlined in MOWER BLADE CLUTCH AND BELT section. Loosen drive belt finger on idler pulley and swing clear of belt. Loosen engine pulley drive belt guard securing nuts and slide belt guard away from drive pulley. Release tension on traction belt idler as needed, then slip drive belt off pulleys.

Inspect drive pulleys for excessive wear, cracks, looseness, burrs or any other damage and renew parts as needed. Renew drive belt, then reassemble in reverse order of disassembly.

TRANSMISSION

All Models

Transmission used on Models Scamp 5 and 405 are equipped with two forward gears and one reverse. The transmission used on Models Scamp 8 and 508 is equipped with two forward gears. No reverse gear is used in this transmission as forward-reverse drive belt arrangement allows for forward or reverse operation in either gear. All units are of the transaxle type with the transmission gears and shafts, differential and axle shafts contained in one case.

Transmission used on Models 526, 830 and 1036 is equipped with five forward gears and one reverse. Power is transmitted from transmission assembly to differential assembly by use of a chain and sprockets.

NOTE: On models so equipped, make certain that safety interlock switches are connected and are in good operating condition before returning mower to service.

Models Scamp 5 and 405

REMOVE AND REINSTALL. The transmission shafts, shifter shafts and left axle shaft extend through the rear frame. The following procedures will outline the removal of components necessary for removal of the transmission and differential. Actual removal of the shafts, gears and differential is outlined in the Simplicity portion of the TRANSMISSION REPAIR section of this manual.

Unbolt and remove engine hood on all models so equipped. On electric start models, disconnect battery cables and remove battery. On all models, disconnect spark plug wire. Unbolt and remove the pulley and belt guard from left side, then remove the belt guide. Place blade clutch control lever in disengaged position and remove the pto drive belt from engine pulley. Depress the clutch-brake pedal and remove the main drive belt. Unbolt and remove transmission input pulley, then remove the rear hitch plate. Loosen the set screw and remove brake drum and key from brake shaft. Support rear of unit and remove rear wheel assemblies. Remove shift links, retaining rings and springs from shifter shafts and the transmission case. All gears, shafts and differential assembly can now be removed.

Reassemble by reversing the disassembly procedure. Lubricate with general purpose lithium grease.

Models Scamp 8 and 508

REMOVE AND REINSTALL. To remove the transaxle assembly, disconnect spark plug wire, open rear frame cover and on electric start models, remove the battery. Push primary drive belt idler forward until primary belt can be removed from engine pulley. Remove primary belt from left side of trans-

mission pulley. Remove reverse drive and forward drive belts from transmission pulley. Unbolt and remove brake band, loosen set screw and remove brake drum. Attach a hoist to rear frame, unbolt transmission and left axle housing from frame, then raise rear of unit to clear shift lever. Roll transaxle assembly from chassis.

Reinstall transaxle by reversing the removal procedure. Adjust clutch and brake as required.

Models 526, 830 and 1036

REMOVE AND REINSTALL. Remove mower deck and drive belt as outlined in MOWER BLADE CLUTCH AND BELT section. Raise hood covering engine and transmission assembly. On models equipped with electric start disconnect battery cables from battery. Unhook engine spark plug wire, then unplug transmission interlock switch. Remove transmission input pulley (7–Fig. AC13) retaining ring (9), then withdraw pulley from transmission input shaft. Remove Woodruff key (8) from input shaft groove and save. Remove hairpin clips (3) from shaft rod (2), then slide rod out of pivot lever (4). Loosen traction drive chain tensioner, then slide drive chain off transmission output shaft sprocket. Inspect and remove any part that will obstruct in removal of transmission. Remove transmission mounting bolts, then lift transmission assembly (6) out of chassis and set to the side for inspection and repair.

Reinstall transmission in reverse order of removal. Adjust traction drive chain and reconnect electrical components.

All Models

OVERHAUL. For Models Scamp 5, Scamp 8, 405 and 508 refer to the Simplicity paragraphs in the TRANSMISSION REPAIR section of this manual.

For Models 526, 830 and 1036 refer to the Peerless Series 700 paragraphs in the TRANSMISSION REPAIR section of this manual.

DIFFERENTIAL

Models Scamp 5, Scamp 8, 405 and 508

R&R AND OVERHAUL. Transmission gears, shafts, differential and axle shafts are contained in one case. To remove the transaxle assembly, refer to the R&R procedures outlined earlier in TRANSMISSION paragraphs.

For differential overhaul procedures, refer to the Simplicity paragraphs in the TRANSMISSION REPAIR section of this manual.

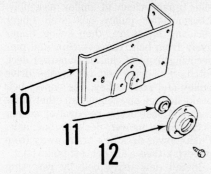

Fig. AC14—View showing axle support (10), bearing (11) and flange (12) used for differential mounting on Models 526, 830 and 1036.

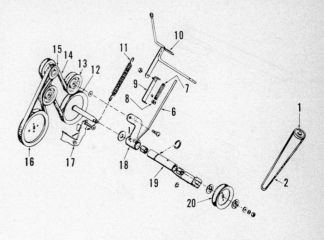

Fig. AC16— View of blade clutch (pto) linkage, pulleys and drive belts used on Models Scamp 5 and 405.

1. Mower pulley
2. Secondary mower belt
6. Clutch rod
7. Tension spring
8. Set collar
9. Rod bracket
10. Blade clutch (pto) control lever
11. Spring
12. Pulley & jackshaft
13. Blade clutch idler pulley
14. Mower clutch belt
15. Engine pulley
16. Transmission input pulley
17. Pto brake
18. Clutch idler arm
19. Jackshaft pulley bracket
20. Jackshaft pulley

Models 526, 830 and 1036

REMOVE AND REINSTALL. Remove mower deck and drive belt as outlined in MOWER BLADE CLUTCH AND BELT section. Raise rear of chassis until rear wheels are clear of ground. Remove E-clips retaining rear wheel assemblies on axle shaft, then remove wheel assemblies. Release tension on drive chain, then slip chain off transmission output sprocket. Unhook ground brake actuating lever from brake band assembly. Remove screws securing axle shaft bearing flanges (12 – Fig. AC14) to chassis axle support (10); left and right, then slide axle shafts out of support slots. Place differential assembly to the side for inspection and repair.

For differential overhaul procedures, refer to the Stewart paragraphs in the DIFFERENTIAL section of this manual.

MOWER BLADE CLUTCH AND BELT

All Models

Belt idler type blade clutch is used on all models. If the mower drive belt slips during normal operation, check and adjust the clutch tension on belt. If belt cannot be adjusted, due to excessive wear or stretching, renew the belt. On models equipped with two belts, R&R procedures will be given for both belts.

SET COLLAR

1/2-IN.

Fig. AC17— On Models Scamp 8 and 508, when blade clutch control lever is in engaged position, clearance between set collar and rod bracket should be ¾-inch.

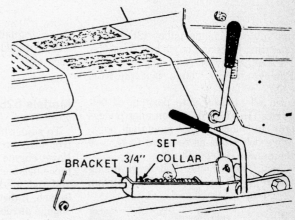

BRACKET   3/4"   SET COLLAR

Models Scamp 5 and 405

To adjust the blade clutch, move the control lever to fully engaged position. Clearance between the set collar and bracket should be ½-inch as shown in Fig. AC15. If clearance is incorrect, disengage blade clutch, loosen set screw and reposition set collar. Engage clutch and recheck clearance.

To remove the mower clutch belt (14 – Fig. AC16), unbolt and remove engine hood on models so equipped. Disconnect spark plug wire, then unbolt and remove pulley and belt guard from left side. Remove belt guide and place blade clutch lever (10) in disengaged position. On models so equipped, pull the pto brake (17) away from belt. Remove the mower drive clutch belt. Pull jackshaft pulley bracket (19) forward and remove belt

from pulley (20). Remove belt from mower.

Install new belts by reversing the removal procedure and adjust blade clutch as required.

Models Scamp 8 and 508

To adjust the blade clutch, move the control lever to fully engaged position. Clearance between the set collar and rod bracket should be ¾-inch measured as shown in Fig. AC17. If clearance is not correct, disengage blade clutch, loosen set screw and reposition set collar on rod. Engage clutch and recheck clearance.

To remove the mower drive belt (3 – Fig. AC18), first remove mower unit as follows: Disconnect spark plug wire and on electric start models, open rear frame cover and remove battery. Place blade clutch lever in disengaged position. Disconnect the mower lift link and blade clutch rod, then unpin front of mower from frame. Unbolt pulley and belt cover from mower housing. Remove mower drive belt from mower pulley and slide mower unit out from under left side.

Refer to Fig. AC11 and push primary drive belt idler (1) forward until primary (transmission drive) belt (3) can be re-

Fig. AC15—On Models Scamp 5 and 405, when blade clutch control lever is in engaged position, clearance between set collar and rod bracket should be ½-inch.

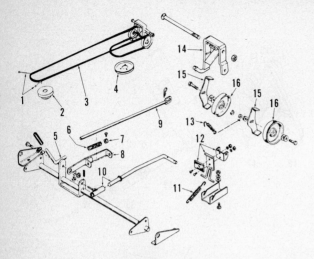

Fig. AC18 — View of blade clutch linkage, pulleys and mower drive belt used on Models Scamp 8 and 508.

1. Belt stops
2. Engine pulley
3. Mower drive belt
4. Mower pulley
5. Blade clutch control lever
6. Tension spring
7. Set collar
8. Rod bracket
9. Clutch rod
10. Blade brake rod
11. Brake spring
12. Blade brake
13. Spring
14. Blade clutch idler arm
15. Belt guides
16. Clutch idler pulleys

moved from engine pulley (2). Loosen mounting bolts and move belt stops (1 – Fig. AC18) away from engine pulley. Remove mower clutch belt (3) from engine pulley.

On all models, note position of belt guides (15 – Fig. AC18) on idler pulleys (16) and loosen the mounting bolts. Move belt stops as required to remove the mower clutch belt (3).

Install new mower belt by reversing the removal procedure. Make certain

Fig. AC19 — View showing procedure for adjusting pto control linkage. Adjust by loosening set screw (3) and sliding set collar (1). Distance is measured between set collar (1) and rod guide (2).

that belt guides and belt stops are properly installed. Adjust blade clutch as necessary.

Models 526, 830 and 1036

To adjust blade clutch, move pto control lever to fully engaged position. Raise engine cover, then measure distance between set collar (1 – Fig. AC19) and rod guide (2). Distance should be $\frac{1}{2}$-inch. If distance is incorrect, then disengage pto control lever. Loosen set collar (1) securing setscrew (3). Slide set collar on control rod, then retighten setscrew. Engage control lever, then recheck distance. Continue adjustment procedure until correct distance is attained.

To remove the mower drive belt, first remove mower deck unit as follows: Open engine cover, then disconnect spark plug wire and on electric start models, disconnect battery cable. Place pto control lever in disengaged position. Place mower height control lever in lowest cutting position. Unbolt rear mower bracket (6 – Fig. AC20) on 26 and 30-inch models and (7 – Fig. AC21) on 36-inch models. Loosen cap screws securing engine pulley belt guard, then

slide guard clear of pulley assembly. Loosen idler pulley belt stop finger securing hardware, then swing finger away from belt. Remove clips and pins securing front of chassis to mower deck hitch. Slip drive belt off engine drive pulley and idler pulley. Disconnect belt brake linkage and remove all other parts that will obstruct in removal of mower deck. Slide mower deck to one side. Push mower deck belt brake away from pulley(s), then slide belt off pulley(s).

Install new drive belt by reversing removal procedures. Reinstall mower deck in reverse order of removal. Clearance between belt stop and drive belt should be $\frac{1}{8}$-inch when pto control lever is engaged. Clearance between engine pulley belt guard and lower pulley should be $\frac{1}{8}$-inch.

MOWER BLADES AND SPINDLES

All Models

Models Scamp 5, 405, Scamp 8, 508, 526 and 830 are equipped with single blade rotary mowers. A twin blade rotary mower is used on Model 1036.

CAUTION: Always disconnect spark plug wire and on electric start models disconnect battery cable before performing any inspection, adjustment or other service on the mower.

Make certain that safety starting switches are connected and in good operating condition before returning unit to service.

Models Scamp 5 and 405

REMOVE AND REINSTALL. To remove the blade, spindle and bearings, remove mower unit as follows: Place blade clutch control lever in disengaged position. Pull jackshaft pulley housing

Fig. AC21 — View showing 36-inch mower deck assembly used on Model 1036.

1. Brake rod	
2. Lift rod	5. Cotter pin
3. Jam nut	6. Hitch
4. Leveling rod	7. Mower bracket

Fig. AC20 — View showing 26 and 30-inch mower deck assembly used on Models 526 and 830.

1. Hitch
2. Lift rod flange
3. Belt brake
4. Pulley belt guard
5. Spacer
6. Mower bracket
7. Brake rod

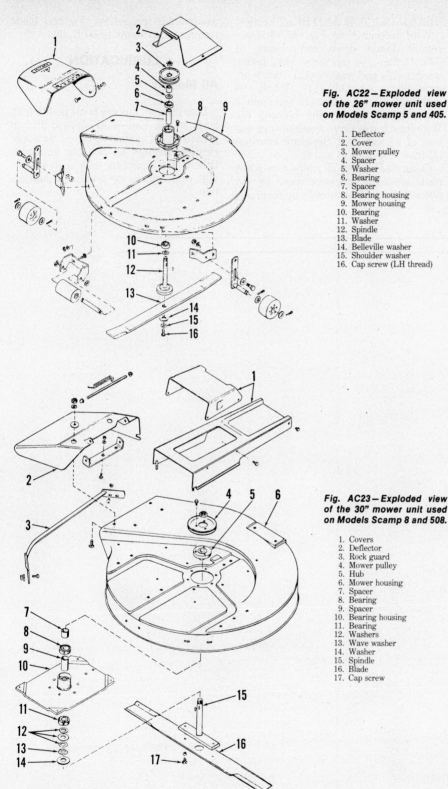

Fig. AC22—Exploded view of the 26" mower unit used on Models Scamp 5 and 405.

1. Deflector
2. Cover
3. Mower pulley
4. Spacer
5. Washer
6. Bearing
7. Spacer
8. Bearing housing
9. Mower housing
10. Bearing
11. Washer
12. Spindle
13. Blade
14. Belleville washer
15. Shoulder washer
16. Cap screw (LH thread)

Fig. AC23—Exploded view of the 30" mower unit used on Models Scamp 8 and 508.

1. Covers
2. Deflector
3. Rock guard
4. Mower pulley
5. Hub
6. Mower housing
7. Spacer
8. Bearing
9. Spacer
10. Bearing housing
11. Bearing
12. Washers
13. Wave washer
14. Washer
15. Spindle
16. Blade
17. Cap screw

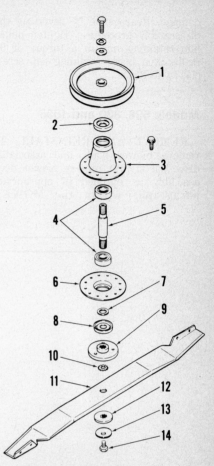

Fig. AC24—Exploded view showing mower blade and associated drive parts on 526 and 830 models. Model 1036 is similar.

1. Pulley	8. Shield
2. Shield	9. Blade adaptor
3. Top housing	10. Ring
4. Bearing	11. Blade
5. Shaft	12. Washer
6. Bottom housing	13. Spring washer
7. Shim	14. Cap screw

ft.-lbs. and left hand thread cap screw (16) to a torque of 80 ft.-lbs.

Models Scamp 8 and 508

REMOVE AND REINSTALL. To remove the blade, spindle and bearings, remove mower unit as follows: Place blade clutch control lever in disengaged position. Disconnect blade clutch rod, then disconnect lift link and unpin front of mower from frame. Unbolt and remove covers (1–Fig. AC23) and remove mower drive belt from pulley (4). Slide mower unit from under left side. Remove cap screws (17) and blade (16). Unbolt mower pulley (4), then unbolt and remove bearing housing (10) from mower housing (6). Remove nut, hub (5) and spacer (7) from top of bearing housing. Remove spindle (15) and washers (12, 13 and 14). Separate bearings (8 and 11) and spacer (9) from bearing housing.

Clean and inspect all parts and renew any showing excessive wear or other

(19–Fig. AC16) forward and remove mower belt (2) from pulley (20). Unhook rear of mower and unpin front of mower, then slide mower unit from under left side. Unbolt cover (2–Fig. AC22) and remove mower belt. Remove left hand thread cap screw (16), washers (14 and 15) and blade (13). Remove nut, pulley (3), spacer (4) and washer (5) from above and spindle (12) and washer (11) from below. Unbolt bearing housing (8) and separate bearings (6 and 10) and spacer (7) from housing.

Clean and inspect all parts and renew any showing excessive wear or other damage. Reassemble by reversing disassembly procedure. Tighten mower pulley retaining nut to a torque of 70

damage. Reassemble by reversing the disassembly procedure. Tighten pulley hub retaining nut to a torque of 95 ft.-lbs. and blade retaining cap screw (17) to 45 ft.-lbs.

Models 526, 830 and 1036

REMOVE AND REINSTALL. To remove mower blade and associated drive parts, first remove mower deck assembly as outlined in appropriate paragraphs within the MOWER BLADE CLUTCH AND BELT section.

With reference to Fig. AC24 disassemble blade drive components as follows. Remove cap screw (14), spring washer (13) and washer (12), then withdraw blade assembly (11). Unbolt and withdraw mower pulley (1), then unbolt and remove top bearing housing (3). Withdraw lower drive assembly out bottom of mower unit. Separate bearings (4) from housings (3 and 6).

Clean and inspect all parts, renew any parts showing excessive wear or any other damage. Reassemble by reversing disassembly procedure. Tighten blade retaining cap screw to 50 ft.-lbs.

LUBRICATION

All Models

Lubricate all linkage pivot points with SAE 30 oil. Use multi-purpose lithium grease on all models equipped with lubrication fittings on bearing housings. Others are equipped with sealed bearings and require no additional lubrication.

ARIENS

ARIENS COMPANY
655 West Ryan St.
Brillion, WI 54110

Model	Make	Engine		Cutting
		Model	Horsepower	Width, In.
EMPEROR 925000 & 927000	Tecumseh	V60	6	26
	Tecumseh	V70	7	28
	Tecumseh	VM80	8	28-32
	B&S	191000	8	30-32
	Tecumseh	VM100	10	30-38

FRONT AXLE

All Models

R&R AND OVERHAUL. Axle main member is center-pivoted in channel at front of chassis. Repair procedure will depend upon extent of service required, however, reference to Fig. AR1 will indicate correct assembly. Axle pivot nut (19 – Fig. AR1) should be tightened to 35 ft.-lbs. torque.

STEERING GEAR

All Models

R&R, OVERHAUL AND ADJUST. If equipped with a rear service bar, drain fuel and operate engine until all fuel is removed from carburetor. Disconnect spark plug wire, remove battery, air cleaner and rear grass catcher. Raise front enough to tip unit back onto service bar. Be careful not to knock unit over while servicing.

If not equipped with service bar at rear, it is necessary to raise unit using sufficient blocks or hoist to permit access to steering gear from below. Be extremely careful to prevent unit from falling over while servicing. If unit is raised only at front remove fluids as described in preceding paragraph. Always disconnect spark plug wire when servicing to help prevent injury.

The steering gear assembly (Fig. AR2) can be removed after detaching link (15 – Fig. AR1) from steering gear (12) and steering shaft (20) from pinion (14).

Adjustment to compensate for wear of pinion and steering gear teeth is possible without removing steering gear assembly. To adjust, loosen nut (1 – Fig. AR1 or Fig. AR2) then tighten nut (2) slightly. Tighten nut (1), then turn steering to be sure that steering does not bind anywhere in operating range. If binding is observed, loosen nut (1), loosen nut (2) retighten nut (1), then recheck steering for smoothness.

ENGINE

All Models

Briggs & Stratton and Tecumseh engines are originally installed as listed in the Condensed Service Data at the beginning of this section. The engine must be accurately identified by the complete model number in order to obtain correct repair parts.

Briggs & Stratton engines have the model number stamped into the cooling fan shroud and will usually be a six digit number such as 190702. The first three digits indicate that the engine is an aluminum block, 8 hp. engine which has 19.44 cubic inch displacement. The last three digits indicate further variations

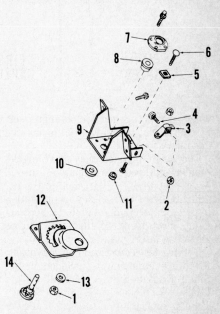

Fig. AR2 – Exploded view of steering gear. Refer to Fig. AR1 for cross sectional view.

1. Nut	8. Bushing
2. Nut	9. Bracket
3. Adjuster block	10. Washer
4. Screw	11. Bushing
5. Pivot spacer	12. Steering gear
6. Screw	13. Washer
7. Retainer	14. Pinion

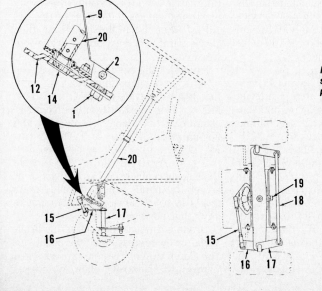

Fig. AR1 – Views of steering system and front axle. Axle pivot nut (19) should be tightened to 35 ft.-lbs. torque.

1. Nut
2. Nut
9. Bracket
12. Steering gear
14. Pinion
15. Steering link
16. Arm
17. Spindle
18. Tie rod
19. Nut
20. Steering shaft

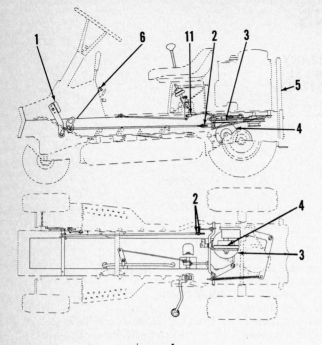

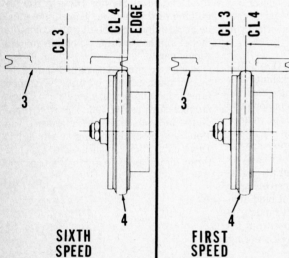

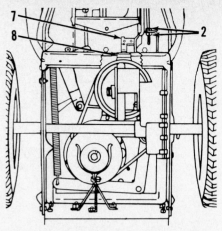

Fig. AR3 — Drawing of clutch operating system. Refer to text for adjustment procedures.

1. Pedal
2. Adjusting nuts
3. Drive disc
4. Friction wheel
5. Rear service bar
6. Parking brake control
11. Link

Fig. AR4 — Refer to text for adjustment of traction drive clutch and friction drive. Linkage is interactive and must be adjusted correctly.

Fig. AR3A — Views of drive disc (3) and friction wheel (4) showing relative location in first speed and sixth speed.

TRACTION DRIVE
TRACTION DRIVE BELT

All Models

REMOVE AND REINSTALL. To remove traction drive belt, it is first necessary to remove the mower drive belt, then remove friction wheel as described in FRICTION DRIVE, R&R AND OVERHAUL paragraphs. Remove belt, then locate new belt in grooves of engine pulley, drive disc and idler pulley. Make sure belt is not twisted, then install and adjust friction wheel. Install and adjust mower drive belt as described in MOWER DRIVE BELT paragraphs.

TRACTION DRIVE CLUTCH

All Models

ADJUSTMENT. Clutch action is accomplished by raising drive disc (3 – Fig. AR3) vertically away from the friction wheel (4). The clutch and brake are interactive and depressing the brake pedal also causes the clutch to release. Adjust the clutch linkage first, then check and if necessary adjust brake controls.

To adjust the clutch, tip unit back on to rear service bar (5) or raise unit sufficiently to provide access from below. Move speed control lever to "Neutral" position, depress clutch pedal fully and engage parking brake (6). Turn nuts (2 – Fig. AR4) until clearance between neutral stop (7) carrier yoke (8) is 1/8-1/4 inch. Release parking brake and rotate rear wheels by hand. Wheels should rotate freely in "Neutral" position, but not in any other position.

Brake is engaged by depressing brake pedal or by depressing clutch pedal past the clutch range. To adjust, move speed control to "Neutral", then loosen nut

which will be needed for obtaining service parts but which will not usually be necessary for servicing.

Tecumseh engines may have model number stamped into cooling fan shroud or on a plate or tag attached to the engine. Be sure to transfer identification tags from original engine to replacement short block so that unit can be identified later.

Service procedures and specifications for the engine unit is included in the Small Engine Service Manual or the Large Air Cooled Engine Service Manual depending on the size of the engine.

Engines which have less than 16 cubic inches of displacement are included in the Small Engine Service Manual. Briggs & Stratton and Tecumseh engines with 6 horsepower and less are included in this manual.

Engines which have 16 cubic inches of displacement or more are included in the Large Air Cooled Engine Service Manual. Briggs & Stratton and Tecumseh engines with 7 horsepower and more are included in this manual.

REMOVE AND REINSTALL. The engine can be removed after removing the drive belts, detaching wires and control cables from engine, then unbolting engine from mounting deck.

Crankcase should be filled with approximately 2-2 1/4 pints of oil. SAE 30W oil should be used when ambient temperatures are above 30 degrees F., SAE 10W oil should be used between 32 degrees and 0 degrees F. and multi-viscosity SAE 5W20 or 5W30 oil should be used at temperatures below 0 degrees F. Multi-viscosity SAE 10W30 or 10W40 oil can be used for summer.

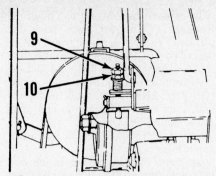

Fig. AR5 — Refer to text for adjustment of brake linkage.

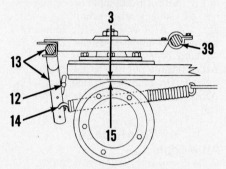

Fig. AR6 — Refer to text for adjustment of friction drive.

Fig. AR7 — Exploded view of drive and die cast carrier control parts. Refer to Fig. AR8 for stamped transfer frame used on some models.

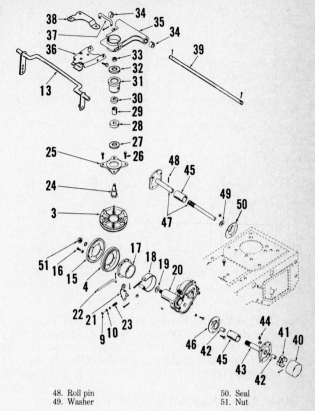

3. Drive disc
4. Friction wheel
9. Nut
10. Nut
13. Clutch shaft
15. Guard
16. Screws
17. Hub
18. Brake band
19. Washer
20. Gear case
21. Lever
22. Link
23. Spring
24. Spindle
25. Adapter
26. Screws
27. Shim
28. Bearing
29. Bushing
30. Bearing
31. Housing
32. Washer
33. Nut
34. Bushings
35. Carrier
36. Carrier yoke
37. Link
38. Bellcrank
39. Transfer shaft
40. Hub cap
41. Retainer
42. Bushing
43. Left axle
44. Grease fitting
45. Spacers
46. Seal
47. Right axle
48. Roll pin
49. Washer
50. Seal
51. Nut

(9 – Fig. AR5) using two ½-inch wrenches. Be careful not to distort the brake band when loosening nuts. Tighten nut (10) until wheels just start to bind, then loosen 1½ turns and tighten locknut (9). Recheck brake action using clutch pedal. The drive disc should just move away from friction wheel (clutch action) when clutch pedal is approximately ¾-inch from limit of travel. The final ¾-inch should be sufficient to engage brake. Be sure that brake band is not twisted.

FRICTION DRIVE

All Models

ADJUSTMENT. Disengagement of the clutch is accomplished by raising the drive disc (3 – Fig. AR3) vertically away from the friction wheel (4). Refer to preceding TRACTION DRIVE CLUTCH paragraphs for adjustment procedures.

Speed control is accomplished by moving the drive disc (3) laterally across the friction surface of the friction wheel (4). Locating the center of the drive disc over the friction wheel provides a neutral position.

To adjust shift positions, proceed as follows: Move carrier yoke (8 – Fig. AR4) until it is centered on neutral stop (7). Detach link (11 – Fig. AR3) from speed control lever. Change length of link (11) if necessary so that it can be re-

Fig. AR8 — Exploded view of stamped transfer frame and related parts used on some models. Refer to Fig. AR7 for legend except the following.

27W. Washer
32. Spacer
35. Cover
36. Transfer frame

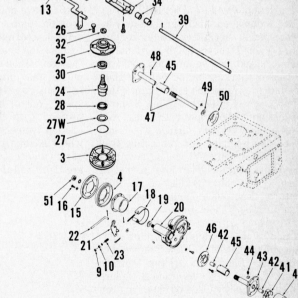

attached with shift lever in neutral and carrier yoke centered. Connect link and shift to sixth speed. Check to be sure that rubber friction wheel (3 – Fig. AR3A) is still on drive disc (3). Shift to first and reverse speeds and check to be sure that the carrier yoke (8 – Fig. AR4)

moves off center in both directions. Distance between centerline of drive disc (CL3) and centerline of friction wheel (CL4) should be 11/16-¾ inch in first speed.

The friction wheel must be removed to measure clearance between guard (15 –

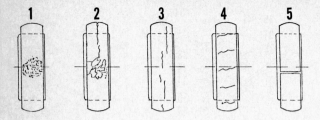

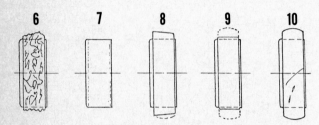

Fig. AR9 — Drawings of some friction wheel failures. Refer to text for description of conditions shown.

Fig. AR6) and drive disc (3) when adjusting position of stop (12). Refer to the following R&R AND OVERHAUL paragraphs for removal of friction wheel and adjusting stop.

R&R AND OVERHAUL. The friction drive consists of friction wheel (4 – Fig. AR7), drive disc (3) and related parts of spindle carrier (24 through 36). Some models are equipped with die cast carrier as shown in Fig. AR7 while other models have stamped frame for carrier as shown in Fig. AR8.

If friction wheel (4 – Fig. AR7 or Fig. AR8) is damaged, analysis of oil may indicate repair necessary. Refer to Fig. AR9. View 1 shows small scuff marks and localized flat spot which could be caused by parking brake not releasing completely or by drive disc contacting friction wheel in neutral. View 2 shows larger chunked out spot with cracks running around friction surface which can be caused by drive disc contacting friction wheel in neutral. View 3 shows normal deterioration after long period of normal use, but if this occurs prematurely, cause is too much pressure on friction roller. View 4 shows shiny friction surface with cracks across surface caused by not enough pressure or operator riding the clutch. View 5 shows split indicating failure of friction surface seam. View 6 shows radical failure which is usually caused by slippage even though pressure is too much (or at least enough). Overloading mower or operating on long inclines for extended time can result in damage shown in View 6. Improper bond may result in missing rubber friction surface as shown in View 7. View 8 shows friction surface worn at angle caused by bent or loose spindle or carrier frame. Rubber friction surface may be prematurely worn and rough-

ened as shown in View 9 if the drive disc surface is not smooth. Cuts in the friction surface as shown in View 10 are usually caused by foreign objects caught in drive assembly.

To disassemble and adjust, proceed as follows:

Remove battery, drain fuel, then raise unit or tip unit back onto service bar if so equipped. Remove attaching screws (16 – Fig. AR7 or Fig. AR8), remove guard (15) and friction wheel (4). When reinstalling, assemble guard (15) without wheel (4), then adjust stop (12 – Fig. AR6). Move speed control to a forward position, then check clearance between drive disc (3) and guard (15). Loosen the two nuts on clutch shaft stop screw (12), then reposition stop screw in slot so that only a small amount of clearance exists between guard (15) and disc (3). Move speed control back to "Neutral", then proceed with disassembly or reinstall friction wheel. Be sure that friction wheel is correctly located over shoulder of hub and that all five retaining screws are tightened upon final assembly.

To remove the carrier assembly (24 through 36 – Fig. AR7) and drive disc (3), first remove friction wheel (4) as described in previous paragraph. Remove cotter pin from link (37) and detach from lever (38). Disconnect spring (14 – Fig. AR6) from shaft and lever (13), then remove cotter pin and withdraw shaft (39). Disengage yoke from clutch shaft (13) and withdraw carrier assembly.

On some models, carrier is die cast (35) as shown in Fig. AR7, while other models use stamped assembly as shown in Fig. AR8. Disassembly procedure will be self evident after examining applicable Fig. AR7 or Fig. AR8. Shims (27) are used to adjust bearings (28 and 30) on both types. On die cast type (Fig. AR7), tighten nut (33) to 250-275 in.-lbs. torque. On stamped frame type (Fig. AR8), tighten nut (33) to 45 ft.-lbs. torque.

GEAR CASE
All Models

R&R AND OVERHAUL. To remove traction drive gear case (20 – Fig. AR7), first remove hub cap retaining cotter pin, hub cap (40) and retainer (41) from the short lift side axle (43). Drive roll pin (48) from long right side axle, then withdraw axle. Remove washer (49) from inside bearing. Remove cotter pin, then detach rod (22) from lever (21). Unbolt and remove seal (46). Remove screws attaching gear case (20) to frame, then withdraw gear case assembly.

To disassemble gear case, remove nut (51), then slide friction wheel (4), guard (15) and hub (17) from shaft as an assembly. Remove Woodruff key (4 – Fig. AR11) from shaft and screws retaining cover (15) to housing (3). Insert screwdrivers in slots provided and pry cover from housing. Breather (16) can be removed by pressing it out from the inside toward outside. Install breather by pressing in from the outside of cover.

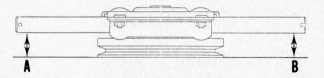

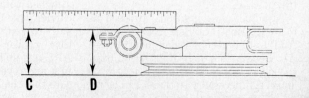

Fig. AR10 — Check for bent parts by measuring as shown. Dimensions (A and B) and (C and D) should be within 1/32-inch.

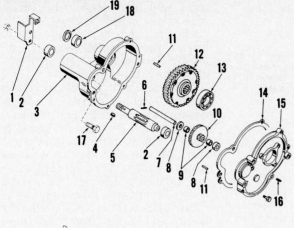

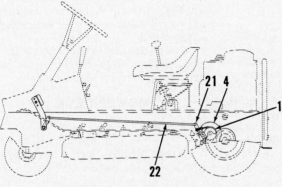

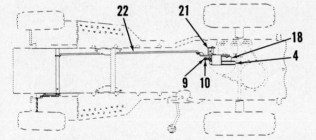

Fig. AR11—Exploded view of gear case.

1. Brake bracket
2. Ball bearing
3. Housing
4. Woodruff key
5. Pinion shaft
6. Roll pin
7. Idler shaft
8. Washer
9. Needle bearings
10. Cluster gear
11. Groove pin
12. Differential assy.
13. Ball bearing
14. Gasket
15. Cover
16. Breather

Fig. AR12—Views of brake controls refer to text for adjustment procedures.

4. Friction wheel
9. Nut
10. Nut
18. Brake band
21. Lever
22. Link

ing the brake pedal causes the clutch to release. Adjust clutch linkage first, then adjust brake controls. Refer to TRACTION DRIVE CLUTCH Adjustment paragraphs for procedures.

R&R AND OVERHAUL. The brake band (18–Fig. AR7) operates by contracting around hub (17). Refer to GEAR CASE, R&R AND OVERHAUL paragraphs to remove the gear case and disassemble brake. When assembling, tighten hub retaining nut (51) to 70 ft.-lbs. torque. Adjust brake as described in TRACTION DRIVE CLUTCH Adjustment paragraphs.

MOWER DRIVE BELT

All Models

ADJUSTMENT. During first hours of operation, new mower drive belt will stretch and will require adjustment as follows.

Set mower height in the middle notch. Hold mower clutch lever (2–Fig. AR13) so that front edge of lever is aligned with rear edge of forward notch (detent) of quadrant as viewed through slot in left side of cowl. Turn adjustment screw (3) with ¾-inch socket. Press socket against spring clip when turning screw (3), then be sure that spring clip engages adjustment screw when released. The mower clutch lever will pull rearward slightly as tension is increased sufficiently. Mower clutch should begin to engage when front edge of lever is aligned with rear edge of forward notch in quadrant and should be fully locked in when lever is completely engaging notch.

Over-tightening will cause premature belt and bearing wear. If mower is usually operated in lowest cutting position, readjust belt tension with height set in this (lowest) position. On models equipped with belt finger retaining belt in idler pulley, belt slippage can also be caused by finger adjusted against belt. Correct adjustment is ⅛-inch clearance between retaining finger and back edge of belt. Blade attaching nut should be tightened to 50-55 ft.-lbs. torque.

Differential assembly (12) should be installed with small inside diameter spline toward inside of housing (3). If renewal is necessary, pinion (5) and cluster gear (10) are available only as matched set. Numbered side of bearings (9) should be facing out and should be flush with face of gear. Bushing (18) is removed by pressing from outside. Press bushing into housing until bottomed against shoulder in bore. Seal (19) should be flush with face of housing. Special ribbed screws (17) are pressed into housing to attach brake bracket (1). Gear case cavity should be filled with lubricant before installing cover.

Reinstall by reversing removal procedure. Tighten nut (51) to 70 ft.-lbs. torque. Refer to TRACTION DRIVE CLUTCH Adjustment paragraphs and to FRICTION DRIVE Adjustment paragraphs while installing and assembling gear case.

GROUND DRIVE BRAKES

All Models

ADJUSTMENT. Refer to Fig. AR12 for views of brake controls. The clutch and brake are interactive and depress-

Fig. AR13—View of mower clutch adjustment screw (3), mower lever (2) and associated controls.

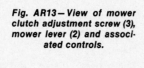

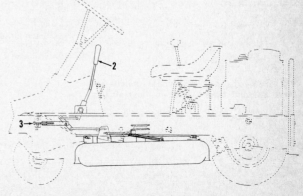

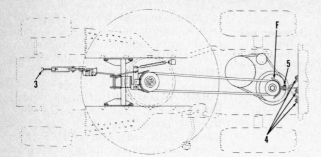

Fig. AR14—View of drive belt controls and belt guard (finger) for mower drive. Refer to text.

REMOVE AND REINSTALL. If equipped with a rear service bar, drain fuel, remove battery and tip unit up onto rear service bar. If not equipped with service bar, drain fuel, remove battery, then raise unit sufficiently to provide easy access to underside. On all models, use all possible precautions to prevent unit from falling while servicing. Move mower clutch lever to "OUT" (disengaged) position. Loosen three nuts (4–Fig. AR14) and move rear finger out of the way. Roll belt out of pulley groove. Renew drive belt, then reinstall in reverse order of disassembly.

MOWER SPINDLE

All Models

LUBRICATION. On all models, mower spindle bearings (7 and 11–Fig. AR15 and Fig. AR16) should be packed with grease when assembling and should not require additional lubrication until disassembled for other service. On

38-inch mowers with two blades (3–Fig. AR16) and spindles (8), bearings (24) for drive pulley (13) can be lubricated through grease fitting (33) about every 25 hours of operation.

R&R AND OVERHAUL. To remove spindle (8–Fig. AR15 or Fig. AR16), first remove mower deck as outlined in appropriate following paragraph. Remove blade retaining nut (1), lockwasher (2) and blade (3 or 17). On models with one cutting blade, remove nut (15–Fig. AR15), washer (14), pulley (13) and top Woodruff key (9). On all models, remove hub (5–Fig. AR15 or Fig. AR16), slinger (6) and lower Wood-

ruff key (9). On models with two blades, remove belt guards (34 and 35–Fig. AR16) and belt to mower spindle pulleys. On all models, bump spindle (8–Fig. AR15 or Fig. AR16) out of spindle housing (10).

On models with two blades, center drive pulley (13–Fig. AR16) can be removed after removing cap (26), adjustment nut (15). Bearings (24) should be cleaned, inspected and renewed if necessary. Be sure that bearing cups are seated in pulley. Pack bearing cones before installing and be sure that seal (23) is installed properly. Adjust bearings by tightening nut (15) until pulley rolls freely with no end play, then install cotter pin. Grease fitting (33) permits bearing lubrication after mower is assembled and installed.

On all models, blade retaining nut (1–Fig. AR15 or Fig. AR16) should be tightened to 50-55 ft.-lbs. torque.

MOWER DECK

All Models

ADJUSTMENT. Before attempting any adjustment, mower unit should be

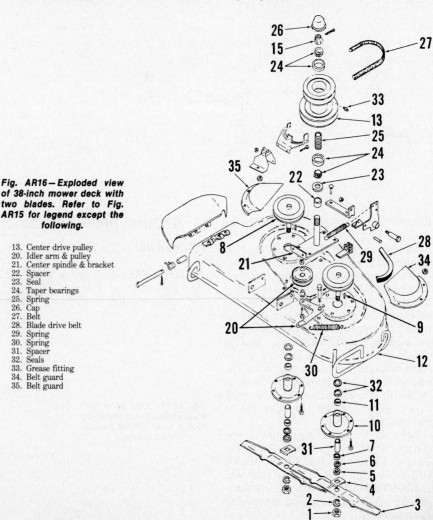

Fig. AR16—Exploded view of 38-inch mower deck with two blades. Refer to Fig. AR15 for legend except the following.

13. Center drive pulley
20. Idler arm & pulley
21. Center spindle & bracket
22. Spacer
23. Seal
24. Taper bearings
25. Spring
26. Cap
27. Belt
28. Blade drive belt
29. Spring
30. Spring
31. Spacer
32. Seals
33. Grease fitting
34. Belt guard
35. Belt guard

Fig. AR15—Views of mower deck, spindle and three different blades used on most models.

1. Nut
2. Lockwasher
3. Blade
4. Blade tray
5. Retainer hub
6. Bearing slinger
7. Bearing
8. Spindle
9. Woodruff keys
10. Spindle housing
11. Bearing
12. Mower deck
13. Pulley
14. Washer
15. Nut
16. Vanes
17. Blade
18. Blade tray (late type)

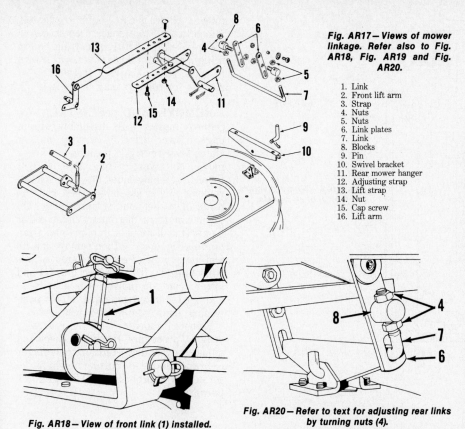

1. Link
2. Front lift arm
3. Strap
4. Nuts
5. Nuts
6. Link plates
7. Link
8. Blocks
9. Pin
10. Swivel bracket
11. Rear mower hanger
12. Adjusting strap
13. Lift strap
14. Nut
15. Cap screw
16. Lift arm

Fig. AR18—View of front link (1) installed.

Fig. AR20—Refer to text for adjusting rear links by turning nuts (4).

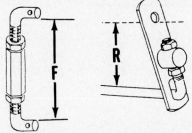

Fig. AR19—Refer to text for standard adjustment lengths for front link (F) and link rod (R).

located on a flat level surface and be sure that tires are correctly inflated to 12 psi. Since measurements are made at the blade cutting surface, it is important to prevent engine from starting. Turn ignition OFF and remove key. Disconnect wire from engine spark plug and ground wire to engine in such a way that wire will not come loose and swing back up to the spark plug.

Cutting height is adjustable in the range of 1-4 inches as measured from ground to cutting edge of mower blade. Cutting height may change slightly when mower clutch is engaged, so measurements should be made with clutch engaged.

Rotate blade so that blade is positioned straight from left to right, then measure distance from cutting edge of blade to ground on both sides. This distance should be the same and should be very close to the desired cutting height.

Rotate blade so that blade is positioned straight from front to rear, then measure distance from cutting edge of blade to ground at front and at rear. Blade cutting edge should be 1/4-3/8 inch closer to ground at front than at rear.

Shortening link (1—Fig. AR17) will raise front of mower and lengthening will lower front. The link is also shown at (1—Fig. AR18). To change link, loosen locknuts, then turn center coupling of link. Tighten both locknuts when length is correct and link ends are both pointing in same direction. Adjustment may be more easily accomplished after link is removed. Normal distance between ends of link (F—Fig. AR19) depend upon mower width as follows:

Mower width	Length of link
26 inches	2-11/16 inches
28 inches	3 inches
30 inches	2-13/16 inches
32 inches	3¼ inches

Height of blade at rear is changed by turning nuts (4 and 5—Fig. AR17) on rear link assembly (4, 5, 6, 7 and 8). Turning nuts (4 and 5) down on link (7) effectively shortens distance between top ends of link plates (6) and cross bar of link (7) and raises rear of mower. Moving nuts only on one side (either 4 or 5), will tip mower blade either to left or right. Nuts should be tight against blocks (8). Normal distance between holes in link plates and cross bar of link rod (R—Fig. AR19) should be 2⅛ inches for mowers which are 26 inches wide, 2½ inches for mowers which are 28, 30 and 32 inches wide. Be sure that distance between cutting edges of blade and ground is the same when blade is positioned across mower (left to right). Difference between left and right requires adjustment of only one side (either nuts 4 or nuts 5—Fig. AR17).

If it is not possible to adjust the cutting height so that front is 1/4-3/8 inch lower in the front, it is possible that lift strap (13 and 12) are not adjusted correctly. To adjust, remove the mower assembly, loosen nut (14) on carriage bolt and remove screw (15). Slide lift strap halves (12 and 13) either longer (to lower rear) or shorter (to raise rear), then tighten nut (14). Install screw (15) in threaded hole of strap (13) which is aligned with hole in strap (12). For 28 inch wide mowers, normal location for screw (15) is in hole closest to hole for carriage bolt. Center hole (as shown) is normal position for screw (15) on other width mowers.

After cutting height and forward tilt of mower blade is correctly set as described in preceding paragraphs, runners should be spaced to correspond to the cutting height. If cutting height of

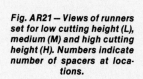

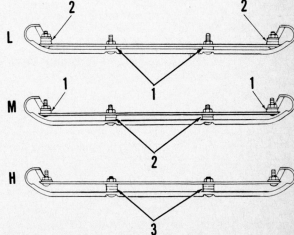

Fig. AR21—Views of runners set for low cutting height (L), medium (M) and high cutting height (H). Numbers indicate number of spacers at locations.

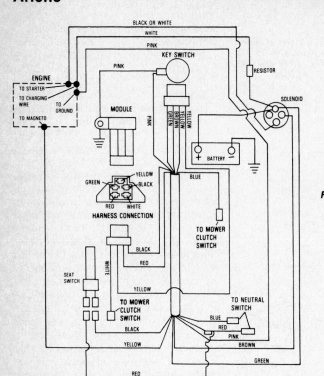

BLACK OR WHITE
WHITE
PINK
KEY SWITCH
RESISTOR
PINK
ENGINE
TO STARTER
TO CHARGING WIRE
TO MAGNETO
TO GROUND
MODULE
SOLENOID
PINK
YELLOW
BROWN
GREEN
YELLOW
BATTERY
+ −
GREEN
YELLOW
BLACK
RED   WHITE
HARNESS CONNECTION
BLUE
TO MOWER CLUTCH SWITCH
BLACK
RED
SEAT SWITCH
WHITE
YELLOW
TO MOWER CLUTCH SWITCH
TO NEUTRAL SWITCH
BLUE
RED
PINK
BLACK
BROWN
YELLOW
GREEN
RED

Fig. AR22 — Wiring diagram typical of all models.

mower blade is set to approximately 1-inch, install two spacers as shown in view "L" – Fig. AR21. If cutting height is high, install all three spacers as shown in view "H". Incorrect installation of spacers for runners may cause scalping.

REMOVE AND REINSTALL. To remove mower deck assembly, disengage mower drive by moving mower clutch lever to "OUT" position and lower mower to lowest setting. Remove belt finger attaching nuts (4 – Fig. AR14), remove finger assembly, then roll the blade drive belt out of rear pulley. On 32-inch mowers, disconnect clutch rod at front. On all models, remove clip from pin (9 – Fig. AR17), then remove pin (9) and lower rear of mower. Remove clips from link (1), then remove link (1) and lower front of mower. On models so equipped, remove pins and detach positioning arms from mower deck.

Reinstall mower deck by reversing removal procedure. Be sure to check and adjust mower drive tension, belt finger clearance and blade cutting height as described in appropriate ADJUSTMENT paragraphs.

BOLENS

BOLENS INTERNATIONAL
215 So. Park
Port Washington, WI 53074

Model	Make	Engine Model	Horsepower	Cutting Width, In.
728	Tecumseh	V70	7	28
828	Tecumseh	VM80	8	28
829	Tecumseh	VM80	8	28

For service information and procedures regarding the following models, refer to the appropriate GILSON section of this manual.

Bolens Models	Gilson Models
1011GM	52073
1011HM	52061

FRONT AXLE AND STEERING SYSTEM

All Models

REMOVE AND REINSTALL. Axle main member (13–Fig. B1) mounts within a box section at front of main frame and is center-pivoted on its mounting bolt (14). To separate front axle from unit, block under frame with front wheels clear. Unbolt outer ends of tie rods (12) from arm sections of steering spindles (20); and after removing a retainer ring (24) with spacer (23), remove either of the front wheels (22). After wheel is removed, axle limiter (16) must be pulled off so that roll pin (19) can be driven from steering spindle (20). Spindle can now be lowered out of axle end for clean-up, evaluation and renewal if necessary. Take care not to lose flat washers (17). When an end of axle main member is so stripped of interfering parts, remove pivot bolt (14) and spacer (15) at center and withdraw axle from box section of frame. Complete dismantling, if necessary, by removal of remaining wheel and spindle from axle. If service is to be limited to axle and spindles, reverse this order to reassemble, then lubricate spindle at grease fitting (18) using a grease gun charged with #16020 Multi-purpose grease or a good grade of lithium-base automotive chassis grease.

NOTE: If additional work on steering system is planned, consider the following:

CAUTION: Frequently, the performance of maintenance, adjustment or repair operations on a riding mower is more convenient if mower is standing on end and these units are provided with a stand for this purpose. This procedure can be con-sidered a recommended practice providing the following safety recommendations are performed:

1. **Drain fuel tank or make certain that fuel level is low enough so that fuel will not drain out.**
2. **Close fuel shut-off valve if so equipped.**
3. **Remove battery on models so equipped.**
4. **Disconnect spark plug wire and tie out of way.**
5. **Although not absolutely essential, it is recommended that crankcase oil be drained to avoid flooding the combustion chamber with oil when engine is tilted.**
6. **Secure mower from tipping by lashing unit to a nearby post or overhead beam.**

To remove steering gear, carefully drive out roll pin (5–Fig. B1) and pull wheel from steering shaft (6). Unbolt inner ends of non-adjustable tie rods (12) from steering arm (11), then unbolt retainer (9) from column base (7) and parts (8, 9, 10 and 11) will come away as a unit. Set these aside, remove retainer (2) at top end of steering shaft (6) and pull shaft down and out of column (7). Do not lose thrust washers (3) and take care not to damage shaft bearing (4) during shaft

removal. Separate gear sector (8) from steering lever arm (11) by carefully driving out roll pin (10). Thoroughly clean all parts and carefully evaluate shaft bearing (4), pinion portion of shaft (6) and sector (8) for possible renewal. To reassemble, proceed in reverse order. Use oil to lubricate bearing (4), shaft (6) and gear sector (8).

ENGINE

All Models

For overhaul and repair procedures on engines listed in Specification Table refer to Large Air Cooled Engines Service Manual.

Refer to the following paragraphs for removal and installation procedures.

REMOVE AND REINSTALL. To remove engine, first remove spark plug lead for safety, then, reach under and release mower drive belt from mower spindle sheave. Remove idler arm tension spring or pull idler away from belt to relieve pressure. Slip belt off mower drive clutch sheave under engine. Disconnect lift chains (remove klik pins) from mower deck. Chains are relaxed when mower lift lever is set in lowest

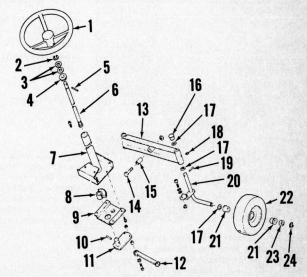

Fig. B1–Exploded view of front axle and steering system for Models 728, 828 and 829.

1. Steering wheel
2. Retainer
3. Thrust washer (2)
4. Shaft bearing
5. Roll pin
6. Shaft & pinion
7. Column & base
8. Gear sector
9. Retainer
10. Roll pin
11. Steering arm
12. Tie rod (2)
13. Front axle
14. Pivot bolt
15. Spacer
16. Axle limiter (2)
17. Flat washer
18. Grease fitting
19. Roll pin
20. Spindle (2)
21. Flange bearing (2)
22. Wheel & tire
23. Spacer
24. Retaining ring

cutting position. Remove four spring cotter pins (hairpin style), and pull out four clevis pins which attach mower lift linkages to main frame, then pull mowing unit out at either side.

For convenience, set mower chassis up on its rear stand, as previously outlined, relieve tension on main drive belt idler pulley and remove drive belt. Fit a pin punch or small bar into clutch extension shaft as in Fig. B2, then carefully back out clutch cap screw to remove mower drive clutch assembly. Clutch unit is spring-loaded (See MOWER CLUTCH section) and must be handled with care to prevent injury or loss of parts. To remove clutch extension,

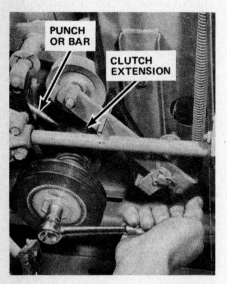

Fig. B2—View to show technique for removal of mower clutch-brake assembly. Unit is spring-loaded. Observe cautions in text.

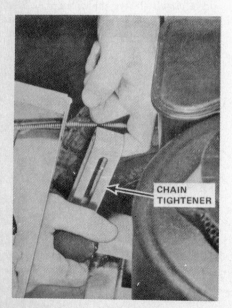

Fig. B3—Remove slack from drive chain, as shown, using hand pressure only. Some units may have a T-handle on retaining nut instead of knob.

back out **both** socket head set screws using a 7/32-inch Allen wrench. Screws are located in upper pulley hub directly over the keyway. An ordinary gear puller may be used to pull extension shaft and main drive pulley from engine crankshaft lower end, however, be sure to replace clutch cap screw engaging five or six threads so that puller will not damage shaft end. Lower mower chassis to rest on its wheels.

If not previously done on electric start models, raise operator's seat and disconnect ground cable from battery.

For engine access, on Model 728, raise operator's seat, disengage spring wire retainer from groove in seat pivot shaft, then shift seat from side to side and separate from its brackets. Engine cover can now be unlatched at rear and tilted forward on its hinges or hinge pins can be pulled for complete removal. On Models 828 and 829, remove operator's seat, then remove two wing latches inside center panel assembly so that engine cover/fender assembly can be lifted off. If desired, remove rear stand.

On all models, disconnect yellow-insulated ground wire from engine. This wire runs from seat safety switch. On electric start models, disconnect starter cable (red) from starter motor. Disconnect throttle control bowden cable at carburetor. Check for possible interference of other parts not mentioned here such as a special muffler kit, then unbolt and remove all engine base bolts so that engine can be lifted and set aside.

Reinstall engine by reversing this procedure.

TRACTION DRIVE CLUTCH AND BRAKE

All Models

ADJUSTMENT. Because of interaction of control linkages of main drive clutch, transmission disc brake and automatic disengagement of mower drive clutch/brake, adjustment of each element should be checked and performed, **when necessary,** in this sequence:

DRIVE BELT AND CHAIN. Drive belt tension is correct when belt runs deep enough to remain in drive pulley of engine crankshaft and in driven pulley on transmission input shaft if belt idler (clutch) is disengaged. If main drive belt is too loose on sheaves, loosen drive chain tensioner (Fig. B3) and transmission to chassis bolts and move transmission forward in slots to eliminate excessive slack in main drive belt. A correctly adjusted belt will remain in sheaves when tension on clutch idler is released and transmission pulley can rotate freely. If drive belt is worn or stretched to a point that such adjustment is not possible, renew belt. When belt is correctly adjusted, readjust chain tightener **by hand only** as shown in Fig. B3 to remove slack from drive chain. If chain is damaged or defective, unbolt forward end of chain guard and swing guard out of the way. Disconnect master link and remove chain for repair or renewal.

TRANSMISSION BRAKE. A properly adjusted disc brake should keep rear wheels from turning except when drive pedal is depressed one inch or more. When drive pedal is released during operation, there is controlled service braking for normal stopping. When BRAKE pedal is sharply applied, mower blade clutch will disengage, and drive wheels will lock up for emergency stop. Brake lock holds pedal down for parking.

NOTE: If drive belt tension has been adjusted by shifting transmission as covered under DRIVE BELT AND CHAIN, all linkages for transmission brake, drive pedal and brake pedal must be checked and readjusted as necessary.

Adjust transmission brake by these steps: Depress DRIVE pedal one inch or more so that brake is released and brake lever shown in Fig. B4 is vertical. Move stop against brake lever and lock, then turn adjusting nut in or out so that brake pads are just clear of brake disc. Now, allow DRIVE pedal to retract so that

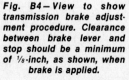

Fig. B4—View to show transmission brake adjustment procedure. Clearance between brake lever and stop should be a minimum of 1/8-inch, as shown, when brake is applied.

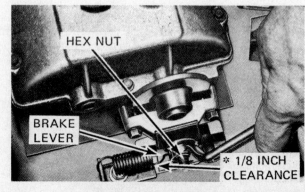

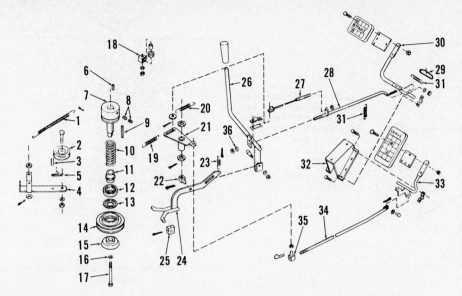

Fig. B5 — Exploded view of clutch, brake and drive control linkages. Refer to text for adjustment sequence and procedure.

1. Extension spring
2. Idler pulley
3. Drive pin
4. Clutch idler arm
5. Belt guide
6. Crankshaft key (¼X1)
7. Pulley and shaft
8. Set screws (5/16-18)
9. Key (3/16X¼X2¼)
10. Clutch spring
11. Retainer
12. Ball bearing
13. Slinger
14. Clutch sheave
15. Clutch cone
16. Pyramid washer
17. Cap screw (⅜X4½-24)
18. Cable guide assy.
19. Clutch spring
20. Brake spring
21. Cam
22. Pivot stud
23. Extension spring
24. Clutch-brake lever
25. Brake block (2)
26. Blade control lever
27. Brake cable
28. Brake rod
29. Link
30. Brake lever
31. Extension spring (2)
32. Linkage support
33. Clutch-drive lever
34. Clutch rod
35. Adjustment block
36. Pivot washer

brake is applied, and reset stop for minimum ⅛-inch clearance from brake lever. If brake does not just release when DRIVE pedal is depressed one inch (brake disc should rotate freely), then recheck adjustment. Brake spring (20–Fig. B5) should be hooked into whichever hole in brake lever will hold drive pin (3) of clutch idler arm (4) 5/16-inch clear of rear axle. Remove clevis pin at forward end of clutch rod (34) while making this adjustment.

DRIVE PEDAL. Adjustment of drive pedal should always coincide with transmission brake adjustment. Proceed as follows:

When transmission brake adjustment has been checked or performed, remove clevis pin at forward end of drive pedal (clutch) rod (34). Now, rotate rod (34) in threaded adjustment block (35) as necessary for clevis pin to be slipped back easily to reconnect with lever (33). Be sure that 5/16-inch clearance between pin (3) and rear axle is not affected and there is no interference or binding of parts. Damaged or defective springs, keepers or pins should be renewed if performance is unsatisfactory or adjustments cannot be made.

BRAKE PEDAL. As previously indicated, service braking, for routine slowing or stopping of mower is a function of DRIVE pedal. BRAKE pedal is

applied for emergency stops. When properly adjusted, its action will lock mower drive wheels, disengage mower drive clutch and halt blade rotation simultaneously.

When correctly adjusted, BRAKE pedal should lock rear wheels when depressed half way. Adjust by use of locknut on brake cable (27–Fig. B5). To do so, tighten nut against welded tab on brake rod (28) until brake engages solidly with pedal pushed down half of complete stroke. Be sure not to over-tighten. If brake cable is drawn up too tight, transmission brake will not be allowed to release. When brake cable adjustment is satisfactorily completed, press pedal down hard and engage brake lock. With mower blade clutch disengaged (handle to rear in "OFF" position) pivot washer (36) on brake rod (28) should contact welded tab on blade control lever (26). Turn locknut behind pivot washer (36) to adjust if necessary. Adjustment of mowing unit clutch-brake is covered in MOWER CLUTCH section which follows.

REAR AXLE AND DIFFERENTIAL

All Models

REMOVE AND REINSTALL. Rear axle, differential and wheels are removed together. Easiest procedure is to

set mower chassis up on its rear stand as previously outlined. With unit so set up, unbolt and remove chain guard to uncover axle sprocket and remove master link and drive chain. Remove cotter pin from pivot (22–Fig. B5) and shift clutch-brake lever (24) out of the way. Remove two bolts from each of the axle bearing blocks and lift axle assembly complete from frame. Wheel hubs are cross-bolted to outer ends of axle shafts. When wheels are removed, spacers, thrust washers and bearing blocks can be slipped off axle ends. Four cap screws are used to hold chain sprocket to differential.

Reverse procedure to reinstall after renewing all parts which are damaged or defective.

OVERHAUL. The differentials used on all models are manufactured by Indus Wheel (Mast Foos). For differential overhaul procedures refer to the DIFFERENTIAL REPAIR section of this manual.

TRANSMISSION

All Models

REMOVE AND REINSTALL. To remove transmission, remove mower deck and set unit up on rear stand as previously outlined. Remove main drive belt, then remove snap ring from transmission input shaft. Using a properly fitted puller, remove transmission pulley. Take care not to lose Woodruff key. Remove chain guard, relieve chain tension, then pull master link and remove drive chain. Disconnect brake spring from disc brake lever. Loosen mounting screws. Lower unit to rest on its wheels; then raise seat, back out cap screw and remove shift lever at right side of chassis. Reach under frame and back out four pre-loosened flanged cap screws then lift out transmission.

Reinstall transmission by reversal of removal sequence.

OVERHAUL. Peerless Model 503 transmissions are used on all models. For transmission overhaul procedure, refer to the TRANSMISSION REPAIR section of this manual.

MOWER CLUTCH

All Models

ADJUSTMENT. For correct operation of mower clutch-brake, adjustment specifications shown in Fig. B6 should be used. Follow these steps:

Place control lever in rear "OFF" position. Remove cotter pin (4–Fig. B6) from adjuster (3) and slip lever off pivot

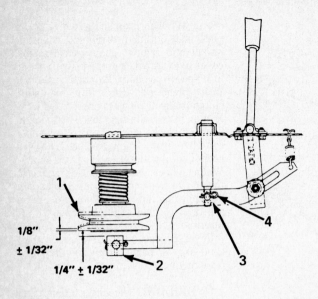

Fig. B6 — View of adjustment points for mowing unit clutch-brake. Note specifications. Refer to text for procedure.

1. Sheave
2. Brake block
3. Adjuster block
4. Hairpin cotter

1/8"
± 1/32"

1/4" ± 1/32"

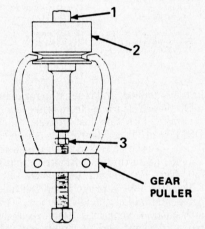

Fig. B7 — Sectional view of mower clutch-brake. See text.

1. Engine crankshaft	7. Clutch cap screw
2. Crankshaft key	8. Pyramid washer
3. Clutch key	9. Clutch facing
4. Retainer	10. Slinger
5. Sheave	11. Ball bearing
6. Cone clutch	12. Clutch spring

Fig. B8 — Use puller as shown to remove main drive pulley and extension shaft from engine crankshaft. See text.

1. Crankshaft
2. Pulley hub
3. Clutch cap screw

stud. Turn adjuster block in or out on threaded mounting stud to set ¼-inch clearance between brake blocks (2) and friction surface of sheave (1). Move lever forward to "ON" position and measure upward movement of sheave when lifted by brake blocks. This measurement should be ⅛-inch, plus or minus 1/32-inch. If not, cone clutch inside sheave will not disengage. If adjustment cannot be made by raising or lowering adjusting block (3), remove cotter pins from brake blocks (2) and rotate each block one-quarter turn to provide a new, unworn contact surface. If these procedures do not succeed, it may be necessary to dismantle and overhaul clutch-brake assembly. See following section.

NOTE: Brake blocks should track on outer margin of pulley friction surface. If they are misaligned, loosen control handle mounting bolts and shift assembly from side to side to gain correct alignment.

R&R AND OVERHAUL. If mower clutch-brake is worn, damaged or defective, remove as follows:

Remove mowing unit drive belt and disconnect mower lift chains, then remove four clevis pins which attach mower lift linkage to frame and remove mower deck from under unit. Set chassis up on its rear stand as previously outlined. Refer to Fig. B2, insert a pin punch or bar into clutch extension shaft and remove clutch retainer cap screw as shown.

CAUTION: Clutch spring (12 – Fig. B7) is under compression, and when threads of cap screw (7) are released from engine crankshaft (1) a pad or cushion should be used to prevent accidental injury.

Parts of assembly are shown in exploded view in Fig. B5 with sectional view in Fig. B7 to show order of parts

arrangement. All parts should be thoroughly cleaned in non-petroleum base solvent and carefully evaluated for wear or damage. Renew parts as needed. Cone clutch (6 – Fig. B7 or 15 – Fig. B5) may be renewed completely or new clutch facing (9) can be installed. Follow this procedure: Remove compression spring (12) and retainer (4). Carefully pull or press bearing (11) from hub of sheave (5) so as not to damage slinger (10). Place adhesive side of new clutch facing (9) on well-cleaned surface of clutch cone (6), then assemble cone into cup portion of sheave (5) holding parts firmly together with a bolt and nut arrangement and large fitted washers. Place clutch in an oven, which has been preheated to 400°F, for 20 minutes.

When reassembling cone clutch, keep these points in mind: Retainer (4) must slip freely on shaft extension. Renew if doubtful. Key (3) which locks clutch cone (6) to extension shaft must not ride up ramp in keyway during installation. Renew pyramid washer (8) if questionable. Torque cap screw (7) to 15 ft.-lbs. Reassemble and install clutch-brake assembly in reverse of order used to dismantle.

NOTE: Extension shaft and main drive pulley may need to be removed from engine crankshaft due to keyway or pulley damage or defect. To do so, use a conventional gear puller, being sure to reinstall clutch cap screw (3) by five or six threads for puller screw to bear upon as shown in Fig. B8. Do not overlook stacked set screws (8 – Fig. B5) installed in pulley hub above keyway.

MOWER BLADE SPINDLE

All Models

R&R AND OVERHAUL. When looseness, noise, binding or unsatisfac-

tory operation indicate trouble in spindle of cutting deck, proceed as follows:

With mower deck removed from under chassis, block mower blade from turning by use of a scrap of lumber, remove grease fitting (15 – Fig. B9), then back off locknut (1). Using an Allen wrench, back out set screw in pulley hub over Woodruff key and remove sheave (14). Remove Woodruff key (2) and top spacer (3) then pull spindle shaft (8) down and out of housing (12). If desired, separate housing from mower deck (4) by removal of flange screws (13), though bearings (5) can be tapped out of housing without removing from deck. Note that bearing spacer (11) prevents use of a puller for bearing removal. After cleanup and determination of which parts will

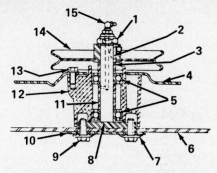

Fig. B9—Sectional view of mower spindle and bearing assembly.

1. Locknut
2. Woodruff key
3. Top spacer
4. Mower deck
5. Ball bearings
6. Blade
7. Pyramid washer
8. Spindle
9. Cap screw
10. Plate stiffener
11. Bearing spacer
12. Spindle housing
13. Flange set screw
14. Sheave
15. Grease fitting

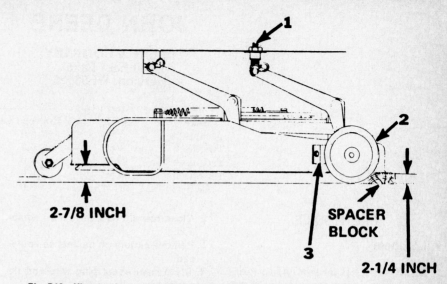

Fig. B10— View to show mower deck leveling technique and measurement points. See text.

1. Leveling stud & nut   2. Gage wheel   3. Gage wheel bracket

require renewal, reassemble and reinstall in reverse order. Press bearings into housing with shielded sides outward, and torque locknut (1) to 90 ft.-lbs. Use a hand grease gun with #16020 Multipurpose grease to lubricate spindle before return to service.

MOWER DECK LEVELING

All Models

ADJUSTMENT. Park unit on a good, level surface such as a garage floor and set spacers 2¼ inches thick under rear mower skirt as shown in Fig. B10. Check condition of leveling linkages and pins and be sure tires are inflated evenly to 20 psi. Move lift lever down to its lowest position and loosen nuts of gage wheel brackets (3). Use leveling nuts and studs (1) to adjust height of forward edge of cutting deck to 2⅞ inches above floor surface as shown. Install square stud of gage wheel axle in middle hole of gage wheel mounting brackets, and with gage wheels in contact with floor, tighten both gage wheel brackets. Pull out spacers from under rear edge of mower deck and set lift handle in fourth notch from bottom. Raise mower seat and adjust lift chain wing nuts to remove slack from chain. When mower cutting height is changed, square stud on gage wheel axle should always be reset in proper hole (1 thru 5) of gage wheel bracket to correspond with whichever of five lever positions is chosen.

COLUMBIA

Model	Make	Engine Model	Horsepower	Cutting Width, In.
380	B&S	130000	5	25
400	B&S	130000	5	25
455	B&S	190000	8	34
480	B&S	190000	8	34
485	B&S	190000	8	34
495	B&S	190000	8	38
497	B&S	230000	10	38
525	B&S	190000	8	26

For service information and procedures, use the following model cross reference and refer to the MTD section of this manual.

Columbia Models	MTD Models
380	380
400	400
455	445
480	480
485	485
495	495
497	497
525	525

JOHN DEERE

DEERE & COMPANY
220 East Lake
Horicon, WI 53032

Model	Engine			Cutting Width, In.
	Make	Model	Horsepower	
55	Tecumseh	V50	5.0	26
56	Tecumseh	V60	6.0	28
57	Tecumseh	V70	7.0	34
66	Tecumseh	V60	6.0	30
68	B&S	191702	8.0	30/34

FRONT AXLE

All Models

REMOVE AND REINSTALL. Axle main member is center-pivoted in cross channel at forward end of chassis frame. Mowing unit is fitted to a clevis or to a draft rod (11 – Fig. JD2) which is part of axle assembly. If removal of steering gear or front axle is planned, mower deck should first be separated from chassis. Removal of under-chassis items may be done with all wheels on the ground, however, these mowers are provided with a rear-mounted stand (optional on Models 66 and 68) designed for placing mower upright on its rear end. If this is done, observe the following:

CAUTION: Frequently, the performance of maintenance, adjustment or repair operations on a riding mower is more convenient if mower is standing on end. This procedure can be considered a recommended practice providing the following safety recommendations are performed:

1. Drain fuel tank or make certain that fuel level is low enough so that fuel will not drain out.

2. Close fuel shut-off valve if so equipped.

3. Remove battery on models so equipped.

4. Disconnect spark plug wire and tie out of way.

5. Although not absolutely essential, it it recommended that crankcase oil be drained to avoid flooding the combustion chamber with oil when engine is tilted.

6. Secure mower from tipping by lashing unit to a nearby post or overhead beam.

If Models 66 and 68 are not equipped with a stand, cradle blocks made of scrap lumber can be fitted under rear wheels so that each tire is lifted about five inches clear of floor.

Remove mower unit as outlined in MOWER DECK paragraphs. Set machine up on rear stand as previously outlined.

Before disassembly of steering gear parts, check carefully for looseness, damage or wear through entire system with special attention to condition of non-adjustable tie rods (5 – Fig. JD1) or gear sector (9 – Fig. JD2) and steering shaft pinion.

To remove front axle, first remove wheels for weight reduction, then disconnect tie rods from steering spindles and remove pivot bolt from center of frame channel. With axle assembly separated from frame, remove snap rings (6 – Fig. JD1 or 18 – Fig. JD2) and remove spindles from axle ends.

STEERING GEAR

Model 55

REMOVE AND REINSTALL. To remove steering gear from Model 55, disconnect tie rods (5 – Fig. JD1) from steering arm, carefully drive out pin (4) at hub of handlebar (1), pull handlebar off steering shaft (3) and lower steering shaft out of control console.

All Other Models

REMOVE AND REINSTALL. To remove steering gear from other models, disconnect tie rods (10 – Fig. JD2) and remove gear sector (9) at lower end of steering shaft. Carefully drive out groove pin (5) from steering wheel hub,

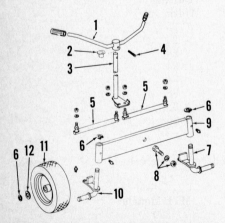

Fig. JD1 – Exploded view of front axle and steering gear parts for Model 55.

1. Handlebar	7. Spindle, LH
2. Plastic bushing	8. Axle pivot bolt
3. Steering shaft &	(½x1½)
lever	9. Axle assy.
4. Drive pin	10. Spindle RH
5. Tie rod (2)	11. Wheel & tire
6. Snap ring (4)	12. Spindle washer

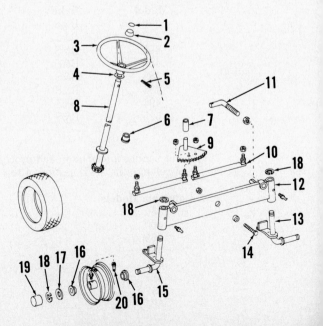

Fig. JD2 – Exploded view of typical steering gear and front axle parts for Models 56, 57, 66 and 68. Model 68 shown.

1. Logo
2. Cap
3. Wheel
4. Shim washer
5. Groove pin
6. Plastic bushing
7. Bronze bushing
8. Steering shaft & pinion
9. Gear sector
10. Tie rod
11. Draft rod
12. Axle
13. Spindle, LH
14. Pivot bolt
15. Spindle, RH
16. Wheel bearing (4)
17. Special washer (2)
18. Snap ring (4)
19. Cap
20. Valve stem

remove steering wheel and note number of shim washers (4) used to take up vertical play of steering shaft. Pull steering shaft (3) down and out of control console taking care not to damage bushing (6). Reassemble in reverse order. Lubricate at fittings using John Deere Multipurpose Lubricant No. TY2098.

ENGINE

All Models

For overhaul and repair procedures on engines listed in Specification Table, refer to Small Air Cooled Engines Service Manual.

Refer to following paragraphs for removal and installation procedure.

REMOVE AND REINSTALL. To remove engine, tip unit up to rest on its rear stand as previously outlined, and remove mowing deck and mower drive belt after first unbolting hitch and belt guide on Models 55, 56 and 57 or belt guide only on Models 66 and 68. Back out cap screw from end of engine crankshaft and use a puller to remove mower drive and traction drive pulleys. While chassis is up-ended, it is advisable to partially loosen seat pedestal cap screws (Models 55, 56 and 57) and engine mounting bolts from under side for convenience in later removal. Lower unit to rest on all wheels.

On Model 55, removal of seat pedestal to expose engine is optional. On Models 56 and 57, engine shroud and seat assembly are hinged to pedestal and may be tilted forward out of way. Access to engine is improved by complete removal of pedestal. Models 66 and 68 require only that seat assembly and shroud be tipped forward to uncover engine enclosure housing. Remove fuel tank vent tube at carburetor end, slide engine enclosure box forward to disengage latch pins at rear and lift front of enclosure simultaneously to expose engine.

On models equipped for electric starting, disconnect cable from starter terminal post. Disconnect key switch wire and safety interlock wire at engine end. Disconnect throttle control bowden cable at carburetor and on Models 66 and 68 disconnect fuel line at inlet fitting on carburetor. Remove mounting bolts and lift engine from frame.

Reverse removal sequence to reinstall engine.

NOTE: On Models 66 and 68, to prevent damage to muffler during engine removal, it may be advisable to back out manifold cap screws at valve cage in cylinder block so that muffler and exhaust pipe can remain in place.

TRACTION DRIVE CLUTCH AND DRIVE BELT

All Models

ADJUSTMENT. The traction drive clutch consists of a spring-loaded idler with no special adjustment feature. Figs. JD3 through JD6 show linkages and parts arrangement of traction controls.

If performance is marginal or inadequate, with slippage an apparent problem, inspection of clutch idler, belt and linkage will probably disclose that belt is damaged, stretched or glazed, idler pulley or its bearing may be worn or defective, driving pulley (engine) or drive pulley (transmission) may be damaged or loose on shafts, or linkage rods, springs or cotters may be broken, bent, missing or disconnected. Refer to appropriate figures for parts identification and placement in reassembly.

Models 55-56-57

REMOVE AND REINSTALL. To change drive belt, remove mower deck as outlined in MOWER DECK section. Set unit up on its rear stand as previously outlined. Mower drive belt and rear hitch will have been removed as a part of mowing deck removal task, which will leave underside open for full access to traction drive system.

Proceed as follows: On older units with a common pedal for drive clutch-brake, depress pedal and set brake lock. With newer production models, which have a separate clutch pedal, depress clutch pedal and wedge a piece of wood

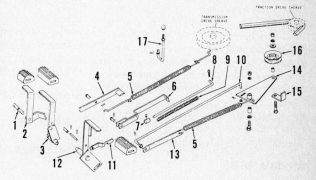

Fig. JD3—Exploded view of traction clutch-brake and mower clutch linkages used on early Model 55 (to Ser. No. 20950) and on early Model 56 (to Ser. No. 21750).

1. Clutch shaft
2. Mower drive throw-out pedal
3. Mower drive clutch pedal
4. Latch lever
5. Tension spring (2)
6. Clutch arm
7. Locking collar
8. Mower clutch rod
9. Mower clutch spring
10. Traction clutch rod
11. Brake lock
12. Clutch-brake pedal
13. Link
14. Traction clutch arm
15. Belt guide
16. Traction clutch idler
17. Brake rod assy.

Fig. JD4—Exploded view of mower and traction control linkages used on Model 55 (Ser. No. 20951 and after), Model 56 (to Ser. No. 44000) and Model 57 (to Ser. No. 14000). Refer to Fig. JD3 for parts identification legend.

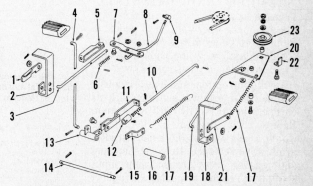

Fig. JD5—Exploded view of mower and traction control linkages as used on Model 56 (Ser. No. 44001 and after) and on Model 57 (Ser. No. 14001 and after).

1. Brake lock
2. Brake pedal
3. Brake rod (front)
4. Mower clutch rod
5. Retainer
6. Spring
7. Brake link
8. Brake rod (rear)
9. Brake adjuster
10. Mower clutch rod
11. Mower clutch arm
12. Locking collar
13. Throwout pivot
14. Pedal shaft
15. Spring link
16. Spacer
17. Clutch spring (2)
18. Clutch pedal
19. Clutch rod
20. Traction clutch arm
21. Spring link
22. Belt guide
23. Traction clutch idler

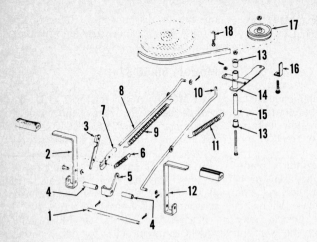

Fig. JD6 — Exploded view of traction clutch and brake linkages used on Models 66 and 68.

1. Shaft
2. Brake pedal
3. Brake lock
4. Spacer (2)
5. Throwout pivot
6. Return spring
7. Brake tube
8. Brake rod
9. Brake spring
10. Clutch rod
11. Clutch spring
12. Clutch pedal
13. Nylon bushing (2)
14. Idler arm
15. Bushing
16. Belt guide
17. Clutch idler pulley
18. Belt guide

scrap between clutch arm and frame to hold clutch disengaged. Next, remove idler pulley from clutch arm. Belt will be completely loose, but it will be necessary to bend belt guides away from transmission pulley so belt can be lifted out of pulley. Slip belt to rear and between axle and driving pulley for removal. Proceed in reverse order to reinstall, and be sure to bend belt guides back toward transmission pulley, leaving ¼-inch clearance. After mower is reinstalled, adjust belt guides on hitch plate for 1/16 to ⅛-inch clearance from driving pulley.

Models 66-68

REMOVE AND REINSTALL. Remove mower deck as outlined in MOWER DECK section. Set unit up on its rear stand as previously outlined. Unbolt and remove frame reinforcement located next to transmission pulley, then remove pulley belt guide. Remove idler pulley and slip belt out between pulleys and rear axle. Reverse procedure to install new belt. Adjust belt guide to ⅛-inch clearance between guide and belt.

TRANSMISSION

Models 55-56-57

REMOVE AND REINSTALL. Transmissions on these models are located under operator's seat pedestal. Removal of pedestal is a matter of convenience and is a mechanic's option. Other removal steps are as follows:

Tip unit up on rear stand as previously outlined and remove mowing deck as outlined in MOWER DECK section. Relieve belt tension by depressing clutch. Use a wooden block against idler arm for convenience. Remove snap ring and washer from transmission input shaft and remove transmission pulley. Do not lose key from shaft or pulley hub. Disconnect return spring for disc brake and remove brake rod from brake lever.

Disconnect leads from neutral-start safety switch when so equipped. Loosen drive chain tensioner and remove master link and drive chain. Back out the four mounting cap screws and remove transmission. Reverse this procedure to reinstall and reset tension on drive chain as outlined in DRIVE CHAIN paragraphs.

Models 66-68

REMOVE AND REINSTALL. To remove transmission, first remove mower deck as outlined in MOWER

DECK section. Now, block idler arm in released position to relieve belt tension and remove snap ring, washer, input pulley and key from transmission input shaft. Slightly loosen each of the transmission mounting cap screws working from underside of unit, then set mower down on all four wheels. Raise and unbolt engine shroud at hinge pivots and remove shroud, then disconnect fuel line and remove fuel tank and its mounting bracket by unbolting from frame. With brake pedal relaxed, remove cotter pin and disconnect brake rod from lever of transmission disc brake. Disconnect wires from neutral-start safety switch. Raise and block under frame at right rear so wheel is clear, then remove wheel. Remove cotter pin and collar from inner (small diameter) axle shaft, then pull axle by its flange out of chain case. Note that inner end of axle will be disengaged from differential side gear inside chain case. See Fig. JD8. Wrap this end with a paper shop towel to keep it clean for reassembly. At upper end of chain case, disconnect lift arm spring and carefully remove groove pin which attaches hub of lift arm (bellcrank) to cross shaft, then pull arm off shaft end. Now, unbolt and remove outer cover, taking care not to damage gasket if re-use is planned. When chain case is thus

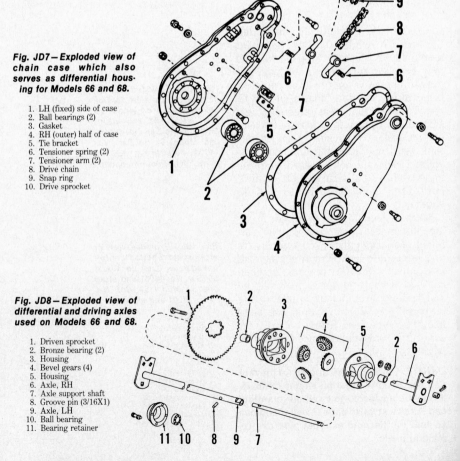

Fig. JD7 — Exploded view of chain case which also serves as differential housing for Models 66 and 68.

1. LH (fixed) side of case
2. Ball bearings (2)
3. Gasket
4. RH (outer) half of case
5. Tie bracket
6. Tensioner spring (2)
7. Tensioner arm (2)
8. Drive chain
9. Snap ring
10. Drive sprocket

Fig. JD8 — Exploded view of differential and driving axles used on Models 66 and 68.

1. Driven sprocket
2. Bronze bearing (2)
3. Housing
4. Bevel gears (4)
5. Housing
6. Axle, RH
7. Axle support shaft
8. Groove pin (3/16X1)
9. Axle, LH
10. Ball bearing
11. Bearing retainer

opened, remove snap ring from transmission input shaft and remove input sprocket and Woodruff key. Remove transmission mounting cap screws then move transmission inward so shaft end clears chain case half and lift out of mower.

NOTE: Take care not to lose or damage nylon bushing which supports lift shaft in flanges of chain case.

Reverse this procedure to reinstall transmission.

All Models

OVERHAUL. For transmission overhaul service, see TRANSMISSION REPAIR section in this manual. Transmissions used in these models are identified in the following table:

Riding Mower Model	Foote Transmission Model
55, 56, 57 (Two-speed)	2240-2 (Deere No. AM32670)
56, 57 (Three-speed)	2500-6 (Deere No. AM34464)
66, 68 (Five-speed)	2600-2 (Deere No. AM35523)

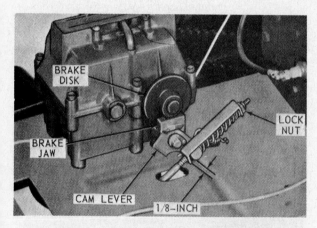

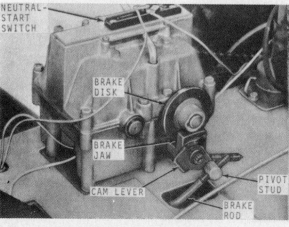

DRIVE CHAIN

Models 55-56-57

R&R AND ADJUSTMENT. Adjustment of chain tension is performed with seat and shroud assembly raised. There is no need for mower removal or other disassembly. Proceed as follows: Loosen axle bolt for idler sprocket and move idler bolt in its slot under transmission platform. Tension is correct when chain can be flexed ½-inch at a mid-point between drive and driven sprocket in top strand. If chain is defective, loosen adjusting sprocket and pull master link. Chain can then be removed for cleaning, lubrication, repair or renewal.

Models 66-68

R&R AND ADJUSTMENT. These models are without external adjustment means for drive chain. If drive chain becomes noisy due to looseness within chain case, it is likely that spring-loaded automatic chain tensioner assembly (see Fig. JD7) is defective. Refer to DIFFERENTIAL AND REAR AXLE paragraphs.

Fig. JD9 — View of brake adjustment point for transmission disc brake on Foote two-speed model. Refer to text.

Fig. JD10 — View of disc brake used on three-speed Foote transmission for Models 56 and 57 in 1973-74. See text for adjustment procedure.

DIFFERENTIAL AND REAR AXLE

Model 55

REMOVE AND REINSTALL. Set mower up on rear stand as previously outlined, then release drive chain tensioner, disconnect master link and remove drive chain. Unbolt bearing blocks at each end of rear axle and withdraw entire rear axle, differential and rear wheels as a unit.

Models 56-57

NOTE: Some earlier production Models 56 and 57 (pre-1972) were equipped with same rear axle bearing block and mounting as used on Model 55.

REMOVE AND REINSTALL. Remove mower deck as outlined in MOWER DECK section. Set mower up on its rear stand as previously outlined. Remove cap screw from engine crankshaft which retains mower drive pulley. Use a suitable puller to remove pulley from shaft.

Remove rear wheels. In earlier production, wheel hubs are directly bolted by through-bolt to axle shaft. Later, wheel is attached by lug bolts to a flanged hub which is cross-bolted to end of axle. Whichever design is used, strip all bushings, spacers, keys and washers from outer end of axle and unbolt axle bearing retainer from frame. Release tension on drive chain, then remove master link and lift chain off sprockets. Shift axle assembly from side to side to clear frame and lift out. Reverse these steps to reinstall.

Models 66-68

REMOVE AND REINSTALL. Remove mower deck as outlined in MOWER DECK paragraphs, and set unit on rear stand as previously outlined.

NOTE: If service to differential only is planned, remove only right rear wheel. If entire axle is to be serviced remove both rear wheels.

Mower lift linkage must be removed from lift shaft which passes through chain case flange. Disconnect lift arm spring, remove groove pin which attaches lift arm hub to shaft and pull arm from shaft. Remove cotter pin and retainer collar from axle support shaft (7 – Fig. JD8) and pull short right hand axle (6) out by its flanged end. Unbolt outer case half (4 – Fig. JD7) and carefully separate from inner half of case (1). Remove chain tensioners (7) with springs (6) so chain will relax, then remove snap ring (9), transmission output shaft sprocket (10) and drive chain.

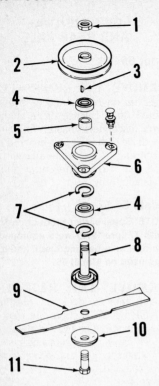

Fig. JD11—Exploded view to show parts arrangement of typical spindle assembly. Spindle for Model 68, 34 inch mower deck, is shown.

1. Locknut, ¾-inch
2. Sheave
3. Woodruff key (3/16x¾)
4. Bearing
5. Spacer
6. Hub flange
7. Snap ring (2)
8. Spindle
9. Blade
10. Drive washer
11. Cap screw (7/16X1)

At this point, entire assembly, driven sprocket (1–Fig. JD8) and differential (3, 4 and 5) can be slipped from left hand axle shaft (9) for further disassembly and overhaul as needed. Ball bearing sets (2–Fig. JD7) should be checked and evaluated for possible renewal. It is unlikely that inner half of chain case (1) will need to be unbolted from frame; procedure is apparent if this becomes necessary. Left side axle shaft (9–Fig. JD8) may readily be pulled out for inspection and evaluation of groove pin (8) and support shaft (7). Unbolt and remove bearing retainer (11) for access to left side axle bearing (10). Reinstall all parts in reverse of this sequence.

OVERHAUL. For overhaul service, see DIFFERENTIAL REPAIR section in this manual. Models 55, 56 and 57 may be equipped with Peerless or Mast-Foos differentials. Models 66 and 68 use a John Deere differential.

GROUND DRIVE BRAKE

All Models

Disc brake which serves both as a service brake and as a parking brake when locked in applied position is cam-actuated with its disc (rotor) assembly keyed to inboard end of transmission output shaft.

ADJUSTMENT. Design of brake mechanisms varies somewhat dependent upon transmission model used. Brake shown in Fig. JD9 is found on two-speed transmissions on Models 55, 56 and 57. To adjust, set parking brake lock, tilt seat forward on pedestal and take up locknut on end of brake rod to set ⅛-inch gap as shown between spring retainer strap and threaded adjuster stud. If adjustment is unsatisfactory, check condition of brake pads and operating linkage, renewing parts which prove defective.

Brake shown in Fig. JD10 is used on mower Models 56 and 57 with three-speed transmissions in 1973 production and after. To adjust, tilt engine shroud and seat forward for access to transmission and brake. On these models, spring tension on brake rod must be maintained so that cam lever just engages cam ramp on brake jaw, without causing brake pads to drag on disc, but providing good response to brake pedal. Pivot stud, threaded on brake rod, rotates inward to tighten, outward to release tension. If adjustment is ineffective, check condition of brake pads and linkage and renew damaged or defective parts.

On Models 66 and 68 which have a five-speed transmission, a set screw adjustment in brake jaw is provided to take up wear on brake pads. Tilt engine shroud and seat forward, then use a ⅛-inch Allen wrench on No. 8 socket head set screw to set clearance between brake pad and rotor disc at 0.022 inch working through opening in fuel tank support with brake pedal in relaxed position. Pads (friction pucks) and all brake parts are renewable, and should be checked for condition whenever adjustment is needed or when service to transmission is being performed.

All Models

R&R AND OVERHAUL. Tilt seat and engine shroud forward for access to transmission and brake. Disconnect brake rod from cam lever. Remove shoulder bolt and withdraw brake jaw with outer pad. Slide brake disc from transmission shaft and remove inner brake pad from transmission case. Clean and inspect parts for wear or other damage. Renew parts as required and reassemble by reversing removal procedure. Brake disc must slide freely on transmission shaft. Torque shoulder bolt to 275 in.-lbs. Adjust brake as outlined in previous paragraphs.

MOWER DRIVE BELTS

Models 55-56

REMOVE AND REINSTALL. Disconnect spark plug wire and remove mower as outlined in MOWER DECK paragraphs. Loosen and move belt guides as needed. Remove idler arm retaining pin and lift idler off pivot post.

Fig. JD12—View of mower deck level adjustment points typical for Models 55, 56 and 57. See text.

1. Draft arm adjuster bolt
2. Transfer rod

Fig. JD13—View of mower deck leveling points on Models 66 and 68. Refer to text for procedure.

A. Blade tip
B. Block for measurement
C. Lift link adjuster nuts

Unbolt and remove idler pulley from arm and remove belt. Renew drive belt in reverse order of removal. Adjust belt guides to ⅛-inch clearance from pulley.

NOTE: Be sure belt is installed with back side against idler pulley and idler is centered with blade pulley.

Model 57

REMOVE AND REINSTALL. To remove mower drive belt, disconnect spark plug wire and remove mower as outlined in MOWER DECK paragraphs. Remove belt shields from mower deck. Loosen idler pulley belt guide and slip belt out of pulley and remove from mower. Renew belt by reversing removal procedure.

Mower deck belt can be removed without removing drive belt. Remove belt shields from mower deck. Remove cotter pin from idler arm stud and slide idler arms upward on stud. Flex idler arm inward and slip belt from idler pulley and remove from mower. Renew belt by reversing removal procedure.

Models 66-68 With 30 Inch Mower

REMOVE AND REINSTALL. Disconnect spark plug wire and remove mower as outlined in MOWER DECK paragraphs. Loosen belt guides and remove idler pulleys. Remove belt from mower. Renew belt by reversing removal procedure.

Model 68 With 34 Inch Mower

REMOVE AND REINSTALL. Disconnect spark plug wire and remove mower as outlined in MOWER DECK paragraphs. Remove shields from mower deck. Remove bolt and spacer from idler arms and remove belt from pulleys. Renew belt by reversing removal procedure.

MOWER SPINDLE

All Models

REMOVE AND REINSTALL. To remove blade spindle from mower deck, first separate mower from chassis as outlined in MOWER DECK paragraphs. Fit a scrap of wood under mower deck to prevent blade from turning, then remove locknut (1 – Fig. JD11) at top of drive spindle. Use a carefully fitted puller and remove pulley (2).

NOTE: Sheaves are less likely to be crushed or distorted if a three-jaw gear puller is used. Remove Woodruff key (3) and pull spindle and blade down and out of hub/flange (6). Disassemble blade from spindle. Ball bearing (4) may be pulled or driven from hub/flange (6). However, note

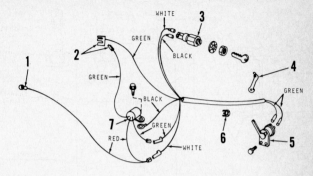

Fig. JD14 – Pictorial view of ignition and safety start circuits for Models 66 and 68 with manual start.

1. Engine terminal
2. Transmission terminal
3. Ignition switch/lock
4. Harness retainer (2)
5. Neutral-start switch
6. Wire loom bushing
7. Ignition interlock

that snap ring (7) at bottom must be removed before lower bearing can be pulled or tapped from bearing bore.

IMPORTANT: Do not force, and use heat sparingly if its use becomes necessary.

Thoroughly clean all parts and renew any showing excessive wear or other damage. If hub flange has been removed from deck, torque bolts to 20-25 ft.-lbs. Reassemble spindle and torque locknut (1) to 60-70 ft.-lbs. Torque cap screw (11) to 80 ft.-lbs.

MOWER DECK

Models 55-56-57

LEVEL ADJUSTMENT. Unit should be parked on a smooth, level surface such as a driveway or concrete garage floor. Tire pressures should be adjusted to specification shown in Operator's Manual for each particular model. Remove spark plug wire and disengage mower clutch. Set mower height control in lowest position, then measure height of blade tip above floor at four points, 90 degrees apart, by rotating blade a quarter-turn for each measurement. All

measurements should be equal within ⅛-inch. Check of opposite end of blade will show if blade is bent, how much, and direction of bend. Damaged blade should be repaired or renewed.

For side-to-side leveling, loosen bolt (1 – Fig. JD12) in right front draft link and tilt mower deck up or down laterally until blade is level; retighten adjuster.

To level deck from front-to-rear, remove cotter pin (2) at forward end of rod and rotate threaded end of rod inward to raise rear of mower deck or out to lower. Recheck level, reset cotter pin.

Models 66-68

LEVEL ADJUSTMENT. Check tires for proper inflation and park mower on level, smooth surface. Remove spark plug lead and disengage mower clutch. Lower mower height control lever, then rotate blade(s) by hand to point from side-to-side as shown in Fig. JD13. Then, measure height of each blade tip above surface. Make side-to-side adjustment by turning adjusting nuts (C – Fig. JD13) to raise or lower lift link on each side to equalize blade height.

To make front-to-rear adjustment, rotate mower blade to point straight ahead. Blade tip at rear should be

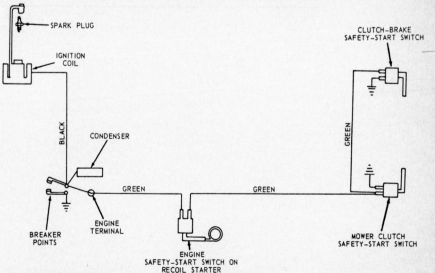

Fig. JD15 – Typical schematic for ignition and safety-start wiring for recoil-start Models 55, 56 and 57 in pre-1972 production.

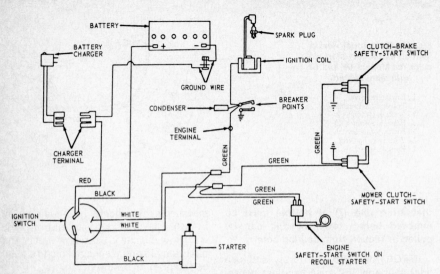

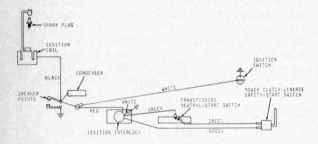

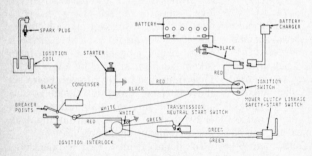

Fig. JD16 — Typical schematic for ignition and safety-start systems for electric-start Models 56 and 57 in pre-1972 production.

Fig. JD17 — Typical schematic for ignition and safety-start wiring for recoil-start Models 56 (Ser. No. 44,001 and after) and 57 (Ser. No. 14,001 and after).

Fig. JD18 — Typical schematic for ignition and safety-start systems for electric-start Models 56 (Ser. No. 44,001 and after) and 57 (Ser. No. 14,001 and after).

⅛-inch higher than in front. Adjust by turning nuts on draft rod (11 – Fig. JD2) at each side of front axle.

Gage wheels should be set to just clear surface with mower at preferred cutting height.

All Models

REMOVE AND INSTALL. Disconnect spark plug wire and place mower in lowest position. Pull spring pins and disconnect clutch rod and lift linkage. Unbolt rear hitch and slip blade drive belt off engine pulley. Remove lock pin from rear draft link. Disconnect lift helper spring on Models 66 and 68 only. On Models 55, 56 and 57, pull spring pins and disconnect front draft links at front axle. Remove mower deck from under chassis.

Reverse removal procedure to install mower. Adjust belt guide to 1/16-⅛inch clearance from engine pulley.

ELECTRICAL

All Models

Units are equipped with safety interlock starting systems to prevent starting engine with drives engaged. Refer to wiring diagrams JD14 through JD18.

DIXON ZTR

DIXON INDUSTRIES, INC.

P.O. Box 494
Airport Industrial Park
Coffeyville, KS 67337

Model	Make	Engine Model	Horsepower	Cutting Width, In.
ZTR-1	B&S	190000*	8	42
ZTR-2	B&S	190000*	8	42
ZTR-3	B&S	190000	8	42
ZTR-308	B&S	190000*	8	30
ZTR-424	B&S	250000*	11	42
ZTR-425	B&S	250000*	11	42
ZTR-426	B&S	250000*	11	42

*Synchro-balanced

STEERING SYSTEM

These units do not have a conventional front axle or steering system; instead, heavy-duty, rubber-tired caster wheels mounted on front corners of frame respond to steering effort of mower transaxle by swivel action in desired direction of turn. See TRANSAXLE section for details of steering adjustment and control.

ENGINE

For overhaul and repair procedures on engines listed in Specification Table refer to Small or Large Air Cooled Engines Service Manual.

All Models

REMOVE AND REINSTALL. To remove engine, first disengage mower drive then disconnect throttle control cable which extends from throttle hand control at right front of operator's seat and separate bowden cable from engine at carburetor.

On electric start models, disconnect battery ground cable and cable from starter solenoid to starter motor. Disconnect ignition switch (primary) wires.

Loosen engine mounting bolts (four) holding engine base to its platform on mower frame, loosen locknut of engine tensioner bolt, back off tensioner and remove mower deck and transaxle drive belts from pulleys under engine. Remove belt keeper (7–Fig. D1) if necessary. Completely remove engine mounting bolts from slotted openings in platform and lift engine from frame. Reverse procedure to reinstall engine.

CLUTCH

All Models

Clutch function is incorporated in drive unit of transaxle. There is no con-

1. Brake arms
2. Brake rod
3. Brake pivot
4. Transaxle assy.
5. Drive pulleys
6. Key
7. Belt keeper
8. Drive belt
9. Starter solenoid
10. Drive chains
11. Hub & sprocket
12. Shims
13. Control rod assemblies
14. Control lever assemblies
15. Lift lever
16. Lift tie rods
17. Lift links
18. Brake rod
19. Pivot pin
20. Brake lever
21. Pedal
22. Caster assy.

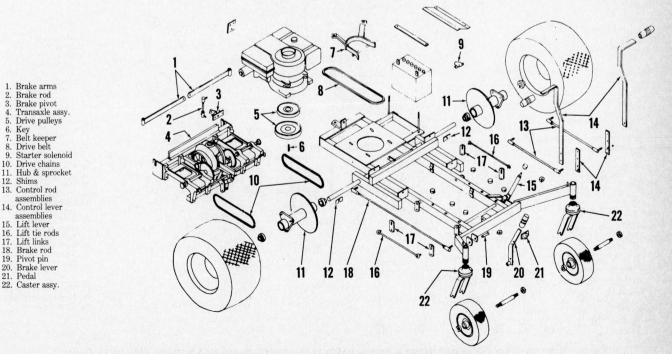

Fig. D1 — Exploded view of mower chassis and frame assemblies.

ventional, separate, belt-type master clutch. See TRANSAXLE section for clutch functions and adjustments.

BRAKES

All Models

These units do not have a service brake as such. Braking is accomplished by moving control levers opposite to direction of movement. Proper adjustment of transaxle will insure that braking is effective for safe operation.

Parking brake, mounted at rear of transaxle frame, is an over-center, linkage design used to hold unit after motion is halted. If adjustment of parking brake becomes necessary, unbolt ball joint of linkage rod (18–Fig. D1) and shorten rod working length to tighten brake. If brake arms (1) exert excessive pressure against inner sidewalls of tires, readjust to decrease pressure. Test for holding by pushing mower by hand with parking brake applied.

TRANSAXLE

All Models

The friction-drive transaxle provides for all functions of clutch, brakes, steering, variable speed transmission and differential. Each of the rear (Driving) wheels rotates independently on its own axle shaft. If proper adjustments are

maintained among working parts, especially the friction drive components and control linkages, over-all operation will be satisfactory.

ADJUST. In order to gain sufficient access to make adjustments, body assembly should be removed from frame. To do so, proceed as follows:

Disconnect throttle control bowden cable from engine. Separate interfering electrical wires by disconnecting at looms and multiconnectors at front and rear of body pan. Pay particular attention to key switch on Model ZTR-1 and to seat safety switch wire on all models. Unbolt upper (handle) portion of control levers (14–Fig. D1) from lower section and set aside. Remove control handle (1–Fig. D3) by removing through-bolt which attaches handle to cutting deck control rod (2). Remove two bolts which attach brake pedal (21–Fig. D1) to brake lever arm (20). Now, remove two acorn style nuts at front of body floor and sheet metal screws which secure body to transaxle mount at rear. Note that brake arm clips are released at the same time. Body assembly can now be lifted clear of frame. Reverse these steps to reinstall body.

After body has been removed, a check for correct adjustments to drive train involves the following:

ENGINE TO TRANSAXLE DRIVE BELT. For operation without undue

slippage or excess loading of shafts due to tightness, main drive belt should take ½-inch deflection by hand between engine and transaxle pulley centers. To adjust, slightly loosen four engine mounting bolts and the locknut on tensioner bolt; screw tensioner in or out to gain proper belt flex. Retighten mounting bolts and locknut.

NOTE: Drive pulley for mower deck is also keyed to engine crankshaft.

DRIVE CHAINS. Two sets of drive chains should be checked for correct adjustment. Primary chain (20–Fig. D2) driven by 9-tooth sprocket (18) of cradle shaft (11) which is fitted directly to discup (3) drives 24-tooth intermediate sprocket (19) on support shaft (15). This chain must run loose so that cradle (10) can pivot freely. There is no provision for take-up of excessive slack on this chain; therefore, if extreme wear of sprockets or chain links develops, renewal of chain and/or sprockets will be necessary. Chains are fitted with master links for convenience.

Final drive chains (10–Fig. D1) drive 60-tooth axle sprockets (11) and are driven by 9-tooth output sprocket (17–Fig. D2) on support shaft (15). To adjust these chains, proceed as follows: Loosen four bolts which attach transaxle to main frame after providing support under rear of frame to relieve weight on drive wheels. Insert additional shims (12–Fig. D1) as needed to

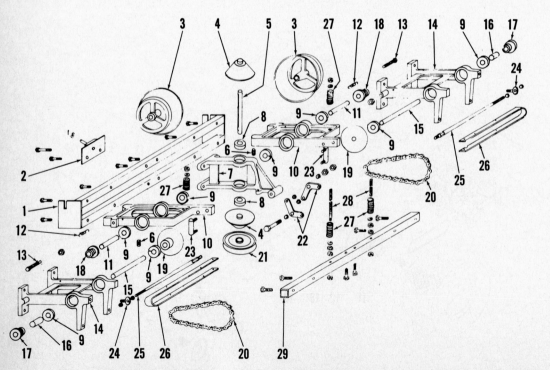

1. Main frame
2. Brake mount
3. Discup (2)
4. Drive cone (2)
5. Frame (input) shaft
6. Cradle spring (2)
7. Center frame
8. Frame shaft bearing (2)
9. Shaft bearing (8)
10. Cradle (2)
11. Cradle shaft (2)
12. Pivot spring (2)
13. Pivot bolt (2)
14. Support (2)
15. Support shaft (2)
16. Spacer (2)
17. Sprocket (2)
18. Sprocket (2)
19. Intermediate sprocket (2)
20. Primary chain (2)
21. Input pulley
22. Bellcrank (2)
23. Control ball joint (2)
24. Thumb screw mount (2)
25. Torque rod (2)
26. Stiffener
27. Mounting springs (4)
28. Mounting studs (2)
29. Front rail

Fig. D2 — Exploded view of friction-drive transaxle. Refer to text for service procedures and parts functions. Shaft roll pin (12 used) are not shown.

overcome wear and slack between drive sprockets and chain. It is not necessary to remove master link from chain when fitting shims. When correctly adjusted, a small deflection (3/16 to ¼-inch) in chain at mid-point between sprocket centers will provide adequate slack to minimize both noise and tension strain and to prevent rapid wear. it is important that deflection be equal on each side of drive.

NOTE: When shims have been added, check and readjust engine to transaxle drive belt as well as control rod linkage which will have been affected.

POSITIVE NEUTRAL ADJUSTMENT. When in neutral, control levers should align, side by side, approximately 10 degrees (4 inches) to the rear of vertical, in easy reach of operator. Lever action should allow for free movement forward and back by operator and engagement (drive application) should be smooth and positive. There should be little or no side play in controls. All slack and looseness should be eliminated by careful check and tightening of all control linkage beginning with levers (14 – Fig. D1) through lever pivots and control rods (13) to bell cranks (22 – Fig. D2) and to ball joints (23) which deflect cradle assemblies (10). When control lever and linkage adjustments are correct, mower should not creep with controls in neutral. Because controls are spring-loaded for return to neutral whenever released, the term "positive neutral" is used.

To properly adjust for positive neutral, pressure must be equalized on springs (27) which are fitted above and below cradle (10) on mounting stud (28). Tighten or loosen adjustment nuts so that clearance between friction face of drive cone (4) and contact zone of steel discup (3) is as nearly equal as possible. Use feeler gage stock of 0.020 inch thickness. As a preliminary check of neutral adjustment, move levers forward and back until contact between cones and discups can be felt, noting if range of movement for each lever is not very nearly equal. After locknuts are drawn up, recheck with feeler gage to insure that adjustment was not disturbed. Levers should spring back to neutral when released and mower should not creep after engine is started.

CRADLE/DISCUP ADJUSTMENT. This adjustment is made with transaxle controls in neutral. A preliminary check to determine if inner edge of discup (3 – Fig. D2) is exactly parallel to adjacent edge of center frame (7) is important. If discup does not run true in relation to center frame, further adjustment is ineffective. This is normally a factory setting performed with production jigs and gages; however, satisfactory measurements can be achieved by using a good Vernier caliper, particularly one equipped with a "T" head. If mower is blocked under frame so that drive wheels can be rotated freely, objectionable excess runout is easily observed and measured while discup turns. If bad alignment is found, carefully check condition of cradle shaft (11) and shaft bearings (9). Damaged or defective parts must be renewed. When discups are parallel, us a 0.020-inch feeler gage to check clearance between discup and friction surface of upper and lower driving cones (4). Check at all four contact areas. If need for clearance adjustment is apparent, before proceeding, check condition of input (frame) shaft (5) and its bearings (8). Looseness due to wear or bearing failure can be corrected only by renewal of defective parts. To proceed with clearance adjustment, follow this sequence:

1. Determine direction in which discup must be moved to set required clearance (0.020 inch) between discup and drive cone.

2. To INCREASE clearance (that is, to move discup AWAY from cone) loosen nut on threaded torque rod (25) at inboard side of thumb screw mount (24), then tighten outer nut by same number of turns.

3. In order to keep cradle (10) and discup (3) parallel to center frame (7), pivot bolt (13) must be adjusted to correspond with changes in length of torque rod (25). This is accomplished by loosening locknut on pivot bolt without allowing bolt to turn, then screwing bolt in COUNTERCLOCKWISE direction to move discup AWAY from cone.

NOTE: Use opposite procedures and rotation to move discup TOWARD drive cone or to DECREASE clearance. Whenever torque rod length is adjusted, it must be kept in mind that pivot bolt will require simultaneous adjustment in order to keep discup parallel to center frame. Because of thread difference, the turn ratio of torque rod nuts to pivot bolt is 2:1. When adjusting, it should be noted that two turns of torque rod nuts will equal one turn of pivot bolt.

4. When clearance is correct, recheck positive neutral adjustment along with control levers and linkages.

TRANSAXLE TROUBLESHOOTING. Some problems which may be encountered during mower operation are listed with recommended corrective action.

Unit will not travel in straight line. Check for loose control lever, then for loose or damaged control rod to include all linkage to transaxle. Tighten, adjust or renew parts as necessary. If trouble is not corrected, remove body and check clearance adjustments between discups and drive cones. Be sure that all cradle springs (6, 12 and 27 – Fig. D2) are in place and fully functional. If all items checked appear normal, it is possible that friction surfaces of drive cones are damaged or contaminated with foreign material such as oil. Repair or renew as necessary. Both cones must be renewed though only one may be defective.

Torque rods break during operation. This condition may be caused by excessive clearance between discup and drive cone. Check and correct as necessary. It is possible that some older models may not have torque rod stiffeners (26) installed. If this is the case, install the parts needed.

Mower travels only in a circle. Check for broken torque rod first, then check for possible broken pivot or broken shaft. If there is no evident problem here, block under frame so that wheels are clear and check carefully for possible broken roll pin. There is a total of twelve roll or spirol pins used to secure sprockets to support shafts, sprockets and discups to cradle shafts and bearings and drive cones to frame shaft. Any problem with a drive chain will be apparent – disengagement or breakage. It is advisable to check condition of sprocket teeth and for proper chain tension. Repair or adjust as necessary.

Mower creeps in neutral. Check for looseness in linkage first, then check for correct clearance between discups and cones. It is also possible that a cradle and its discup have slipped out of alignment with center frame. This condition can cause discup to contact drive cone away from proper friction zone so as to make correct clearance adjustment nearly impossible. Check measurement and correct condition. Readjust positive neutral.

SPECIAL NOTE: Correct clearance between drive cones and discups has been emphasized as being 0.020 inch. Operation of transaxle will be acceptable at 0.005 more or less than this setting. However, excessive clearance such as 0.035 or 0.040 will cause excessive slippage and result in overheating of cones and discups leading to damage or destruction.

OVERHAUL. If previously covered adjustments and parts renewals do not suffice to restore transaxle performance, disassembly, careful evaluation of all parts, thorough cleaning and careful reassembly with new parts will be neces-

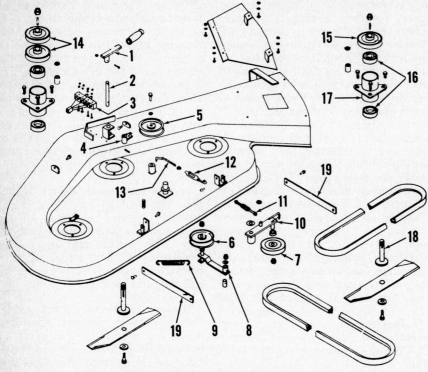

Fig. D3 — Exploded view of mower cutting deck.

1. Control handle			
2. Control rod	7. V-belt idler	12. Turnbuckle	16. Spindle bearings (6 used)
3. Switch terminals	8. Idler bracket	13. Connecting link	17. Hub assy. (3)
4. Interlock switch cam	9. Tension spring	14. Input & drive pulleys	18. Blade spindle (3)
5. Flat idler	10. Engagement bracket	15. Drive pulley	19. Stabilizers (2)
6. Flat idler	11. Engagement spring		

Fig. D4 — Electrical layout diagrams for all models. Wire color code: Green-Solenoid; Red-Battery (+); Black-Battery (-); Brown-Magneto; White-Ground.

1. Engine	6. Alternator lead wire
2. Battery	7. Magneto lead wire
3. Starter motor	8. Microswitch (NC)
4. Solenoid	9. Microswitch (NO)
5. Headlights	10. Switch ground

sary. To do so, first remove engine and body from frame as previously described. Disconnect control rods, then remove master links from drive chains (10 – Fig. D1) and remove chains. Remove brake arms (1). Unbolt transaxle assembly (4) from main frame taking care not to lose adjustment shims (12). Remove the separated transaxle assembly. Refer to Fig. D2 for parts arrangement and identification.

Remove master links and drive chains (20 – Fig. D2). Further disassembly procedures will be apparent. It is suggested that parts not be removed from transaxle unless there is a clear need to do so. Because transaxle, after drive chains are removed, is open to easy access, all shafts, anti-friction bearings, drive sprockets and roll pins can be carefully examined and checked by hand for looseness, wear or damage. Particular care should be taken with evaluation of discups and drive cones. Damage, due to heat of friction, especially if unit has been operated with excessive clearance between friction drive components should not be overlooked.

Reassemble, using necessary renewal parts and reinstall in reverse order of disassembly. Refer to preceding AD-JUSTMENT section.

MOWER

All Models

ADJUST. Unsatisfactory cutting performance by mower unit, streaking or patchy finish on lawn, is frequently caused by slippage of drive belts.

Tension of main drive belt from engine pulley is increased by moving V-belt idler (7 – Fig. D3) farther forward. Note that this idler also serves for clutch function to engage and disengage mower drive with its linkage connected to control handle (1). Adjustment, when needed, is made by rotation of turnbuckle (12) to shorten or lengthen control linkage and to increase or decrease tension on engagement spring (11). Properly adjusted control linkage will exert enough pressure upon the spring-loaded idler (7) to eliminate slippage and will disengage cleanly in response to movement of control handle (1).

NOTE: Be sure that interlock switch cam (4) which operates safety interlock micro switch on later models makes proper contact.

Drive belt for mower spindles is held under tension by spring (9) which is applied to belt flat side by flat idler pulley (6). Flat idler (5) serves mainly as a belt guide. There are no other service adjustments required for mower deck.

R&R AND OVERHAUL. To remove mowing unit from under mower frame, disconnect spark plug wire first to insure that mower cannot be accidentally started, especially if jumpers or by-pass connectors have been used to override interlocks. Position control handle (1 – Fig. D3) to disengage mower drive and relax drive belt. Remove belt keeper under engine platform and remove drive belt from sheaves. Disconnect wire loom at switch connectors and terminals (3). Unbolt control handle (1) from control rod (2), then remove pins from stabilizers (19). Now, remove pins which attach mower pan (deck) to lift links (17 – Fig. D1) and lower cutting deck down away from frame. Be sure that control rod (2 – Fig. D3) is clear of body and frame and slide cutting unit from under mower frame. It may be considered easier to lift front wheels and roll body and frame rearward on the drive wheels. Clean mower deck thoroughly to prepare for inspection of parts and necessary disassembly.

DYNAMARK
(AMF)

DYNAMARK CORP.

165 W. Chicago Avenue
Chicago, IL 60610

Model	Make	Engine Model	Horsepower	Cutting Width, In.
826	B&S	190000	8	26
832	B&S	190000	8	32
1260	B&S	190000	8	26
1261	B&S	190000	8	32
1269	B&S	130000	5	26
1274	B&S	220000	10	36
1281	B&S	190000	8	36
1284	B&S	190000	8	
		220000	10	
		250000	11	36
1288	B&S	220000	10	42
3180	B&S	250000	11	42
3181	B&S	326000	16	42
3184	B&S	220000	10	36
3188	B&S	190000	8	
		220000	10	
		250000	11	36
3189	B&S	190000	8	32
5180	B&S	250000	11	42
5181	B&S	220000	10	
		326000	16	42
5184	B&S	250000	10	36
5188	Tecumseh	V100	10	
	B&S	250000	11	36 or 42
5267	B&S	130000	5	
		190000	8	26 or 32
5282	B&S	190000	8	36
5285	B&S	190000	8	36
5288	B&S	220000	10	
	B&S	250000	11	
	Tecumseh	V100	10	36 or 42
5289	B&S	190000	8	
	Tecumseh	V100	10	32
5294	B&S	130000	5	
		190000	8	32
5296	B&S	220000	10	
		250000	11	36
5297	B&S	220000	10	42

FRONT AXLE STEERING SYSTEM

Early Models

REMOVE AND REINSTALL. To remove axle main member (24 – Fig. DM1), place mower unit in lowest position and unpin mower from mower hanger. Support front of unit and remove front wheels. Disconnect drag link end (20) from steering arm (22) and tie rod ends (27) from spindles. Drive out roll pin (23), remove steering arm (22) and remove spindle (26) from axle. Remove "E" ring (15) and remove spindle (17). Remove pivot bolt (25), then lower axle main member (24) from frame. Inspect spindle bushings (16) for excessive wear and renew as necessary.

Reassemble by reversing disassembly procedure. Check front wheel toe-in and adjust ends (27) on tie rod (28) to obtain a toe-in of 1/8-inch.

To remove steering gears and shafts, drive out roll pin and remove steering wheel (1) and cover (2). Unbolt and remove console from around steering unit. Disconnect drag link end (20) from quadrant arm (18) and remove cotter from steering arm (18) and tie rod ends (24) from spindles. Drive out roll pin, remove steering arm (18) and lower spindle (23) from axle. Remove "E" ring (20) and spindle (26). Remove pivot bolt (17) and lower axle main member (19) from frame. Inspect spindle bushings (21) for excessive wear and renew as necessary. Reassemble by reversing disassembly procedure. Check front wheel toe-in and adjust ends (24) on tie rod (25) as required to obtain 1/8-inch toe-in.

To remove steering gear, drive out roll pin (2) and remove steering wheel (1). Drive roll pin (6) from pinion gear (7) and remove "E" ring (13) from lower end of shaft (5). Withdraw steering shaft (5) and remove pinion gear (7). Disconnect drag link end (14) from quadrant gear (10). Remove lock bolt (11) and eccentric adjuster (9), then lift out quadrant gear (10). Bushings (3 and 8) can be removed after first removing retaining rings (4 and 12). Reassemble by reversing disassembly procedure. To adjust steering gear free play, loosen lock bolt (11) slightly and rotate eccentric adjuster (9). When excessive play is removed, tighten lock bolt (11).

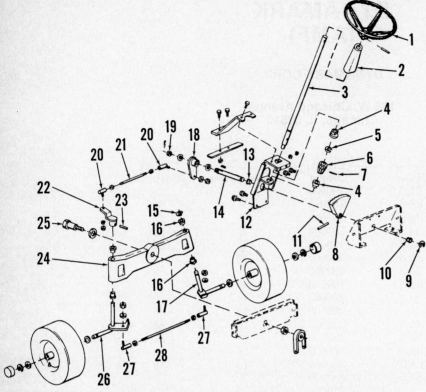

Fig. DM1 — Exploded view of front axle and steering system used on early production models.

1. Steering wheel	8. Quadrant gear	15. "E" ring	22. Steering arm
2. Cover	9. "E" ring	16. Spindle bushings	23. Roll pin
3. Steering shaft	10. Bushing	17. Spindle R.H.	24. Axle main member
4. Bushings	11. Roll pin	18. Quadrant arm	25. Pivot bolt
5. "E" ring	12. Steering support	19. Adjusting nut	26. Spindle L.H.
6. Pinion gear	13. Bushing	20. Drag link ends	27. Tie rod ends
7. Roll pin	14. Quadrant shaft	21. Drag link	28. Tie rod

pin, adjusting nut (19), quadrant arm and washers. Drive out roll pin (11) and remove "E" ring (9). Remove shaft (14) from front and lift out quadrant gear (8). Drive out roll pin (7) and remove "E" ring (5). Withdraw steering shaft (3) and remove pinion gear (6). Bushings (4, 10 and 13) can now be removed. Clean and inspect all parts for wear or other damage and renew as necessary. When reassembling, adjust steering gear free play as follows: Turn adjusting nut (19) clockwise to remove excessive play, then install cotter pin. Free play should be adjusted to a minimum but gears should not bind.

Lubricate all bushings and pivot points with SAE 30 oil. Apply a light coat of lithium grease to pinion gear and quadrant gear teeth.

Late Models

REMOVE AND REINSTALL. To remove axle main member (19 – Fig. DM2), unpin front of mower from mower hanger, then unbolt and remove mower hanger from front support (16). Support front of unit and remove front wheels. Disconnect drag link end (14)

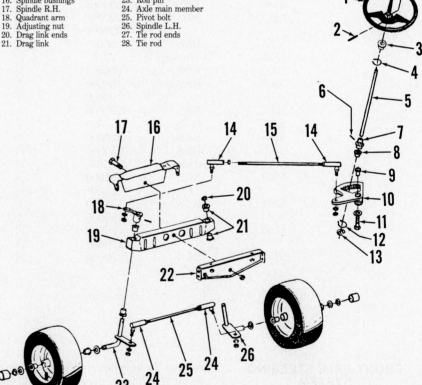

Fig. DM2 — Exploded view of front axle and steering system used on late production models.

1. Steering wheel	8. Bushing	14. Drag link ends	20. "E" ring
2. Roll pin	9. Eccentric adjuster	15. Drag link	21. Spindle bushings
3. Bushing	10. Quadrant gear	16. Front support	22. Axle support
4. Retaining ring	11. Lock bolt	17. Pivot bolt	23. Spindle L.H.
5. Steering shaft	12. Retaining ring	18. Steering arm	24. Tie rod ends
6. Roll pin	13. "E" ring	19. Axle main member	25. Tie rod
7. Pinion gear			26. Spindle R.H.

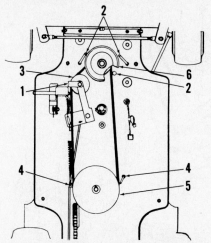

Fig. DM3—Underside view of typical belt idler traction drive clutch, belt and belt guides used on transmission equipped models.

1. Belt retainer
2. Belt guides
3. Clutch idler pulley
4. Belt guides
5. Transmission pulley
6. Engine pulley

Lubricate all bushings and pivot points with SAE 30 oil. Apply a light coat of lithium grease to pinion gear and quadrant gear teeth.

ENGINE

All Models

For overhaul and repair procedures on engines listed in Specification Table refer to Small Air Cooled Engines or Large Air Cooled Engines Service Manual.

REMOVE AND REINSTALL. Disconnect spark plug wire and remove mower as outlined in MOWER DECK paragraphs. Unbolt and remove hood, body panels and protective shields as necessary. Disconnect fuel line and drain fuel tank. On electric start models, disconnect battery cables, starter and alternator wires. On all models, disconnect ignition wiring, throttle cable and all other components as needed. Remove engine pulley belt guides and slip belts off pulley. For assistance in removal of belts refer to TRACTION DRIVE CLUTCH AND DRIVE BELT paragraphs. Remove engine mounting bolts and lift off engine.

Reinstall engine assembly by reversing removal procedure. Adjust belts as outlined in TRACTION DRIVE and MOWER DRIVE BELT paragraphs.

TRACTION DRIVE CLUTCH AND DRIVE BELT

All Models

The traction drive clutch used on all models (except variable speed models) is

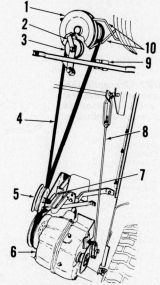

Fig. DM4—Underside view of typical belt idler traction drive clutch, belt and belt guides used on transaxle equipped models.

1. Engine pulley
2. Belt retainer
3. Clutch idler pulley
4. Drive belt
5. Backside idler pulley
6. Transaxle assy.
7. Adjustment slot
8. Brake rod
9. Footrest bracket
10. Belt guides

a belt idler type. When clutch is disengaged, tension is relieved on drive belt and engine pulley turns freely within belt. Two methods are used to adjust traction drive belt depending on how unit is equipped.

On models equipped with variable speed drive, the variable speed pulley is also the clutch idler. The variable speed pulley contains a moveable center section which allows a change in diameter of pulleys for primary and secondary belts. Position of pulleys and belts in disengaged and engaged positions are shown in Fig. DM6.

Models With Transmission-Differential

ADJUSTMENT. Drive belt adjustment is made by moving transmission. To adjust, remove mower as outlined in MOWER DECK paragraphs. Depress clutch-brake pedal to remove belt tension and engage park brake. Loosen transmission mounting bolts and move transmission rearward to tighten belt and retighten bolts. Release pedal and check adjustment, pedal should be approximately straight up and down when adjustment is correct. Check drive chain tension and adjust if needed as outlined in DRIVE CHAIN paragraphs.

REMOVE AND REINSTALL. Remove mower as outlined in MOWER DECK paragraphs. Depress clutch

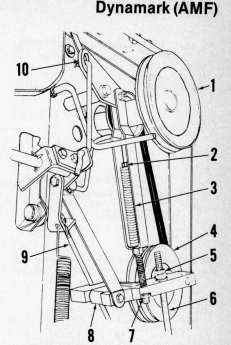

Fig. DM5—Underside view of traction drive variable speed pulley and control linkage used on some models.

1. Engine pulley
2. Adjustment screw
3. Tension spring
4. Variable speed pulley
5. Front jam nut
6. Variable speed adjustment rod
7. Adjustment nut
8. Control arm
9. Brake rod
10. Disengage lever

pedal and engage park brake. Note position of idler pulley belt retainer (1–Fig. DM3), then loosen idler pulley bolt. Loosen and move belt guides (2 and 4) away from pulleys. Slip belt off idler pulley then remove from transmission and engine pulleys.

Renew belt by reversing removal procedure. Hold idler pulley belt retainer in position while tightening idler pulley bolt. Adjust belt guides to 1/16-inch clearance from belt. Adjust belt tension as previously outlined.

Models With Transaxle

ADJUSTMENT. On transaxle equipped models, belt adjustment is made by moving backside idler·(5–Fig. DM4) in its adjusting slot. Adjust idler until clutch-brake pedal is approximately straight up and down.

REMOVE AND REINSTALL. Remove mower as outlined in MOWER DECK paragraphs. Depress brake-clutch pedal and engage park brake. If unit has a footrest bracket (9–Fig. DM4) straddling drive belt, unbolt and remove bracket. Loosen belt guides and move them away from pulleys. Note position of idler pulley belt retainers, then loosen bolts on clutch idler (3) and backside idler (4) pulleys. Remove belt from idlers, engine pulley and transaxle pulley. Remove belt from chassis by pushing it up through gear shift opening.

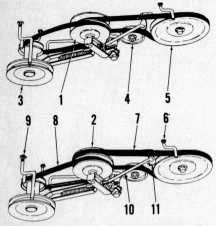

Fig. DM6 — View of variable speed pulley (1) in disengaged position and pulley (2) in fast position.

1. Variable speed pulley (disengaged position)
2. Variable speed pulley (fast position)
3. Engine pulley
4. Backside idler
5. Transmission pulley
6. Belt guide
7. Secondary drive belt
8. Primary drive belt
9. Belt guide
10. Adjustment rod
11. Control lever

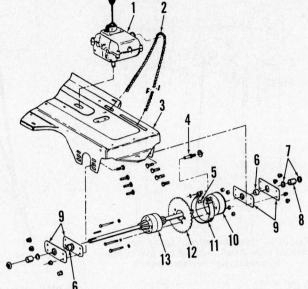

Fig. DM7 — Transmission, drive chain, differential, rear axles and axle bearings used on some models.

1. Transmission
2. Drive chain
3. Frame
4. Brake anchor stud
5. Adjuster pin
6. Axle bearings
7. Washers
8. Spacer
9. Bearing plates
10. Brake drum
11. Brake band
12. Sprocket
13. Differential assy.

Renew belt by reversing removal procedure. Before installing footrest bracket, release brake-clutch pedal. Belt must be under tension to properly align belt and bracket. Adjust belt guides to 1/16-inch clearance from belt. Hold idler pulley belt retainers in place while tightening idler retainer bolts. Adjust belt as previously outlined.

Models With Variable Speed Pulley

ADJUSTMENTS. If unit moves when variable speed lever is in latched position, adjust as follows: Place lever in latched position and remove mower deck. Loosen front jam nut (5 – Fig. DM5) on variable speed adjustment rod (6) two turns, then tighten rear jam nut two turns. Check operation and repeat adjustment as required.

If variable speed lever will not stay in selected speed position, adjust as follows: Place variable speed lever in latched position. Tighten friction adjustment nut, located inside console on right hand side behind engine, one turn. Check operation and repeat adjustment as needed.

To check and adjust brake, proceed as follows: Shift transmission to neutral, then depress brake pedal and engage park brake. Push unit forward or backward. If rear wheels are not locked, tighten adjusting nut on brake rod (9 – Fig. DM5) two turns. Check operation and repeat adjustment as needed.

If variable speed does not return to disengaged position when brake is depressed and held, adjust as follows: With transmission in neutral, start engine and push speed lever to full forward position, then shut off engine. Adjustment nut (7 – Fig. DM5) should be tight against pulley control arm (8) and all connecting parts should have tension against them. To adjust spring (3) hold adjustment screw (2) and tighten adjustment nut (7) until tight against support arm.

REMOVE AND REINSTALL. To remove secondary drive belt (7 – Fig. DM6), first remove mower as outlined in MOWER DECK paragraphs. Unhook adjustment rod (10) from lower end of variable speed lever (11). Loosen belt guide (6) and move away from transmission pulley. Pull backside idler (4) away from belt and slip belt off underside of idler pulley. Roll belt off underside of transmission pulley, then remove belt from variable speed pulley. Belt can be removed over top of pulley by pushing toward front of unit and twisting belt sideways.

To remove primary belt (8 – Fig. DM6) it is necessary to first remove secondary belt as previously outlined. Loosen engine pulley belt guide and move it away from pulley. Remove belt from variable speed pulley by pushing pulley toward front of unit and twisting belt sideways. Remove belt from engine pulley.

Renew belts in reverse order of removal. Adjust belt guides to 1/16-inch clearance from belts. Check and adjust if needed as previously outlined.

TRANSMISSION

All Models

The transmissions used on models so equipped are manufactured by J.B. Foote Foundry Co. or Peerless Division of Tecumseh Products Co. For overhaul and repair procedures refer to Foote or Peerless paragraphs in TRANSMISSION REPAIR section of this manual.

REMOVE AND REINSTALL. Disconnect spark plug wire and remove mower as outlined in MOWER DECK paragraphs. Depress brake-clutch pedal and engage park brake. Loosen belt guides, then slip traction drive belt off transmission pulley. Remove retaining ring and pulley from transmission shaft. Unbolt and remove body panels or protective shields as necessary to provide access to transmission. Disconnect drive chain and transmission safety starting switch wires. If unit is equipped with transmission mounted brake, disconnect brake rod from brake lever. Unbolt and remove shift lever. Unbolt and remove transmission from chassis. Reinstall in reverse order of removal procedure.

TRANSAXLE

All Models

The transaxles used on models so equipped are manufactured by J.B. Foote Foundry Co. or Peerless Division of Tecumseh Products Co. For overhaul and repair procedures, refer to Foote or Peerless paragraphs in TRANSMISSION REPAIR section of this manual.

REMOVE AND REINSTALL. Disconnect spark plug wire and remove mower as outlined in MOWER DECK paragraphs. Depress brake-clutch pedal and engage park brake. Loosen belt guides and roll belt out of idler pulley and transmission pulley. Disconnect brake rod from disc brake lever. Sup-

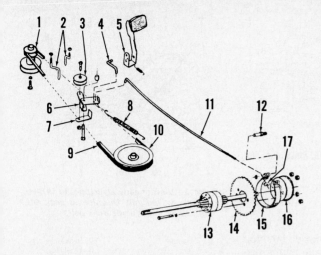

Fig. DM8 — Exploded views of typical main drive clutch and band brake used on some models.

1. Engine pulley
2. Belt guides
3. Clutch idler pulley
4. Clutch link
5. Clutch-brake pedal
6. Idler arm
7. Belt retainer
8. Clutch tension spring
9. Main drive belt
10. Transmission input pulley
11. Brake rod
12. Brake anchor stud
13. Differential assy.
14. Sprocket
15. Brake band
16. Brake drum
17. Adjuster pin

clutch pedal must be fully released. To tighten brake, loosen front nut on brake rod and tighten rear nut.

To adjust brake on models equipped with Peerless Model 639 transaxle, first release brake-clutch pedal. Loosen jam nut on brake actuating lever and turn adjusting nut until brake pads just clear brake disc. Hold adjusting nut and tighten jam nut against it.

To adjust brake on models equipped with all other transaxle models, release brake-clutch pedal. Disconnect and adjust length of brake rod to adjust brake. Do not turn rod more than one turn without reinstalling to check adjustment.

port rear of unit and remove rear wheels. Unbolt and remove transaxle from chassis. Reinstall by reversing removal procedure.

DRIVE CHAIN

All Models

R&R AND ADJUSTMENT. To remove chain, on models so equipped, rotate differential sprocket to locate master link. Disconnect master link and remove chain. Reinstall in reverse order of removal. Lubricate with a light coat of SAE 30 oil.

To adjust chain (2 – Fig. DM7), loosen bolts securing rear axle to frame. On models with band brake, loosen brake anchor stud (4). Move axle rearward (equal distance on both sides) to tighten chain. Chain should deflect approximately ¼-inch. On band brake models, push anchor stud as close as possible to brake band (11) and tighten pin, then adjust brake rod to account for axle movement.

DIFFERENTIAL

All Models

The differential used on models so equipped is manufactured by Peerless Division of Tecumseh Products Co. For overhaul procedure refer to Peerless paragraphs in DIFFERENTIAL REPAIR section of this manual.

REMOVE AND REINSTALL. To remove differential assembly (13 – Fig. DM7), raise rear of unit and support securely. Remove rear wheel assemblies, washers and spacers. On band brake models, disconnect front end of brake rod from clutch idler arm, then unscrew brake rod from adjuster pin (5) and pull brake band (11) from drum (10). Disconnect drive chain. Unbolt bearing

retainers (9) from frame and remove differential assembly. Clean rust, paint and burrs from axle shafts and remove axle bearings (6). Remove nuts inside brake drum and remove drum and sprocket (12).

Reinstall by reversing removal procedure. Adjust drive chain as previously outlined. Lubricate axle bearings with SAE 30 oil.

GROUND DRIVE BRAKE

Models With Band Brake

ADJUSTMENT. To adjust brake, refer to Fig. DM8 and disconnect brake rod (11) from idler arm (6). Turn brake rod into adjuster pin (17) to tighten brake. Reconnect brake rod, fully depress clutch-brake pedal (5) and engage parking brake lock. Main drive belt (9) should be free from engine pulley (1) and rear wheels should be locked. Always make certain that drive belt tension is removed before brake is applied.

Models With Disc Brake

ADJUSTMENT. To adjust brake on transmission equipped models, brake-

Models With Band Brake

R&R AND OVERHAUL. Normal overhaul consists of renewing brake band (15 – Fig. DM8). To remove brake band, disconnect front of brake rod (11) from clutch idler arm (6). Unscrew brake rod adjuster pin (17). Remove band retaining "E" ring from anchor stud (12). Pull brake band from around brake drum (16) and remove band from anchor stud.

Install new brake band by reversing removal procedure and adjust brake as outlined previously.

NOTE: To remove brake drum (16), refer to R&R procedure in DIFFERENTIAL paragraphs.

Models With Disc Brake

R&R AND OVERHAUL. Normal overhaul of disc brake consists of renewing brake pads. To remove brake assembly, disconnect brake rod from brake lever. Unbolt and remove brake carrier, brake disc and brake pads.

Clean and inspect parts and renew as necessary. Reassemble by reversing removal procedure and adjust brake as previously outlined.

Fig. DM9 — Exploded view of typical single spindle mower unit.

1. Safety starting switch
2. Clutch idler arm
3. Blade brake spring
4. Belt retainer
5. Clutch idler pulley
6. Pin
7. Mower mount
7A. Locknut
8. Belt retainers
9. Mower spindle assy.
10. Clutch tension spring
11. Mower belt
12. Blade clutch lever
13. Deflector
14. Lever bracket (upper)
15. Lever bracket (lower)
16. Gage wheels
17. Rear guide bar
18. Mower housing
19. Blade 26"
20. Belleville washers
21. Bolt

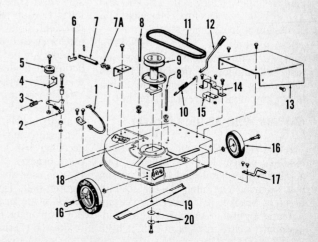

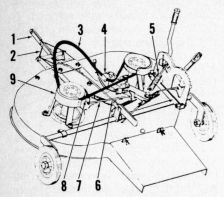

Fig. DM10—View of typical double spindle mower unit used on some models.

1. Bar hitch
2. Pivot bracket
3. Blade drive belt
4. Belt retainer
5. Blade engage spring
6. Lift cable & clevis
7. Blade brake rod
8. Left brake bracket
9. Belt guard

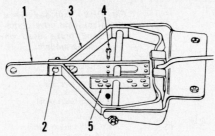

Fig. DM11—View of mower front hitch and pivot bracket.

1. Bar hitch
2. Bolt
3. Pivot bracket
4. Bolt
5. Lock bracket

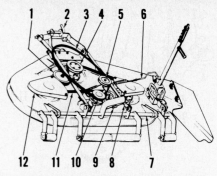

Fig. DM12—View of early style (prior to 1979) triple spindle mower unit. Unit uses a deck drive belt and a blade drive belt.

1. Front idler
2. Hitch pull pin
3. Deck drive belt
4. Center spindle assy.
5. Idler pulley & belt retainer
6. Blade clutch rod
7. Rear roller shaft
8. Eyebolt
9. Deck drive spring
10. Jackshaft pulley
11. Blade drive belt
12. Spindle shield

MOWER DRIVE BELT

Models With 26 Inch Mowers

ADJUSTMENT. On early production models, adjust as follows: Raise and block front of mower deck and remove pull pin from deck front hitch. Loosen jam nut and turn deck hitch in or out to adjust belt tension. Check blade brake action when blade clutch is disengaged. Blade should stop and remain stopped. If blade does not stop, belt adjustment is too tight.

On late production models, adjust as follows: Loosen nut securing hitch bar in pivot bracket. Slide deck forward to loosen belt and rearward to tighten belt. Retighten nut and check blade brake action. If blade does not stop when blade clutch is disengaged, belt adjustment is too tight.

REMOVE AND REINSTALL. Blade drive belt can be replaced without removing mower, but it is easier with mower removed. Note position of idler pulley belt retainer (4—Fig. DM9), then remove idler pulley (5). Loosen and move belt guides for clearance and remove belt from pulleys.

Renew belt in reverse order of removal. With blade clutch engaged, adjust belt guides to provide 1/16-inch clearance between belt and guide. Check belt adjustment as previously outlined.

Models With 32 Inch Mower

ADJUSTMENT. On early style models, adjust belt as follows: Disengage blade clutch and loosen nut on bolt securing bar hitch (1—Fig. DM10) to pivot bracket (2). Loosen front jam nut and tighten rear nut on hitch adjustment screw to tighten belt. Check

blade brake action. If blades do not stop or tend to creep with blade clutch disengaged, adjust nuts on brake rod (7) that connects to left brake bracket (8). There should be approximately 1/16-inch clearance between nut and bracket with clutch disengaged.

On late style models, adjust belt as follows: Disengage blade clutch and remove bolt (4—Fig. DM11) from lock bracket (5). Loosen bolt (2) securing bar hitch (1) to pivot bracket (3) and slide bar hitch forward to tighten belt. Retighten bolt and check blade brake action. If blades do not stop or tend to creep, adjustment is too tight. When adjustment is correct, install bolt in lock bracket.

REMOVE AND REINSTALL. Disconnect spark plug wire and remove mower as outlined in MOWER DECK paragraphs. Note position of idler pulley belt retainer (4—Fig. DM10), and loosen idler pulley mounting bolt. Loosen belt guards (9) and remove belt.

Install new belt by reversing removal procedure. Hold idler pulley belt retainer in position and tighten mounting bolt. Adjust belt guides to 1/16-inch clearance from belt. Check belt adjustment as previously outlined.

Models With 36 Inch Mower

ADJUSTMENT. On early production side discharge mowers, adjust as follows: Loosen jam nut on drive belt adjustment bolt and turn bolt counterclockwise to tighten belt. Hold bolt and tighten jam nut, then check blade brake action. If blades do not stop or tend to creep, belt adjustment is too tight.

On early production rear discharge and intermediate production side discharge models, adjust belt as follows: Disengage mower clutch and loosen bolt securing bar hitch (1—Fig. DM10) to pivot bracket (2), then loosen front and tighten rear nut on adjustment screw to tighten belt. Tighten bar hitch bolt and check blade brake action. If blades do not stop or tend to creep, adjust nuts on brake rod (7) that connects to left brake

bracket (8). There should be approximately 1/16-inch clearance between nuts and bracket with mower clutch disengaged.

On late production side and rear discharge models, adjust belt as follows: Disengage blade clutch and loosen bar hitch bolt (2—Fig. DM11). Remove bolt (4) from lock bracket (5) and slide bar hitch (1) forward to tighten belt. Tighten bar hitch bolt and check blade brake action. If blades do not stop or tend to creep, belt adjustment is too tight. When adjustment is correct, reinstall bolt in lock bracket.

REMOVE AND REINSTALL. Disconnect spark plug wire and remove mower as outlined in MOWER DECK paragraphs. Note position of idler pulley belt retainer (4—Fig. DM10), and loosen idler pulley mounting bolt. Loosen belt guards and remove belt.

Install new belt by reversing removal procedure. Hold idler pulley belt retainer in position and tighten bolt. Adjust belt guides to 1/16-inch clearance from belt. Check belt adjustment as previously outlined.

Models With 42 Inch Mower

ADJUSTMENT. On early production models, two belts are used—a deck drive belt (3—Fig. DM12) and a blade drive belt (11). To adjust deck drive belt, disengage blade clutch and adjust clutch rod (6) to increase or decrease idler spring tension. Deck hanger clevis, mounted on front axle pivot bolt, should be positioned halfway between engine pulley and front axle. To adjust blade drive belt, loosen front idler pulley (1) and move pulley toward center of mower deck to tighten belt.

On late production models, adjust belt as follows: Disengage blade clutch and

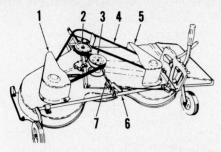

Fig. DM13—View of late style (1979 and on) triple spindle mower unit. Unit uses a single drive belt.

1. Spindle shield
2. Idler pulley & belt retainer
3. Center spindle
4. Drive belt
5. Spindle shield
6. Deck guide bar
7. Lift cable & clevis

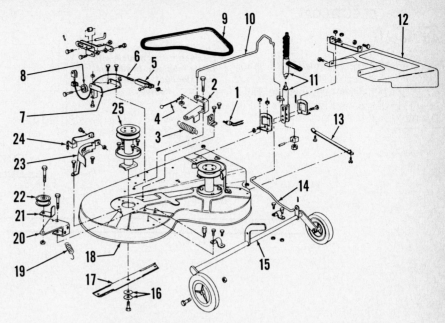

Fig. DM14—Exploded view of typical double spindle mower unit.

1. Safety starting switch
2. Bellcrank
3. Clutch tension spring
4. Brake cable
5. Clevis
6. Lift cable
7. Lift bracket
8. Pulley
9. Mower belt
10. Clutch rod
11. Lift and blade clutch lever
12. Deflector
13. Skid bar
14. Lift rod
15. Gage wheels assy.
16. Belleville washers
17. Blade 18"
18. Mower housing
19. Spring
20. Clutch idler arm
21. Belt retainer
22. Clutch idler pulley
23. Belt guard
24. Blade brake
25. Mower spindle assy.

loosen bar hitch bolt (2–Fig. DM11). Remove bolt from lock bracket (5) and slide bar hitch (1) forward to tighten belt. Tighten bar hitch bolt and check blade brake action. If blades do not stop or tend to creep, belt adjustment is too tight. When adjustment is correct, reinstall bolt in lock bracket.

REMOVE AND REINSTALL. On early production models, remove drive belts as follows: Remove mower as outlined in MOWER DECK paragraphs. Remove belt shields and loosen and move belt guides away from pulleys. Note position of idler pulley belt retainer (5–Fig. DM12), then loosen idler pulley bolt. Note position of deck drive spring (9), then unhook spring. Remove belts from pulleys.

Install new belts by reversing removal procedure. Hold idler pulley belt retainer in position while tightening bolt. Adjust belt guides to 1/16-inch from belt. Adjust belts as previously outlined.

On late production models, remove drive belt as follows: Place mower in lowest position. Loosen and move belt guides away from pulleys. Note position of idler pulley belt retainer (2–Fig. DM13), then loosen retainer. Remove belt from pulleys.

Install new belt by reversing removal procedure. Hold idler pulley belt retainer in position while tightening bolt. Adjust belt guides to 1/16-inch from belt. Adjust belt as previously outlined.

MOWER SPINDLE

All Models

REMOVE AND REINSTALL. The spindle assemblies are similar on all models. The spindle, spindle housing and mower pulley are available only as an assembly. Refer to Fig. DM9 or DM14 and remove mower spindle assembly as follows: Remove mower unit and drive belt. Place wood block between blade and mower housing to prevent blade from turning. Remove retaining cap screw, washers and blade from spindle. Unbolt and remove spindle assembly from top of mower housing. Renew parts as necessary and reassemble by reversing removal procedure. Torque blade mounting cap screw to 30-35 ft.-lbs.

MOWER DECK

All Models

LEVEL ADJUSTMENT. With unit on level surface and tires properly inflated, measure height of blade from ground at front and rear of mower deck. Blade should be approximately ⅛-inch lower in front.

To level mowers equipped with lift cable (6–Fig. DM10), raise and block front of mower deck to remove weight of unit from lift cable. Remove clevis pin from cable rear clevis and turn clevis to raise or lower front of deck. Each 1½ turn of clevis changes front deck height approximately ⅛-inch.

To level 26 inch mowers not equipped with lift cable, loosen front deck hanger mounting bolts and move hanger up or down to desired level position. Retighten hanger mounting bolts.

To level 42 inch mowers not equipped with lift cable, lower mower to lowest position and place block under rear of mower deck. Disconnect lift linkage eye bolts (8–Fig. DM12) from rear roller shaft (7). Turn eye bolts clockwise to lower rear of deck. Both eyebolts must be turned same number of turns. Reconnect eyebolts to roller shaft.

All Models

REMOVE AND REINSTALL. Place mower in lowest position and disengage blade clutch. Turn front wheels full left, and disconnect mower safety switch wires. Remove pull pin securing mower deck to front mounting bracket. Move mower deck forward to free rear deck guide bracket, then loosen belt guides on engine pulley and slip mower drive belt off pulley. Deck can now be slid out from under right side of unit. On some models removal will be easier if right front of unit is raised to provide clearance.

Reinstall deck by reversing removal procedure. Check deck level adjustment and blade drive belt adjustment.

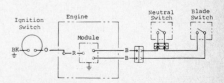

Fig. DM15—Wiring diagram for manual start models with safety interlock starting system.

1. BK-Black
2. O-Orange
3. R-Red
4. B-Brown

ELECTRICAL

All Models

All models are equipped with safety interlock starting systems to prevent starting when drives are engaged. Refer to wiring diagrams Fig. DM15 and Fig. DM16.

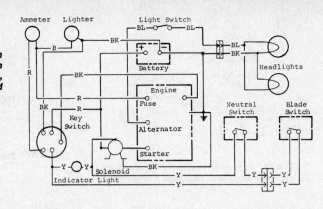

Fig. DM16 — Wiring diagram for electric start models with safety interlock, headlights, indicator light, ammeter and lighter.

1. R-Red
2. B-Brown
3. Y-Yellow
4. BK-Black
5. BL-Blue

FORD

FORD TRACTOR OPERATIONS
2500 East Maple Road
Troy, MI 48084

Model	Make	Engine Model	Horsepower	Cutting Width, In.
51	B&S	130902	5	26
60	B&S	170702	7	30
61	B&S	190702	8	30
65	B&S	170705	7	30
66	B&S	190705	8	30
526	B&S	130902	5	26
830	B&S	190702	8	30
830E	B&S	190705	8	30
1130E	B&S	252707	11	30

For service information and procedures on **FRICTION-DRIVE** Models 830, 830E and 1130E, use the following model cross reference and refer to JACOBSEN section of this manual.

Ford Model	Jacobsen Model
830	RMX
830E	RMX
1130E	RMX

For **BELT-DRIVE** Models 830, 830E and 1130E, refer to the following service procedures.

FRONT AXLE

All Models

Axle main member is center-pivoted and mounts within cross channel at front of main frame. Removal of under-chassis items may be aided by standing mower unit upright on its rear stand. If this is done, observe the following:

CAUTION: Frequently, the performance of maintenance, adjustment or repair operations on a riding mower is more convenient if mower is standing on end and these units are provided with a stand for this purpose. This procedure can be considered a recommended practice providing the following safety recommendations are performed:

1. Drain fuel tank or make certain that fuel level is low enough so that fuel will not drain out.
2. Close fuel shut-off valve if so equipped.
3. Remove battery on models so equipped.
4. Disconnect spark plug wire and tie out of way.

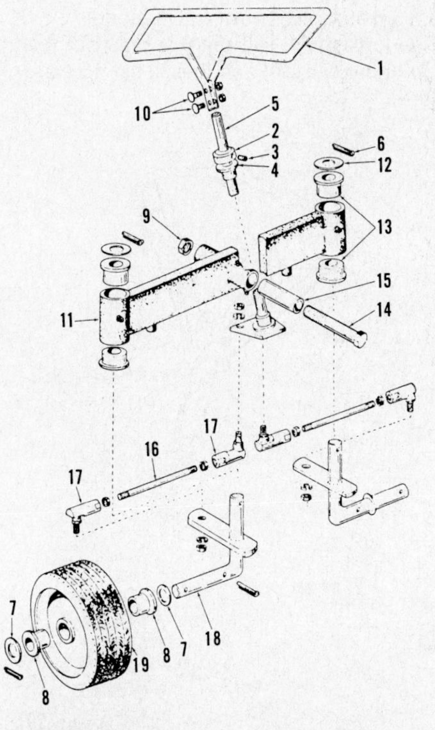

Fig. F1-Exploded view of front axle and steering assembly used on Models 51, 60 and 65.

1. Steering handle
2. Steering shaft collar
3. Set screw
4. Steering shaft bearing
5. Steering shaft
6. Pin
7. Washers
8. Wheel bushings
9. Nut
10. Screws
11. Axle
12. Washer
13. Spindle bushings
14. Axle pivot bolt
15. Axle spacer
16. Steering tie rod
17. Tie rod end
18. Spindle
19. Wheel

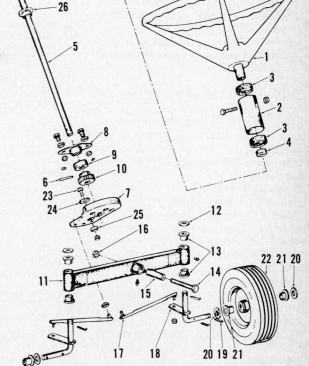

Fig. F2 — Exploded view of front axle and steering gear used on Models 51, 61 and 66.

1. Steering wheel
2. Tube
3. Washers
4. Washer
5. Steering shaft
6. Roll pin
7. Steering gear
8. Bearing
9. Collar
10. Pinion gear
11. Axle
12. Washer
13. Spindle bushings
14. Axle pivot bolt
15. Spacer
16. Nut
17. Tie rod
18. Spindle
19. Pin
20. Washer
21. Wheel bushings
22. Wheel
23. Bolt
24. Washer
25. Spacer
26. Flanged bearing (in steering console)

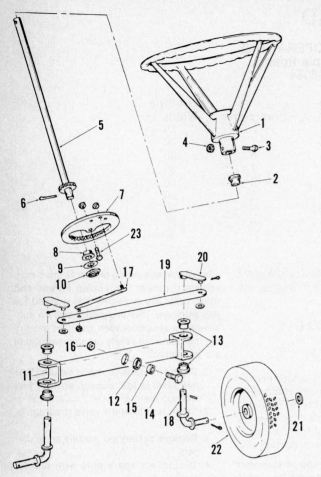

Fig. F3 — Exploded view of front axle and steering gear used on Model 526.

1. Steering wheel
2. Flanged bearing (in steering console)
3. Bolt
4. Nut
5. Steering shaft
6. Pin
7. Steering gear
8. Flanged bearing
9. Washer
10. Washer
11. Axle
12. Bushing
13. Spindle bushings
14. Axle pivot bolt
15. Spacer
16. Nut
17. Tie rod
18. Spindle
19. Steering strap
20. Steering lever
21. Washer
22. Wheel
23. Bolt

REMOVE AND REINSTALL. To remove steering gear, first stand unit on end as previously outlined. Disconnect tie rod ends (17–Fig. F1) from bottom of steering shaft (5). Unbolt steering handle (1) from steering shaft. Loosen set screw (3) in steering shaft collar (2), then lower shaft out of steering support and remove collar (2) and bearing (4).

Reinstall in reverse order of removal. When installing steering shaft, remove all end play, then push collar (2) down against bearing (4) and tighten set screw to hold shaft in position.

Other Models

REMOVE AND REINSTALL. To remove steering gear, stand unit on end as previously outlined. On Model 526 only, remove steering console front cover and frame front plate. Remove steering wheel retaining bolt and pull steering wheel (1–Fig. F2, F3 or F4) from steering shaft (5). All models except Model 526, remove steering tube (2) and washers (3 and 4), then unbolt and remove steering console and shaft upper bearing (26). Disconnect tie rods (17) from steering gear (7). On Models 51, 61 and 66, drive roll pin (6–Fig. F2) from pinion gear (10) and loosen set screws in

5. Although not absolutely essential, it is recommended that crankcase oil be drained to avoid flooding the combustion chamber with oil when engine is tilted.
6. Secure mower from tipping by lashing unit to a nearby post or overhead beam.

REMOVE AND REINSTALL. To remove front axle (11–Fig. F1, F2, F3 or F4), raise and block front of frame or stand unit on end. On all models except Model 526, unbolt outer ends of tie rods (17) from steering spindles (18). On Model 526, unbolt end of tie rod (17) from steering strap (19). On all models, remove front wheels from spindles, then remove pivot bolt (14) and spacer (15) from frame and withdraw axle member from frame channel. With axle assembly separated from frame, remove retaining pins and remove spindles (18) and bushings (13) from axle ends.

Clean and inspect parts for possible renewal. Reinstall in reverse order of removal. Lubricate at fittings using mutli-purpose lithium base grease. use SAE 30 oil at pivot points not having grease fittings.

Fig. F4 — Exploded view of front axle and steering gear used on Models 830, 830E and 1130E.

1. Steering wheel
2. Tube
3. Washers
4. Spring washers
5. Steering shaft
6. Pin
7. Steering gear
8. Flanged bearing
9. Washer
10. Washer
11. Axle
12. Washer
13. Spindle bushings
14. Axle pivot bolt
15. Spacer
16. Nut
17. Tie rod
18. Spindle
19. Washer
20. Felt washer
21. Wheel bushing
22. Wheel
23. Bolt
24. Washer
25. Spacer
26. Flanged bearing (in steering console)

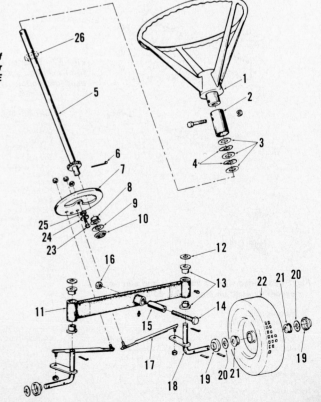

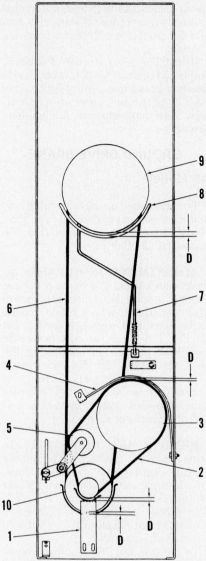

Fig. F5 — Typical traction drive and blade drive belt installation. Note belt guide clearance "D" of 1/16 to ⅛-inch between guide and outside surface of belt or pulley.

1. Engine pulley belt guide
2. Traction drive belt
3. Transmission pulley
4. Transmission pulley belt guide
5. Idler pulley
6. Blade drive belt
7. Blade brake rod
8. Blade drive belt guide
9. Blade pulley
10. Engine pulley

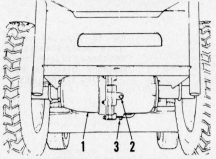

Fig. F6 — Location of oil level check plug and drain plug on transaxle.

1. Transaxle
2. Oil level check plug
3. Drain plug

ENGINE

All Models

For overhaul and repair procedures on engines listed in Specification Table, refer to Small or Large Air Cooled Engines Service Manual.

Refer to following paragraphs for removal and installation procedures.

REMOVE AND REINSTALL. To remove engine, disconnect spark plug wire, battery cables, throttle control cable and wiring harness from engine. Remove hitch plate from rear of frame. Place mower blade clutch control lever in disengaged position and mower height control lever in lowest position. Depress clutch pedal and engage pedal lock. Remove engine pulley belt guide and slip blade drive belt and traction drive belt off engine pulley. Remove

engine mounting bolts and lift engine from frame.

Reinstall engine by reversing removal procedure. Make certain belt guides are reinstalled properly and all safety starting switches are connected and operating properly.

TRACTION DRIVE CLUTCH AND DRIVE BELT

All Models

All models use a pedal-operated, spring-loaded belt idler to apply or relieve tension on drive belt. No adjustment is provided. If belt slippage occurs during normal operation due to belt wear or stretching, belt must be renewed.

REMOVE AND REINSTALL. To remove traction drive belt (2–Fig. F5), disconnect spark plug wire, then place blade clutch control lever in disengaged position and height control lever in lowest position. Depress clutch pedal and engage pedal lock. Stand unit on end as previously outlined. Remove engine pulley belt guide (1) and slip mower drive belt (6) off engine pulley (10). Remove idler pulley (5). Loosen transmission pulley belt guide (8) and move guide away from pulley, then pull belt out between pulleys and frame.

Renew belt in reverse order of removal. Adjust belt guides to provide 1/16 to ⅛-inch clearance between inside of guide and outer surface of belt.

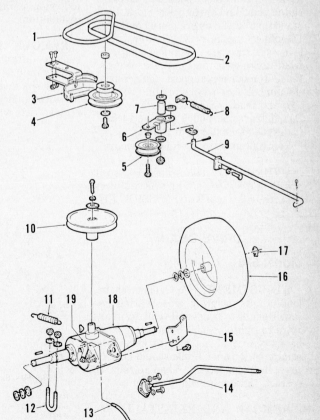

Fig. F7 — Exploded view of typical transaxle assembly and traction drive clutch.

1. Traction drive belt
2. Blade drive belt
3. Engine pulley belt guide
4. Engine pulley
5. Idler pulley
6. Idler pivot bracket
7. Sleeve
8. Idler arm spring
9. Clutch rod
10. Transaxle pulley
11. Brake return spring
12. U-bolt
13. Brake rod
14. Shift lever
15. Mounting bracket
16. Wheel
17. Retaining ring
18. Transaxle
19. Brake lever

collar (9). Unbolt bearing (8) and remove steering shaft, bearing, collar and pinion gear. Remove bolt (23), washer (24) and spacer (25), then lift out steering gear (7). On Models 526, 830, 830E and 1130E, remove pin (6–Fig. F3 or F4) and withdraw steering shaft (5), bearing (8) and washers (9 and 10). Remove retaining bolt (23) and lift out steering gear (7). Clean and inspect parts for wear and renew as necessary.

Reinstall in reverse of removal procedure. Lubricate bushings and pivot points with SAE 30 oil. Apply light coat of multi-purpose grease to pinion gear and steering gear teeth.

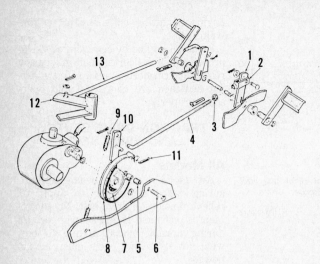

Fig. F8 – Exploded view of typical band brake assembly and clutch and brake control linkages used on early Models 51, 60 and 65.

1. Brake lever
2. Clevis
3. Jam nut
4. Brake rod
5. Spacer
6. Bolt
7. Brake drum
8. Brake band
9. Brake return spring
10. Brake stabilizer
11. Pin
12. Idler bracket
13. Clutch rod

tainer plates and withdraw differential assembly from chassis. Clean paint, rust and dirt from axle ends before removing bearings.

Reinstall in reverse order of removal. Adjust drive chain tension as outlined in previous paragraphs. Apply light coat of SAE 30 oil to chain. Lubricate axle bearings with multi-purpose lithium base grease.

GROUND DRIVE BRAKE

All Models

The brake used on early Models 51, 60 and 65 is a band and drum type (Fig. F8). A caliper type disc brake (Fig. F9) is used on all other models.

ADJUSTMENT. To adjust brake, depress clutch pedal and brake pedal and engage pedal locks, then attempt to push unit. If rear wheels rotate, release brake pedal and tighten adjusting nut (2 – Fig. F9) ½-turn on disc brake. On band brake, loosen jam nut (3 – Fig. F8) on brake rod (4), then disconnect clevis (2) from brake lever (1). Turn clevis ½-turn onto brake rod to tighten brake band. Recheck brake operation and repeat adjustment as required.

REMOVE AND REINSTALL. To remove brake band (8 – Fig. F8), remove pin (11) securing band to brake stabilizer (10). Remove anchor bolt (6) and spacer (5), then slide band off brake drum (7). Renew brake band by reversing removal procedure and adjust brake as outlined in previous paragraph.

To remove disc brake, unhook brake return spring (9 – Fig. F9), then disconnect brake rod (10) from brake cam lever (3). Unscrew cap screws (1), then remove brake parts (2 through 7). Re-

TRANSAXLE

Models Except 526

LUBRICATION. With unit on level surface, check rear axle oil by removing oil level plug (2 – Fig. F6). If lubricant does not run out, add SAE 90 EP oil through oil level hole until it begins to run out. Do not overfill. The transaxle lubricant can be drained by removing drain plug (3) in bottom of housing.

REMOVE AND REINSTALL. To remove transaxle (18 – Fig. F7), disengage mower blade clutch and place mower in lowest position. Depress clutch pedal and engage pedal lock, then stand unit on end as previously outlined. Disconnect brake rod (13) from brake lever (19). Remove retaining ring (17) from axle ends and remove wheels. Place temporary support under axle, then unbolt axle from mounting bracket (15) and frame. Slip traction drive belt (1) off transaxle pulley (10), and lower transaxle assembly from frame.

Reinstall in reverse order of removal.

OVERHAUL. The transaxles used are Peerless models. For overhaul procedures, refer to TRANSMISSION REPAIR service section of this manual.

DRIVE CHAIN

Model 526

ADJUSTMENT. To adjust chain, loosen nuts securing axle bearing retainer plates to frame. Move plates rearward to tighten chain. Both plates must be moved equal distance to keep axle square with frame. Chain should have ⅛ to ¼-inch deflection. When tension is correct, tighten bearing plate nuts.

REMOVE AND REINSTALL. Remove chain guard, then rotate rear

wheel to locate master link at transmission sprocket. Disconnect master link and remove chain. Clean and inspect chain and renew if excessively worn.

Reinstall in reverse order of removal. Adjust chain tension and lubricate with light coat of SAE 30 oil.

DIFFERENTIAL

Model 526

A Peerless model differential is used. For differential overhaul procedures, refer to DIFFERENTIAL REPAIR service section of this manual. Refer to following paragraphs for removal and installation.

REMOVE AND REINSTALL. To remove differential, raise rear of unit and support securely. Disconnect master link and remove drive chain. Remove retaining ring from axle ends and remove rear wheels. Unbolt bearing re-

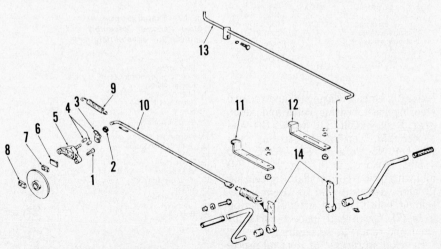

Fig. F9 – Exploded view of typical disc brake and clutch and brake control linkages.

1. Bolt	5. Carrier	9. Brake return spring
2. Adjusting nut	6. Back-up plate	10. Brake rod	13. Clutch rod
3. Brake lever	7. Outer brake pad	11. Brake lock lever	14. Brake and clutch link
4. Actuating pins	8. Inner brake pad	12. Clutch lock lever	15. Brake disc

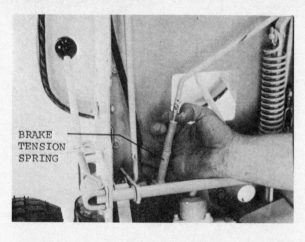

Fig. F10 — View of brake tension spring and holes in brake rod used to adjust length of spring.

BRAKE
TENSION
SPRING

move adjusting nut (2) and separate cam lever (3), actuating pins (4), back-up plate (6) and outer brake pad (7) from carrier (5). Slide brake disc (15) from transmission shaft and remove inner brake pad (8) from holder slot in transmission case.

Clean and inspect all parts for excessive wear or damage. Renew parts as required and reassemble by reversing removal procedure. Adjust brake as previously outlined.

BLADE DRIVE BRAKE

Models 51-60-61-65-66

ADJUSTMENT. Disconnect spark plug wire and disengage blade clutch,

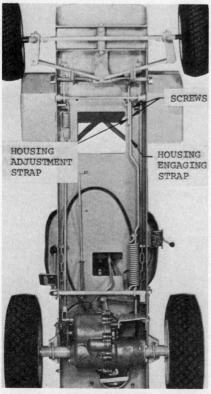

SCREWS

HOUSING
ADJUSTMENT
STRAP

HOUSING
ENGAGING
STRAP

Fig. F11 — View of adjustment strap and engaging strap. Linkage is lengthened to decrease belt tension or shortened to increase belt tension.

then stand unit on rear as previously outlined. Compress brake tension spring (Fig. F10) and reposition retainer pin in blade brake rod to increase spring tension and shorten braking time. An additional adjustment can be made as shown in Fig. F11. Loosen screws holding adjustment strap to engaging strap, and move adjustment strap to lengthen the linkage. Tighten screws in linkage, then recheck braking action and repeat adjustment as necessary.

Models 830-830E-1130E

ADJUSTMENT. Disconnect spark plug wire and disengage blade clutch. Loosen the three screws on both adjustment plates (3 – Fig. F12). Moving adjustment plates rearward, to decrease tension on drive belt (1), will shorten braking time. When adjustment is correct, tighten adjustment plate screws and check mower deck level adjustment.

MOWER DRIVE BELT

Models 51-60-61-65-66

ADJUSTMENT. Disconnect spark plug wire. Disengage blade drive clutch

Fig. F12 — View of mower deck mounting brackets and adjustment plates. Note locating stud on front mounting bracket must be inserted into hole in frame for proper mower deck alignment.

1. Blade drive belt
2. Mounting chain
3. Adjustment plates
4. Cam nuts
5. Mounting straps
6. Pins
7. Support shaft
8. Front mounting bracket
9. Locating stud
10. Rear mounting bracket

and set mower deck in highest position. Loosen screws (Fig. F11) securing housing adjustment strap to housing engaging strap, then move adjustment strap rearward to shorten linkage and increase belt tension. When adjustment is correct, tighten screws. Make sure blade braking action is not slowed by having linkage too short. Refer to blade brake adjustment covered in previous paragraphs.

Model 526

ADJUSTMENT. Disconnect spark plug wire. Disengage blade drive clutch and set mower deck in lowest position. Remove engaging spring (1 – Fig. F13) from idler lever (2) and relocate spring to hole which increases tension on spring.

Models 830-830E-1130E

ADJUSTMENT. Disconnect spark plug wire. Disengage blade drive clutch and set mower deck in lowest position. Loosen the three screws on both adjustment plates (3 – Fig. F12). Move adjustment plates forward to increase tension on drive belt. Tighten adjustment plate screws and check mower deck position. Make sure blade braking action is not slowed by having blade drive belt too tight. Refer to blade brake adjustment covered in previous paragraphs.

All Models

REMOVE AND REINSTALL. Disconnect spark plug wire and remove console tray. Disengage blade clutch and set mower height to lowest position. Stand unit on end as previously outlined. Remove engine pulley belt guide (1 – Fig. F5) and slip belt off engine pulley (10). Move blade clutch control lever to engaged position. Disconnect blade brake rod (7) from brake lever at mower spindle on Models 51, 60, 61, 65 and 66. Remove or tilt remaining belt

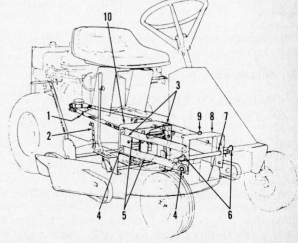
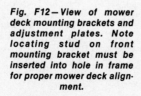

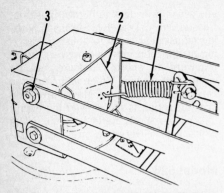

Fig. F13 — View of holes in idler lever used in adjusting blade engage spring tension on Model 526.

1. Engaging spring
2. Idler lever
3. Cam nut

guides as necessary to slip drive belt out of blade pulley and idler pulley, then pull belt free of mower.

Renew belt by reversing removal procedures and adjust to proper tension as previously outlined. Adjust belt guides to provide clearance "D" 1/16 to 1/8-inch between guide and outside surface of belt or pulley.

MOWER SPINDLE

Model 51

REMOVE AND REINSTALL. Remove mower deck as outlined in MOWER DECK paragraphs. Remove blade retaining bolt (1 – Fig. 14) and remove washers (2), blade (3) and blade holder (4) from spindle shaft (7). Remove blade pulley retaining ring (13) and loosen set screws (11), then pull blade pulley (12) off spindle shaft. Unbolt and remove brake mounting bracket (15) and brake lever (14), belt guide (16) and bearing retainer housing (9) from mower deck. Remove snap ring (6) from bottom of spindle shaft and separate shaft and bearings (8 and 10) from retainer housing.

Clean and inspect all parts and renew any showing excessive wear or damage. Reinstall by reversing removal procedure. Spindle bearings are sealed and require no additional lubrication.

Models 60-65

REMOVE AND REINSTALL. Remove mower deck as outlined in MOWER DECK paragraphs. Remove blade retaining bolt (1 – Fig. F15) and remove washers (2), blade (3) and blade holder (4) from spindle shaft (7). Unbolt brake band spacer (15) from brake band (17), then disconnect and remove brake drum from brake band bracket (19). Loosen set screws (11) and remove retaining ring (13) from blade pulley (12),

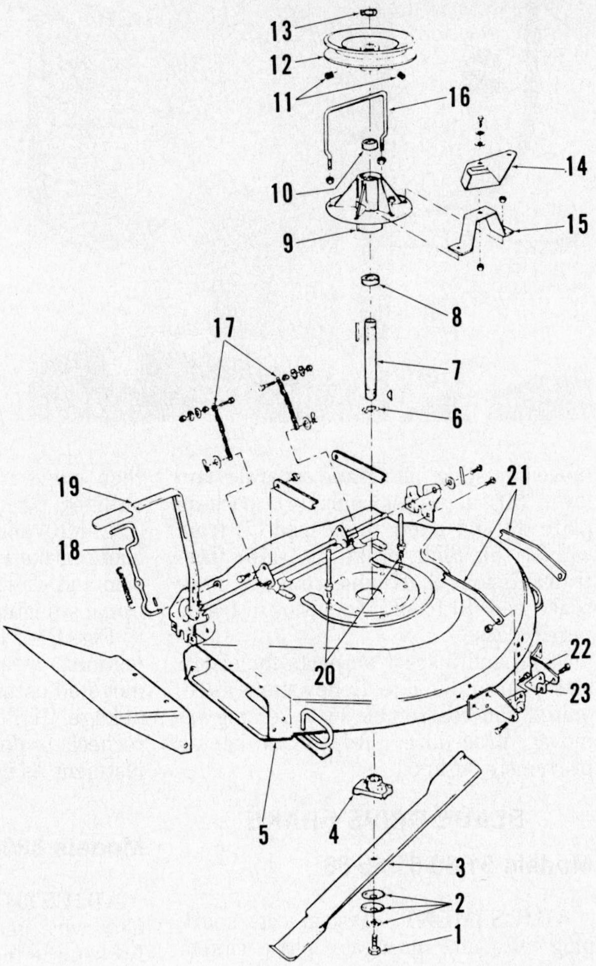

Fig. F14 — Exploded view of mower deck with lift operating controls used on Model 51.

1. Bolt
2. Washers
3. Blade
4. Blade holder
5. Mower deck
6. Retaining ring
7. Spindle shaft
8. Bearing
9. Bearing retainer housing
10. Bearing
11. Set screws
12. Blade pulley
13. Retaining ring
14. Brake lever
15. Brake mounting bracket
16. Belt guide
17. Lift chains
18. Latch slide
19. Height adjusting shaft
20. Belt guides
21. Parallel bars
22. Mounting bracket
23. Pivot bracket

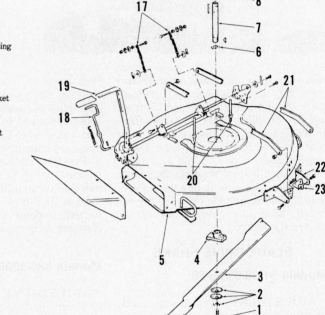

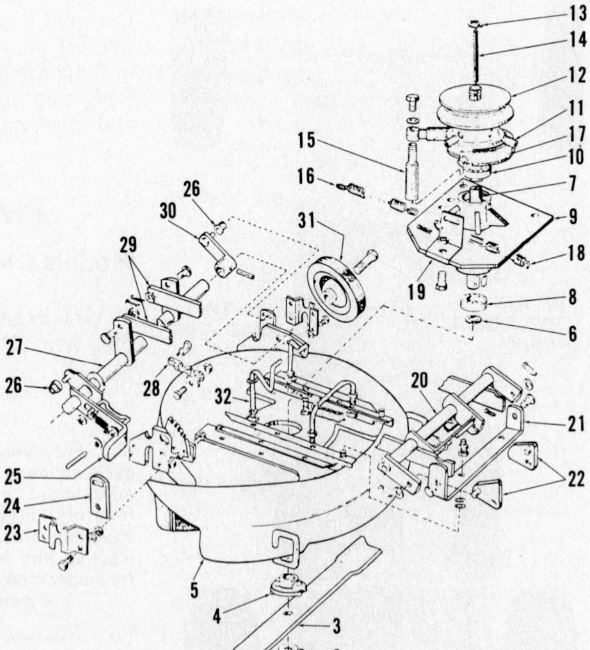

Fig. F15 — Exploded view of mower deck and spindle assembly with lift operating controls used on early Models 60 and 65.

1. Bolt
2. Washers
3. Blade
4. Blade holder
5. Mower deck
6. Retaining ring
7. Spindle shaft
8. Bearing
9. Bearing retainer housing
10. Bearing
11. Set screws
12. Blade pulley
13. Retaining ring
14. Square key
15. Brake band spacer
16. Brake spring
17. Brake band
18. Brake release spring
19. Brake band bracket
20. Pivot shaft
21. Front support
22. Height adjusting plates
23. Height adjustment slide retainer
24. Height adjustment slide
25. Handle grip
26. Height adjustment slide bushings
27. Wheel adjustment shaft
28. Chain
29. Height adjustment straps
30. Height adjustment link
31. Height adjustment wheel
32. Belt guide

then pull pulley off spindle shaft (7). Unbolt and remove bearing retainer housing (9) from mower deck. Remove retaining ring (6) from bottom of spindle

shaft, then separate shaft and bearings (8 and 10) from retainer housing.

Clean and inspect all parts and renew any showing excessive wear or damage.

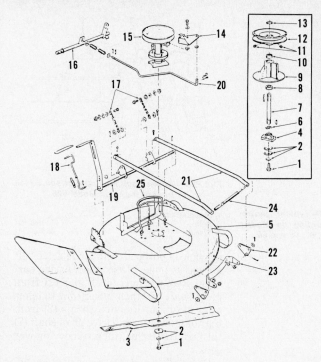

Fig. F16—Exploded view of mower deck and spindle assembly with lift operating controls used on late Models 60 and 65, Models 61 and 66 and early Models 830 and 830E. Note on Models 830 and 830E the spindle is renewed as a complete assembly. All other models can be renewed as shown in inset.

1. Bolt
2. Washers
3. Blade
4. Holder
5. Mower deck
6. Retaining ring
7. Spindle shaft
8. Bearing
9. Bearing retainer housing
10. Bearing
11. Set screws
12. Blade pulley
13. Retaining ring
14. Blade brake lever
15. Blade spindle assy.
16. Clutch lever
17. Chains
18. Slide latch
19. Height adjust shaft
20. Blade brake rod
21. Height adjust straps
22. Height adjust plate
23. Front support
24. Stabilizer
25. Belt guide

MOWER DECK

Models 51-60-61-65-66

LEVEL ADJUSTMENT. Position unit on level surface and disconnect spark plug wire. Measure distance between ground and cutting edge of blade from side to side and front to rear. Blade should be ¼-inch lower in front than at rear. Level mower deck using adjustment slots (Fig. F18) where mounting chains attach to frame. Some models also have slotted holes in front mounting plate for side to side adjustments.

Models 526-830-830E-1130E

LEVEL ADJUSTMENT. Position unit on level surface and disconnect spark plug wire. Measure distance between ground and cutting edge of blade from side to side and front to rear. Blade should be ¼-inch lower in front than at rear. Adjust front to rear height by rotating cam nuts (4—Fig. F12). On Model 526, level deck from side to side by adjusting length of height adjust

Reinstall by reversing removal procedure.

Models 60-61-65-66-830-830E

REMOVE AND REINSTALL. Remove mower deck as outlined in MOWER DECK paragraphs. Remove blade retaining bolt (1—Fig. F16) and remove washers (2), blade (3) and blade holder (4) from spindle shaft (7). Remove retaining ring (13) and loosen set screws (11), then pull blade pulley (12) off spindle shaft (7). Unbolt and remove spindle bearing housing (9) from mower deck. Remove retaining ring (6) from bottom of spindle shaft and separate shaft and bearings (8 and 10) from retainer housing. On Models 830 and 830E, the pulley and spindle assembly (15) is renewed as a complete unit.

Clean and inspect parts and renew any showing excessive wear or damage. Reinstall by reversing removal procedure. Spindle bearings are sealed and require no additional lubrication.

Models 526-830-830E-1130E

REMOVE AND REINSTALL. Remove mower deck as outlined in MOWER DECK paragraphs. Remove blade retaining bolt (1—Fig. F17) and remove washers (2) and blade (3) from spindle. Unbolt and remove idler pulley (9) and idler lever (11) and mounting bracket (13) from mower deck. Unbolt blade spindle assembly (8) and lift off mower deck. Blade spindle assembly is renewed as a complete unit.

Reinstall in reverse order of removal and check level adjustment.

Fig. F17—Exploded view of typical mower deck and lift operating controls used on Model 526 and late Models 830, 830E and 1130E.

1. Bolt
2. Washers
3. Blade
4. Spacer
5. Mower deck
6. Blade engage shaft
7. Blade engage spring
8. Blade spindle assy.
9. Idler pulley
10. Blade brake spring
11. Idler lever
12. Adjustment plates
13. Rear mounting bracket
14. Front mounting bracket
15. Upper mounting strap
16. Lower mounting strap
17. Mounting chains
18. Latch slide
19. Height adjust shaft
20. Cam nuts
21. Support shaft
22. Blade engage lever

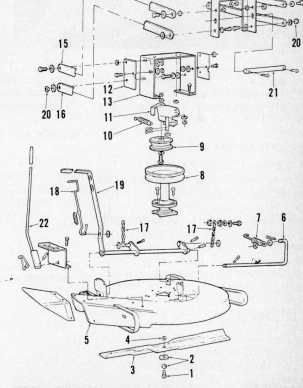

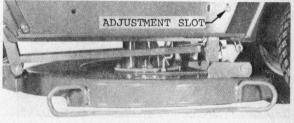

Fig. F18—View of deck mounting chains and adjustment slot.

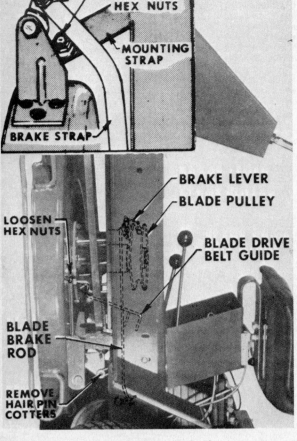

Fig. F19—Typical mower deck mounting on Models 51, 60, 61, 65 and 66.

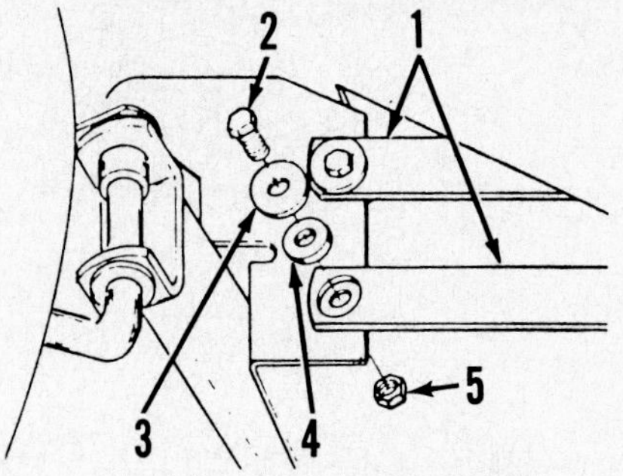

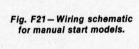

Fig. F20—Mower deck mounting straps attach to front of frame on Model 526.

1. Mounting straps
2. Bolt
3. Washer
4. Bushing
5. Nut

previously outlined. Slip blade drive belt off engine pulley. Disconnect wiring harness. Remove pins (6—Fig. F12) from pivot shaft (7), then lower unit to operating position. Remove chains (2) from rear of mower deck. Drive pivot shaft (7) through the frame, then remove mower deck.

Reinstall in reverse order of removal. Locating stud (9) must be inserted into hole in frame first for proper mower deck alignment. Check mower deck level adjustment.

ELECTRICAL

All Models

All models are equipped with an interlock safety switch system. Safety switches are located on blade clutch control linkage and traction clutch control linkage (except on Model 526, switch is located on transmission shift linkage). The switches are connected in series to a control module mounted on the engine. Refer to wiring schematic Fig. F21 or F22.

cables where cables attach to rear of mower deck. On Models 830, 830E and 1130E use adjustment slot where mounting chains (2—Fig. F12) attach to frame.

Models 51-60-61-65-66

REMOVE AND REINSTALL. Disconnect spark plug wire. Disengage blade drive clutch and set mower height to lowest position. Stand unit on end as previously outlined. Remove blade pulley belt guide (Fig. F19) and disconnect blade brake rod from brake lever, then slip belt off blade pulley. Lower unit to operating position, then disconnect mounting chains from rear of mower deck. Disconnect clutch and brake straps and mounting straps from the height adjust plates on front of mower deck, then remove mower deck.

Reinstall in reverse of removal and check deck level adjustment.

Model 526

REMOVE AND REINSTALL. Disconnect spark plug wire. Disengage blade drive clutch and set mower height to highest position. Stand unit on end as previously outlined. Slip blade drive belt off engine pulley. Disconnect wiring from blade clutch switch. Remove nuts from bottom side of deck which attach height adjust cables. Unbolt deck mount-

ing straps (1—Fig. F20) at front of frame, then remove mower deck.

Reinstall in reverse order of removal. Check mower deck level adjustment.

Models 830-830E-1130E

REMOVE AND REINSTALL. Disconnect spark plug wire. Disengage blade drive clutch and set mower in highest position. Stand unit on end as

Fig. F21—Wiring schematic for manual start models.

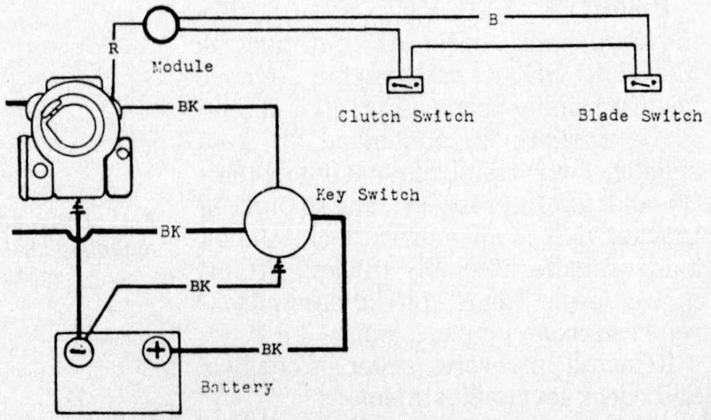

Fig. F22—Wiring schematic for electric start models.

1. R-Red
2. B-Brown
3. BK-Black

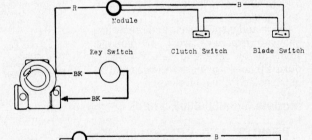

GILSON

GILSON BROTHERS COMPANY
P.O. Box 152
Plymouth, WI 53073

Model	Make	Engine Model	Horsepower	Cutting Width, In.
52013	B&S	130000	5	25
52014	B&S	190000	8	34
52015	B&S	190000	8	34
52031	B&S	130000	5	25
52032	B&S	190000	8	30
52033	B&S	190000	8	30
52038	B&S	130000	5	25
52039	B&S	190000	8	30
52040	B&S	190000	8	30
52051C	B&S	250000	11	36
52060	B&S	250000	11	38
52061	B&S	250000	11	36
52064	B&S	190000	8	38
52065	B&S	250000	11	38
52066	B&S	250000	11	38
52072	B&S	190000	8	38
52073	B&S	250000	11	38
52074	B&S	250000	11	38

FRONT AXLE

Models 52013-52014-52015-52031-52032-52033-52038-52039-52040

R&R AND OVERHAUL. To remove the axle main member (22–Fig. G1), support front of unit and remove front wheels. Disconnect steering link (9) from steering arm (12). Remove pivot pin (11) and lower axle assembly from chassis. On models so equipped, remove stabilizer springs (19). On all models, remove tie rod assembly (13 thru 15). Loosen clamp bolt, remove steering arm (12) and key, then remove spindle (17). Drive out roll pin (21), remove washer (23) and lower spindle (20) out of axle main member (22). Inspect spindle bushings (16) and pivot bushings (24) for excessive wear and renew as necessary. Reassemble by reversing disassembly procedure and adjust tie rod as required for 0 to 1/16-inch toe-in.

To remove steering shaft (2), disconnect steering link (9). Unbolt or unpin and remove steering wheel (1). Remove steering column tube (3), then slide steering shaft (2) downward and out of frame. Clean and inspect all parts and renew any showing excessive wear or other damage. When reassembling, add washers (7) as required to remove excessive end play of steering shaft (2).

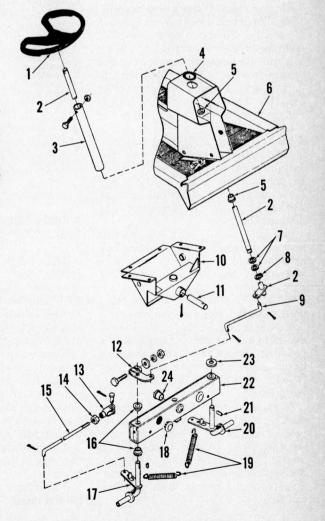

Fig. G1— Exploded view of typical front axle and steering system used on Models 52038, 52039 and 52040. Models 52013, 52014, 52015, 52031, 52032 and 52033 are similar, but are not equipped with stabilizer springs (19).

1. Steering wheel
2. Steering shaft
3. Column tube
4. Grommet
5. Flange bushing (2)
6. Frame
7. Washers
8. Snap ring
9. Steering link
10. Axle support
11. Pivot pin
12. Steering arm
13. Clevis
14. Jam nut
15. Tie rod
16. Flange bushings (4)
17. Spindle R.H.
18. Button plug
19. Stabilizer springs
20. Spindle L.H.
21. Roll pin
22. Axle main member
23. Washer
24. Axle pivot bushings (2)

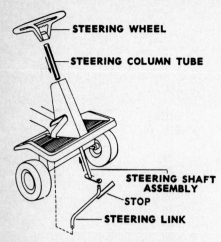

STEERING WHEEL

STEERING COLUMN TUBE

STEERING SHAFT ASSEMBLY

STOP

STEERING LINK

Fig. G2 — On Models 52013, 52014, 52015, 52031, 52032 and 52033, install steering link so that stop is towards steering shaft.

Models 52051C — 52060 — 52061

R&R AND OVERHAUL. Remove mower deck as needed, refer to MOWER BLADE CLUTCH AND BELT section for removal procedures. To remove axle assembly (22 – Fig. G2A) and steering parts, raise and securely block front portion of chassis.

Remove cap (34) and cotter key (30), then slide wheel assembly off spindles (24 and 29). Remove drag link end (14) from spindle steering arm (17) and swing clear of axle assembly. Remove nut and washer from pivot bolt (27). While supporting axle assembly withdraw pivot bolt, then lower complete axle assembly and place to the side for inspection and repair.

With reference to Fig. G2A disassemble and inspect complete axle assembly. Inspect spindles, axle assembly and tie bar for excessive wear, bending or any other damage and renew all parts as needed. Reassemble in reverse order of disassembly.

Inspect steering shaft (4), steering gear (5), bearings (3 and 10), drag link arm (16) and ends (14) for excessive wear or any other damage and renew all parts as needed.

Reinstall and lubricate spindle bushings, steering shaft and wear points in linkage with SAE 30 engine oil. Wipe off excess to prevent dirt accumulation. As provided, grease front wheel bushings using grease zerts on inside of front wheel assemblies with No. 2 wheel bearing grease.

To adjust steering gear mesh between steering shaft (4) and steering gear (5) proceed as follows: Loosen cap screws securing lower bearing flange (10) to frame. Slide bearing flange in slots as needed until gears mesh together evenly and steering is smooth, then retighten cap screws.

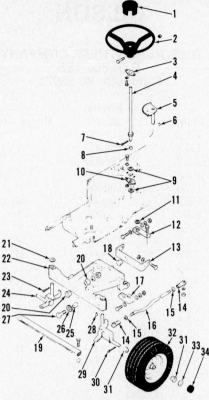

Fig. G2A — Exploded view showing steering components and related parts for Models 52051C, 52061 and 52060.

1. Cap
2. Steering wheel
3. Upper flange bearing
4. Steering shaft
5. Steering gear
6. Woodruff key
7. Cotter key
8. Spacer
9. Spacer washer
10. Lower flange bearing
11. Spacer washer
12. Steering arm
13. Bracket
14. Drag link end
15. Jam nut
16. Drag link
17. Spindle steering arm
18. Bearing
19. Tie bar
20. Bushing
21. Washer
22. Axle assy.
23. Cotter key
24. Right spindle assy.
25. Sleeve
26. Washer
27. Axle pivot bolt
28. Woodruff key
29. Left spindle assy.
30. Cotter key
31. Bushing
32. Wheel assy.
33. Washer
34. Cap

Models 52064-52065-52066-52072-52073-52074

R&R AND OVERHAUL. Remove mower deck as needed, refer to MOWER BLADE CLUTCH AND BELT section for removal procedures. To remove axle assembly (22 – Fig. G2B) and steering parts, raise and securely block front portion of chassis.

Remove cap (34) and cotter key (30), then slide wheel assembly off spindles 24 and 29. Remove drag link end (14) from spindle steering arm (17) and swing clear of axle assembly. Remove nut and washer from pivot bolt (27). While supporting axle assembly withdraw pivot bolt, then lower complete axle assembly and place to the side for inspection and repair.

With reference to Fig. G2B disassemble and inspect complete axle assembly.

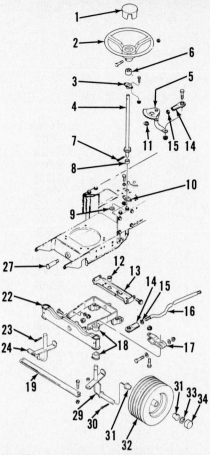

Fig. G2B — Exploded view showing steering components and related parts for Models 52064, 52065, 52066, 52072, 52073 and 52074. Refer to legend at Fig. G2A for identification of parts except for the following.

6. Sleeve
11. "E" Clip
12. Bushing
18. Spindle bushing

Inspect spindles, axle assembly and tie bar for excessive wear, bending or any other damage and renew all parts as needed. Reassemble in reverse order of disassembly.

Inspect steering shaft (4), steering gear (5), bearings (3 and 10), drag link arm (16) and ends (14) for excessive wear or any other damage and renew all parts as needed.

Reinstall and lubricate spindle bushings, steering shaft and wear points in linkage with SAE 30 engine oil. Wipe off excess to prevent dirt accumulation. As provided, grease front wheel bushings using grease zerts on inside of front wheel assembly with No. 2 wheel bearing grease.

To adjust steering gear mesh between steering shaft (4) and steering gear (5) proceed as follows: Loosen cap screws securing lower bearing flange (10) to frame. Slide bearing flange in slots as needed until gears mesh together evenly and steering is smooth, then retighten cap screws.

ENGINE

All Models

For overhaul and repair procedures on engines listed in Specification Table, refer to Small or Large Air Cooled Engines Service Manual. Refer to the following paragraphs for removal and installation procedures.

Models 52013-52014-52015-52031-52032-52033-52038-52039-52040

REMOVE AND REINSTALL. To remove the engine assembly, disconnect spark plug wire. On Models 52038, 52039 and 52040, unbolt and remove engine rear cover and side panels. On electric start models, disconnect battery cables. On all models disconnect ignition wires and throttle control cable. Place mower blade clutch control lever in disengaged position. Depress clutch-brake pedal and engage brake lock. Unbolt and remove engine pulley belt guide and remove mower drive belt from engine pulley. Loosen transmission input pulley belt guide and slip main drive belt from transmission pulley. Remove main drive belt from engine pulley. Remove engine mounting bolts and lift engine assembly from frame.

Reinstall engine by reversing the removal procedure. With main drive clutch engaged (pedal up), there should be 1/16-inch clearance between transmission input pulley belt guide and belt and a clearance of 1/16 to ⅛-inch between engine pulley and the engine pulley belt guide.

Models 52051C-52060-52061-52064-52065-52066-52072-52073-52074

REMOVE AND REINSTALL. Remove mower deck as needed. Disconnect battery ground cable from battery terminal. Disconnect engine spark plug wire. Unbolt and remove engine hood assembly. Remove drive belt from engine drive pulley. Disconnect engine throttle cable and ignition wiring as needed. Inspect and remove all parts that will obstruct in removal of engine assembly. Remove engine mounting bolts and lift engine assembly from frame.

Reinstall engine in reverse order of removal.

TRACTION DRIVE CLUTCH

Models 52013-52014-52015-52031-52032-52033-52038-52039-52040

The traction drive clutch is a belt idler type operated by the clutch-brake pedal

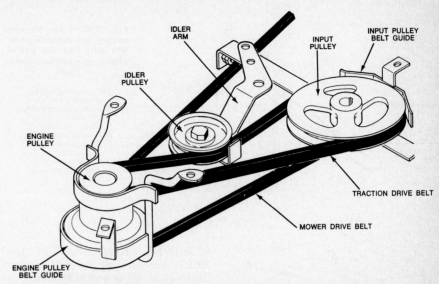

Fig. G3 — View showing traction drive belt, clutch idler and pulley on Models 52013, 52014, 52015, 52031, 52032, 52033, 52038, 52039 and 52040. Note location of belt guides.

on right side. When pedal is depressed, belt tension is removed, allowing engine drive pulley to rotate freely within the drive belt. There is no adjustment on the traction drive clutch. If belt slippage occurs during normal operation due to excessive belt wear or stretching, a new belt should be installed.

REMOVE AND RENEW DRIVE BELT. To remove traction drive belt, depress clutch-brake pedal and engage brake lock. Place mower blade clutch control lever in disengaged position. Unbolt and remove engine pulley belt guide (Fig. G3) and remove mower drive belt from engine pulley. Loosen transmission input pulley belt guide and slip traction drive belt from transmission pulley. Remove traction drive belt from idler and engine pulley.

Install new traction drive belt by reversing removal procedure and adjust belt guide clearances as follows: With main drive clutch engaged (pedal up), there should be 1/16-inch clearance between transmission input pulley belt guide and belt and a clearance of 1/16 – ⅛ inch between engine pulley and

engine pulley belt guide. Adjust brake as outlined in TRACTION DRIVE BRAKE paragraph.

Models 52051C-52060-52064-52065-52066-52072-52073-52074

The traction drive clutch is a belt idler type operated by the clutch-brake pedal on left side. When pedal is depressed, belt tension is removed, allowing engine drive pulley to rotate freely within the drive belt. There is no adjustment on the traction drive clutch. If belt slippage occurs during normal operation due to excessive belt wear or stretching, a new belt should be installed.

REMOVE AND RENEW DRIVE BELT. Remove mower deck as outlined in MOWER BLADE CLUTCH AND BELT section. Disconnect engine spark plug wire and set tractor parking brake. Disconnect battery ground cable from battery terminal. Loosen belt guide (3 – Fig. G3A) mounting bolt. Slide drive belt off input pulley (2) and past belt guide (1). Loosen rear idler pulley (4)

Fig. G3A — View showing traction drive belt, clutch idler and pulley on Models 52051C, 52060, 52064, 52065, 52066, 52072, 52073 and 52074. Note location of belt guides.

1. Belt guide
2. Transaxle input pulley
3. Belt guide
4. Rear idler pulley
5. Front idler pulley
6. Drive belt
7. Belt guide
8. Engine pulley

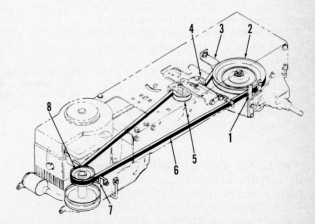

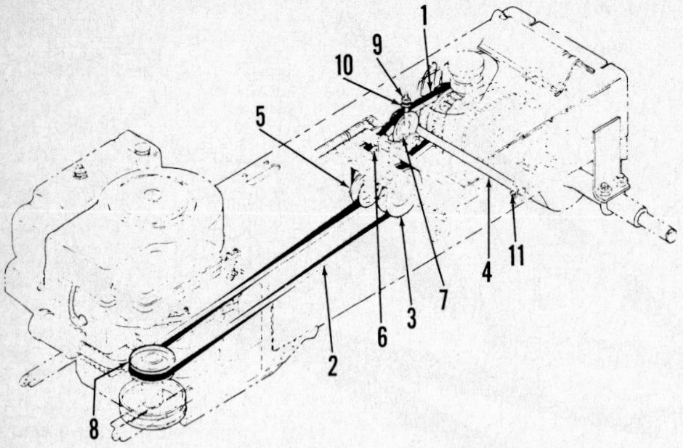

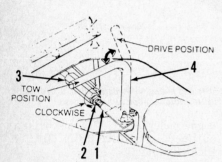

Fig. G3B – View showing primary and secondary traction drive belt and associated components for Model 52061.

1. Secondary drive belt
2. Primary drive belt
3. Inner jackshaft pulley
4. Lift shaft
5. Fixed idler pulley
6. Outer jackshaft pulley
7. Spring loaded idler
8. Engine pulley
9. Frame top nut
10. Frame underside nut
11. Roll pin

Fig. G3C – View showing position of control lever (4) for normal operation and for moving tractor when engine is inoperative. Adjust hydrostatic pump control linkage by loosening locknut (2) and turning turnbuckle (3) on control rod (1).

and belt finger mounting bolt, then slide drive belt off idler pulleys. Remove drive belt retainer (7), then slip drive belt off engine pulley (8). Inspect all drive pulleys for wear and renew as needed. Renew drive belt by reversing removal procedure. To adjust belt guide (3), release parking brake, then position belt guide 1/16 – 1/8 inch from drive belt.

Model 52061

ADJUSTMENT (Primary Drive Belt). Lift up seat support assembly to allow access to frame nut (9 – Fig. G3B). Loosen nut (10) on underside of frame, then turn top nut clockwise 1/2-turn to tighten belt and 1/2-turn counterclockwise to loosen belt. Retighten underside nut, then check drive belt deflection at midpoint of belt. Drive belt deflection should be one inch. Repeat adjustment procedure until correct deflection is attained.

(Secondary Drive Belt). There is no adjustment required on secondary drive belt. Idler pulley is spring loaded, check pulley arm periodically for freedom of movement and for providing correct belt tension.

REMOVE AND RENEW (Primary Drive Belt). Remove bolts and nuts attaching engine pulley belt cover to chassis. Remove mower deck as outlined

in MOWER BLADE CLUTCH AND BELT section. Lift up seat support assembly to allow access to frame nut (9 – Fig. G3B). Loosen nut (9), then remove belt from inner jackshaft pulley (3). Slip belt off engine pulley (8), then withdraw belt. Inspect all drive pulleys for excessive wear and renew as needed. Renew drive belt by reversing removal procedure. Adjust drive belt tension as outlined in ADJUSTMENT paragraph. Complete reassembly in reverse order of disassembly.

(Secondary Drive Belt). Remove bolts and nuts attaching engine pulley belt cover to chassis. Remove mower deck as outlined in MOWER BLADE CLUTCH AND BELT section. Remove roll pin (11 – Fig. G3B), then withdraw lift shaft (4). Push idler pulley (7) down to loosen belt tension, then slip drive belt off outer jackshaft pulley (6). Slip drive belt off transmission pulley, then withdraw drive belt. Inspect all drive pulleys for excessive wear and renew as needed. Renew drive belt by reversing removal procedure. Reassemble in reverse order of disassembly.

HYDROSTATIC PUMP CONTROL LINKAGE

Model 52061

Tractor can be moved a short distance without engine running by lifting seat and turning control lever (4 – Fig. G3C) counterclockwise. Lift lever to lock in place. Do not move tractor faster than walking speed. To release control lever, push lever down and turn clockwise.

Periodically check oil reservoir on top of transmission and fill to indicated level with SAE 20W High Detergent oil as necessary.

LINKAGE ADJUSTMENT. Linkage adjustment is correct when tractor does not move with engine running, clutch engaged and control lever in neutral.

To adjust linkage, raise both rear wheels and securely block. Loosen locknut (2 – Fig. G3C) on control rod (1) and turn turnbuckle (3) clockwise if tractor creeps forward and counterclockwise if tractor creeps backward. Retighten locknut, then recheck tractor for creeping. Continue adjustment procedure until correct adjustment is attained.

TRACTION DRIVE BRAKE

Differential Models

The brake is a band and drum type and is operated by the clutch-brake pedal. The brake drum is secured to the differential sprocket.

ADJUSTMENT. To adjust the brake, refer to Fig. G4 and turn the locknut clockwise to tighten brake. Tighten lock nut about 2½ turns and test brake. If necessary, repeat the adjustment until braking action is satisfactory.

Fig. G4 – View showing brake adjusting nut typical of all models. Refer to text for adjustment procedure.

TURN THIS
HEX. LOCK NUT

BRAKE SHOE
AND LINING

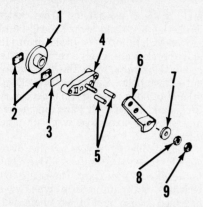

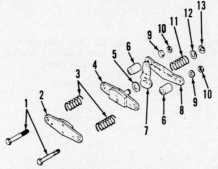

Fig. G4A — Exploded view showing a typical disc brake assembly used on transaxle models.

1. Disc
2. Pads
3. Plate
4. Carrier assy.
5. Dowel pins
6. Lever
7. Flat washer
8. Adjuster nut
9. Locknut

Fig. G4B — Exploded view showing disc brake assembly used on Model 52061.

1. Bolt
2. Pad
3. Spring
4. Pad
5. Flat washer
6. Spacer
7. Lever
8. Plate
9. Lockwasher
10. Nut
11. Spring
12. Flat washer
13. Snap ring

CAUTION: Do not overtighten brake. Make certain that brake does not drag when clutch is engaged.

OVERHAUL. Normal overhaul of the brake consists of renewing the band and lining assembly. The procedure is obvious after examination of the unit and reference to Fig. G7.

Transaxle Models

The brake is a disc type brake that is mounted on the transaxle. Shown in Fig. G4A is a view of a typical disc brake assembly used, other models are similar.

ADJUSTMENT. Loosen locknut (9 – Fig. G4A), then turn adjuster nut (8) so that brake is fully applied when foot pedal is completely engaged. Be sure brake disc (1) revolves freely when foot pedal is in released position. Retighten locknut.

OVERHAUL. Remove brake return spring, then disconnect brake rod from lever (6 – Fig. G4A). Remove mounting cap screws from carrier and pad assembly (4). Withdraw brake disc (1), then remove old brake pads. Renew brake pads, then reassemble in reverse order of disassembly. If brake adjustment is needed, refer to previous paragraph.

Model 52061

The brake is a disc type brake that is mounted on the differential assembly.

ADJUSTMENT. Adjust brake by turning adjuster nut on brake rod. Turn nut clockwise to reduce clearance between brake disc and brake pad. Brake should fully apply when foot pedal is completely engaged. After adjustment and foot pedal released, check to be sure

brake is not overtightened by engaging control lever to tow position and rolling tractor back and forth.

OVERHAUL. Disconnect brake rod from lever (7 – Fig. G4B). Remove nut and lockwasher (10 and 9), then remove pad (4), flat washer (5), lever (7), plate (8), spring (11), flat washer (12) and snap ring (13) as an assembly. Withdraw spacers (6), springs (3) brake disc and inner brake pad (2). Remove snap ring (13) from outer brake pad assembly, then disassemble with reference to Fig. G4B. Renew brake pads, then reassemble in reverse order of disassembly. If brake adjustment is needed, refer to previous paragraph.

TRANSMISSION

All Models So Equipped

The transmission used on Models 52013, 52014 and 52015 is a 3-speed Model 2010-13 manufactured by the J.B. Foote Foundry Co. All other models are equipped with a 3-speed Model 505

Fig. G5 — Underneath view of a typical rear engine rider with mower unit removed.

transmission manufactured by Peerless Division of Tecumseh Products Company.

REMOVE AND REINSTALL. To remove the transmission assembly, disconnect spark plug wire and on electric start models disconnect battery cables. On Models 52038, 52039 and 52040, unbolt and remove right side panel. On all models unbolt and remove drive chain guard. Locate connecting link, then disconnect drive chain. On Models 52013, 52014 and 52015, remove transmission shift handle. Disconnect safety starting switch wires from transmission on Models 52038, 52039 and 52040. On all models, depress clutch-brake pedal and engage brake lock. Unbolt and remove frame cross brace. See Fig. G5. Loosen transmission input pulley belt guide and remove main drive belt from input pulley. Remove the snap ring from transmission input shaft and the two socket head set screws from input pulley. Remove the pulley from transmission input shaft. Unbolt and remove transmission assembly.

Reinstall transmission by reversing the removal procedure. Adjust transmission input pulley belt guide so there is 1/16-inch clearance between belt guide and main drive belt when clutch is engaged. On Models 52038, 52039 and 52040, make certain that safety starting switches are connected and in good condition before returning mower to service.

OVERHAUL. For transmission overhaul procedures, refer to Foote Model 2010-13 on Models 52013, 52014 and 52015 and Peerless Model 505 on all other models in the TRANSMISSION REPAIR section of this manual.

TRANSAXLE

All Models So Equipped

The transaxles used are manufactured by Peerless Division of Tecumseh Prod-

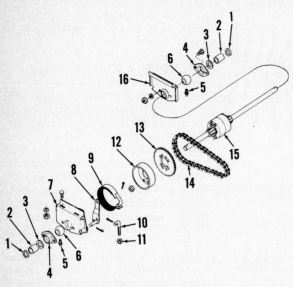

Fig. G6 — Exploded view of rear axle mounting bracket assemblies.

1. Spacer washer
2. Spacer
3. Washer (thin)
4. Bearing clamp
5. Lube fitting
6. Axle bearing
7. Mounting bracket R.H.
8. Brake lever
9. Brake band
10. Link
11. Adjusting nut
12. Brake drum
13. Sprocket
14. Drive chain
15. Axle & differential assy.
16. Mounting bracket L.H.

ucts Company. Models 52051C, 52060, 52064, 52065 and 52066 use a 3-speed Series 600. Models 52072, 52073 and 52074 use a 5-speed Series 800.

REMOVE AND REINSTALL. To remove the transaxle assembly, disconnect spark plug wire and battery ground cable from battery terminal. Loosen belt guide bolts and slide belt guides away from belt. Remove traction drive belt from transaxle input pulley. Disconnect brake linkage. Remove cap screws retaining transaxle to frame and remove "U" bolts securing axle housings to frame. Raise rear of tractor and remove transaxle assembly from tractor.

Reinstall by reversing the removal procedure. Adjust clutch and brake linkage as required.

OVERHAUL. For overhaul procedures, refer to Peerless transaxle Series 600 on Models 52051C, 52060, 52064, 52065 and 52066 and Series 800 on Models 52072, 52073 and 52074 in the TRANSMISSION REPAIR section of this manual.

HYDROSTATIC TRANSMISSION

Model 52061

Model 52061 is equipped with a Model 7 hydrostatic transmission manufactured by Eaton Corporation.

REMOVE AND REINSTALL. To remove transmission and differential, disconnect hydrostatic control rod at transmission. Disconnect brake linkage and remove rear drive belt. Raise rear wheels clear of ground and securely block. Remove axle housing mounting nuts and "U" bolts securing axle housings to frame, then remove transmission and differential assembly from tractor. Separate transmission and differential.

Reinstall by reversing the removal procedure. Adjust linkage rods as needed.

OVERHAUL. For hydrostatic pump overhaul procedures, refer to Eaton Hydrostatic section in the TRANSMISSION REPAIR section of this manual.

DIFFERENTIAL

All Models So Equipped Except 52061

REMOVE AND REINSTALL. To remove the rear axle and differential assembly, raise rear of chassis and support securely. Disconnect brake rod from brake lever (8 – Fig. G6). Locate connecting link and disconnect drive chain. Remove rear wheels, spacer washers (1), spacers (2) and thin washers (3). Unbolt mounting brackets (7 and 16) from frame and remove the rear axle assembly. Clean all rust or paint from axles and slide axle bearing

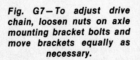

Fig. G7 — To adjust drive chain, loosen nuts on axle mounting bracket bolts and move brackets equally as necessary.

and bracket assemblies from axles. Brake band (9) and lever (8) will be removed with bracket (7). Brake drum (12) and sprocket (13) can be removed after removal of nuts from the four differential thru bolts. Unbolt bearing clamps (4) and remove bearings (6) if necessary. Clean and inspect all parts and renew any showing excessive wear or other damage.

Reassemble by reversing the disassembly procedure. Adjust drive chain and brake as necessary. Lubricate axle bearings with automotive type wheel bearing grease and drive chain with SAE 30 oil.

OVERHAUL. Peerless differentials are used on all models. Model 1523 differential is used on Models 52038, 52039 and 52040. Model 117 differential is used on all other models. For overhaul procedures, refer to the DIFFERENTIAL REPAIR section of this manual.

Model 52061

Refer to Hydrostatic Transmission section for removal of differential assembly. For overhaul procedures, refer to Peerless Series 1300 in the DIFFERENTIAL REPAIR section of this manual.

DRIVE CHAIN

All Models So Equipped

ADJUSTMENT. To adjust the drive chain, refer to Fig. G7 and loosen the hex nuts on bolts securing axle mounting brackets to the frame. Move brackets rearward to tighten chain. Both brackets must be moved the same distance to keep the axle and sprocket square with frame. The chain should have a slight amount of slack, approx-

LOOSEN HEX. NUTS ON THESE TWO BOLTS ON EACH END OF AXLE.

BEARING CLAMP BRACKET

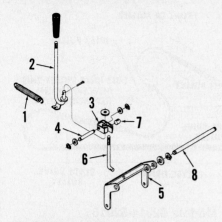

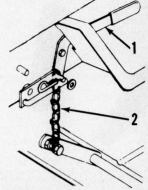

Fig. G9B — View showing lift control lever (1) and chain (2) used in raising and lowering mower deck on Models 52051C, 52060, 52061, 52066 and 52074.

Fig. G8 — Exploded view of blade clutch linkage used on Models 52038, 52039 and 52040.

1. Clutch tension spring
2. Blade clutch control lever
3. Pivot bracket
4. Pivot pin
5. Hanger arm assy.
6. Clutch link
7. Link pivot
8. Hanger pivot shaft

imately ¼ to ⅜-inch. Tighten the hex nuts securely when chain adjustment is correct.

REMOVE AND REINSTALL. To remove the drive chain, remove the right side panel on Models 52038, 52039 and 52040, then unbolt and remove the chain guard from all models. Locate the connecting link, then disconnect and remove the drive chain.

If new chain is to be installed, loosen rear axle mounting brackets (Fig. G7), install chain and adjust chain as outlined previously. Lubricate chain with a light coat of SAE 30 oil.

MOWER BLADE CLUTCH AND BELT

All Models

CAUTION: Always disconnect spark plug wire before performing any inspection, adjustment or other service on the mower unit.

The mower unit on all rear engine models is attached to the hanger arm assembly (5 – Fig. G8 or G9). The blade clutch is operated by control lever (2) and the linkage which moves mower unit forward or rearward, tightening or loosening mower drive belt. When control lever (2) is moved rearward and locked in disengaged position, mower assembly is moved rearward, loosening the belt. The idler spring (10 – Fig. G12, G13 and G14) pulls the idler and brake arm (8) around until brake shoe (9) contacts the mower pulley, stopping blade rotation. At this time, mower drive belt is free from engine crankshaft pulley. When the control lever is moved forward (engaged position), mower is moved forward. This tightens the belt

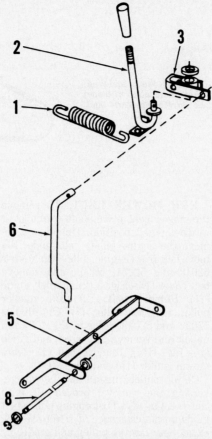

Fig. G9 — Exploded view of blade clutch linkage used on Models 52013, 52014, 52015, 52031, 52032 and 52033. Clutch link (6) is non-adjustable on these models. Refer to Fig. G8 for legend.

and pulls the brake shoe on idler arm free from the mower pulley.

The mower unit on all front engine tractors is attached to the hanger

bracket located at the front of the chassis. On Models 52064, 52065, 52072 and 52073 mower deck is raised and lowered by a chain attached to a lift arm which is operated by a cable attached to a lift control lever located on right-hand side of frame, typical of the one shown in Fig. G9A. On Models 52051C, 52060, 52061, 52066 and 52074 mower deck is raised and lowered by a chain attached to a lift control lever located on right-hand side of frame, typical of the one shown in Fig. G9B. Mower engagement lever is located on left side of instrument panel. When mower control lever is moved to engaged position, idler pulley should tighten tension on drive belt and blade brake should move away from pulleys. When control lever is moved to disengaged position, idler pulley should move away from drive belt and blade brake should come in contact with pulleys. Refer to the following paragraphs for adjustment of mower

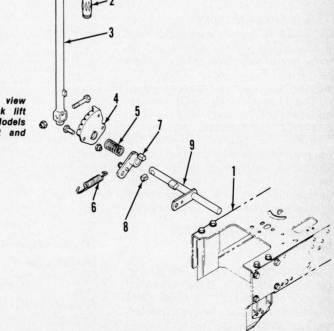

Fig. G9A — Exploded view showing mower deck lift control assembly for Models 52064, 52065, 52072 and 52073.

1. Frame
2. Grip
3. Handle
4. Quadrant
5. Spring
6. Spring
7. Lift arm
8. Spacer
9. Lift shaft

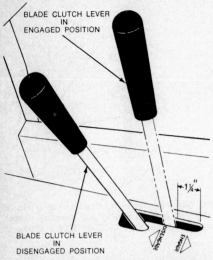

Fig. G10—On Models 52038, 52039 and 52040, adjust blade clutch linkage so that a distance of 1¼ inches exists between clutch lever and end of slot when blade clutch is engaged.

Fig. G11—View showing correct installation of blade belt on Models 52014 and 52015.

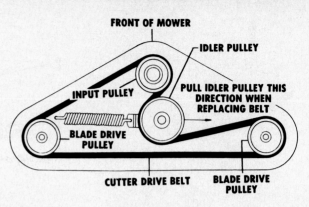

engagement clutch and renewing drive belt.

Models 52038-52039-52040

ADJUSTMENT. If blade clutch does not disengage when blade clutch lever is in disengaged position (Fig. G10), adjust linkage as follows: Disconnect mower from hanger arm assembly. Turn the link clockwise one complete turn into pivot (7–Fig. G8) and reconnect link to hanger arm assembly. Reattach mower and recheck clutch operation. Repeat the adjustment if necessary.

If the clutch control lever contacts front end of slot before the clutch is engaged, adjust link (6) counter-clockwise until clutch lever is approximately 1¼ inches from end of slot when clutch is engaged. See Fig. G10.

R&R MOWER BELT. To remove the mower belt, place blade clutch lever in disengaged position. Unbolt and remove engine pulley belt guide. Remove belt from engine pulley. Unbolt and remove belt cover and guide (15–Fig. G14 or 11 and 13–Fig. G12), then remove belt from mower pulley (16–Fig. G14 or 14–Fig. G12) and idler.

Install new belt by reversing the removal procedure and adjust clutch linkage as outlined in the preceding paragraph. With clutch engaged, adjust engine pulley belt guide to a clearance of 1/16 to ⅛-inch between pulley and guide.

Models 52013-52031-52032-52033

ADJUSTMENT. Clutch link (6–Fig. G9) is non-adjustable. If the mower belt is worn or stretched to a point where blade clutch will not engage, a new belt must be installed.

R&R MOWER BELT. To remove the mower belt, place blade clutch lever in disengaged position. Unbolt and remove the engine pulley belt guide. Remove belt from engine pulley. On Models 52013 and 52031, unbolt and remove belt cover (13–Fig. G12) and belt guide (11). Remove belt (12) from mower pulley (14) and idler (7). On Models 52032 and 52033, refer to Fig. G14 and unbolt and remove belt cover and guide (15–Fig. G14). Remove belt (13) from mower pulley (16) and idler (7).

On all models, install new belt by reversing the removal procedure. With clutch engaged, adjust engine pulley belt guide until a clearance of 1/16 to ⅛-inch exists between the pulley and guide.

Models 52014-52015

ADJUSTMENT. Clutch link (6–Fig. G9) is non-adjustable. If the mower drive belt is worn or stretched to a point where blade clutch will not engage, a new drive belt must be installed.

The blade belt (Fig. G11) is under constant tension of the spring loaded idler and no adjustment is required.

R&R MOWER BELTS. To remove the mower drive belt and the blade belt, place blade clutch lever in disengaged position. Unbolt and remove engine pulley belt guide. Remove mower drive belt from engine pulley. Unbolt and remove belt cover (11–Fig. G13), then

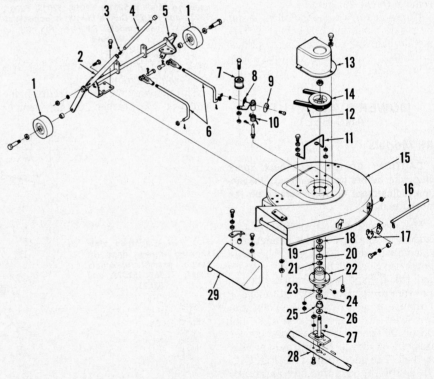

Fig. G12—Exploded view of 25" single blade mower used on Models 52013, 52031 and 52038.

1. Cutting height gage wheels	14. Mower pulley	22. Spindle housing
2. Lever	7. Idler pulley	15. Mower housing	23. Retaining ring
3. Spring	8. Idler & brake arm	16. Mower attaching pin	24. Bearing cup
4. Rod	9. Brake shoe	17. Bellcranks (2)	25. Bearing cone
5. Pivot shaft & support assy.	10. Spring	18. Seal	26. Seal
6. Leveler links	11. Belt guide	19. Bearing cone	27. Spindle
	12. Mower drive belt	20. Bearing cup	28. Blade 25"
	13. Belt cover	21. Retaining ring	29. Deflector

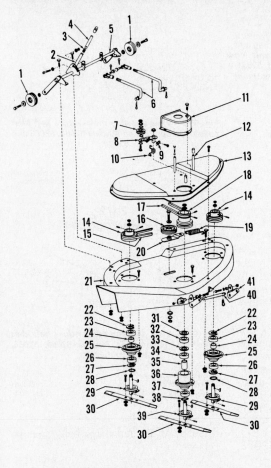

Fig. G13 – Exploded view of 34" 3-blade mower used on Models 52014 and 52015.

1. Cutting height gage wheels
2. Lever
3. Spring
4. Rod
5. Pivot shaft & support assy.
6. Leveler links
7. Idler pulley
8. Idler & brake arm
9. Brake shoe
10. Spring
11. Belt cover
12. Belt guides (2)
13. Blade belt cover
14. Blade pulleys
15. Blade belt
16. Blade belt idler
17. Mower drive belt
18. Input (center spindle) pulley
19. Spring
20. Idler arm
21. Mower housing
22. Retaining ring
23. Bearing
24. Spacer
25. Spindle housing
26. Bearing
27. Retaining ring
28. Spacer
29. Spindle
30. Blades (3)
31. Washer
32. Bearing
33. Washer
34. Bearing
35. Spacer
36. Spindle housing
37. Bearing
38. Spacer
39. Spindle
40. Mower attaching pin
41. Bellcranks (2)

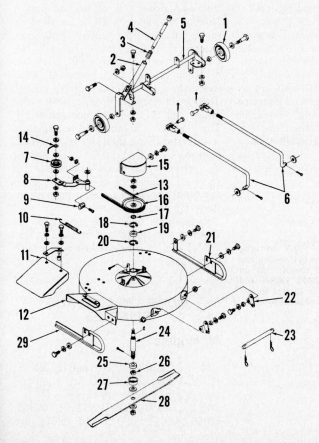

Fig. G14 – Exploded view of 30" single blade mower used on Models 52032, 52033, 52039 and 52040.

1. Cutting height gage wheels
2. Lever
3. Spring
4. Rod
5. Pivot shaft & support assy.
6. Leveler links
7. Idler pulley
8. Idler & brake arm
9. Brake shoe
10. Spring
11. Deflector
12. Mower housing
13. Mower drive belt
14. Belt finger
15. Belt cover & guide
16. Mower pulley
17. Snap ring
18. Retaining ring
19. Bearing
20. Retaining ring
21. Mower runner L.H.
22. Bellcranks (2)
23. Mower attaching pin
24. Spindle
25. Bearing
26. Jam nut
27. Dust cup
28. Blade 30"
29. Mower runner R.H.

remove belt (17) from center spindle pulley (18) and idler (7).

To remove the blade belt (15), unpin front of mower from hanger arm assembly and remove mower from right side of chassis. Unbolt and remove blade belt cover (13). Refer to Fig. G11 and pull idler pulley in direction shown and remove the belt.

Install new belts by reversing the removal procedure. With mower attached, engage the clutch and adjust engine pulley belt guide until a clearance of 1/16 to 1/8-inch exists between the pulley and guide.

Models 52051C-52061

ADJUSTMENT. Move mower engagement lever to engaged position. Turn adjustment nut (1 – Fig. G15) until tension spring (2) measured length is 2⅝ inches as shown in Fig. G15. If correct adjustment can not be attained, then disengage mower engagement lever. Unbolt idler pulley (1 – Fig. G16) from idler arm (5), then reposition idler pulley in back hole (2) on idler arm. Engage mower lever and measure length of tension spring. Readjust adjustment nut (1 – Fig. G15) until correct length is attained. Clearance between belt retainer (3 – Fig. G16) and drive belt should be 1/16-1/8 inch. If drive belt still slips on drive pulleys, then drive belt must be renewed.

R&R MOWER BELT. Move mower engagement lever to disengaged position. Remove engine drive belt cover, then slip drive belt off engine pulley. Unhook mower deck lift chain, then remove pins securing mower deck front hanger to tractor frame hanger. Remove any other components that will obstruct mower deck removal, then withdraw mower deck. Remove mower deck front hanger (6 – Fig. G15), idler pulley (4) and belt guard (7). Slip drive belt off pulleys. Inspect all pulleys for excessive wear or any other damage and renew as needed. Renew drive belt.

Reassemble in reverse order of disassembly. Refer to previous ADJUSTMENT section for drive belt adjustment procedure.

Models 52060-52066-52074

ADJUSTMENT. Move mower engagement lever to engaged position. Turn adjustment nut (1 – Fig. G17) until tension spring (2) measured length is 2⅝ inches as shown in Fig. G18. If correct adjustment can not be attained, then disengage mower engagement lever. Unbolt idler pulley (1) from idler arm (5), then reposition idler pulley in a slot (2) closer to rear of mower deck. There are

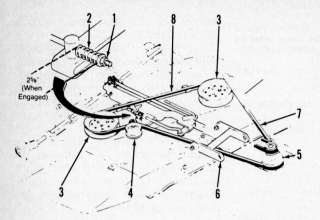

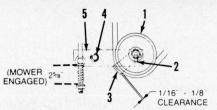

Fig. G15 — View showing mower drive belt routing and associated components on Models 52051C and 52061. Refer to text for adjustment procedure.

1. Adjustment nut
2. Tension spring
3. Spindle drive pulley
4. Idler pulley
5. Engine pulley
6. Front hanger
7. Belt guard
8. Drive belt

Fig. G18 — View showing mower drive belt idler pulley components on Models 52060, 52066 and 52074.

1. Idler pulley
2. Adjustment slots
3. Belt retainer
4. Pivoting pin
5. Idler arm

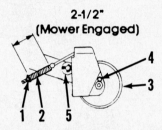

Fig. G19 — View showing mower drive belt idler pulley components on Models 52064, 52065, 52072 and 52073.

1. Adjustment nut
2. Tension spring
3. Idler pulley
4. Front hole
5. Idler arm

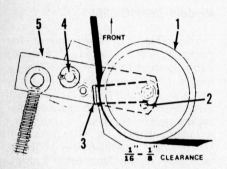

Fig. G16 — View showing mower drive belt idler pulley components on Models 52051C and 52061.

1. Pulley
2. Back hole
3. Belt retainer
4. Pivoting pin
5. Idler arm

four possible adjustment slots. Engage mower lever and measure length of tension spring. Readjust adjustment nut (1 – Fig. G17) until correct length is attained. Clearance between belt retainer (3 – Fig. G18) and drive belt should be 1/16-⅛ inch. If drive belt still slips on drive pulleys and idler pulley has been adjusted as far toward rear of mower as possible, then drive belt must be renewed.

R&R MOWER BELT. Move mower engagement lever to disengaged position. Remove engine drive belt cover,

then slip drive belt off engine pulley. Unhook mower deck lift chain, then remove pins securing mower deck front hanger to tractor frame hanger. Remove any other components that will obstruct mower deck removal, then withdraw mower deck front hanger (6 – Fig. G17), idler pulley (4) and belt guard (7). Slip drive belt off pulleys. Inspect all pulleys for excessive wear or any other damage and renew as needed. Renew drive belt.

Reassemble in reverse order of disassembly. Refer to previous ADJUSTMENT section for drive belt adjustment procedure.

Models 52064-52065-52072-52073

ADJUSTMENT. Move mower engagement lever to engaged position. Turn adjustment nut (1 – Fig. G19) until tension spring (2) measured length is 2½ inches as shown in Fig. G19. If belt still slips after adjustment, then disengage mower engagement lever. Unbolt idler pulley (3) from idler arm (5), then reposition idler in back hole as shown in Fig. G19. Front hole (4) is shown open in Fig. G19. Engage mower lever and measure length of tension spring. Readjust adjustment nut (1) until correct length is attained. Clearance between idler pulley belt retainer and drive belt should be 1/16-⅛ inch. If drive belt still slips on

drive pulleys, then drive belt must be renewed.

R&R MOWER BELT. Move mower engagement lever to disengaged position. Remove engine drive belt guide, then slip drive belt off engine pulley. Unhook mower deck lift chain, then remove pins securing mower deck front hanger to tractor frame hanger. Remove any other components that will obstruct mower deck removal, then withdraw mower deck. Remove mower deck front hanger (6 – Fig. G20), idler pulley belt retainer and belt guard (7). Turn pivot assembly (9) completely counterclockwise, then slip drive belt off pulleys. USE CAUTION when sliding belt between pulleys and brake pads. Inspect all pulleys for excessive wear or any other damage and renew as needed. Renew drive belt.

Reassemble in reverse order of disassembly. Refer to previous ADJUSTMENT section for drive belt adjustment procedure.

MOWER BLADES AND SPINDLES

All Models

Models 52013, 52031, 52032, 52033, 52038, 52039 and 52040 are equipped with single blade rotary mowers. Models 52051C, 52060, 52061, 52064, 52065, 52066, 52072, 52073 and 52074 are

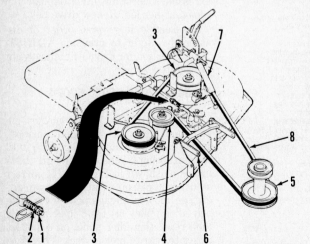

Fig. G17 — View showing mower drive belt routing and associated components on Models 52060, 52066 and 52074. Refer to text for adjustment procedure.

1. Adjustment nut
2. Tension spring
3. Spindle drive pulley
4. Idler pulley
5. Engine pulley
6. Front hanger
7. Belt guard
8. Drive belt

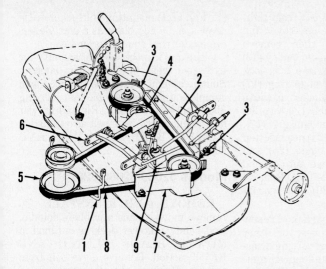

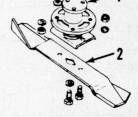

Fig. G20 — View showing mower drive belt routing and associated components on Models 52064, 52065, 52072 and 52073.

1. Belt guide
2. Brake strap
3. Spindle drive pulley
4. Idler pulley
5. Engine pulley
6. Front hanger
7. Belt guard
8. Drive belt
9. Pivot assy.

Fig. G21 — View showing spindle assembly (1) and mower blade (2) used on Models 52051C and 52061.

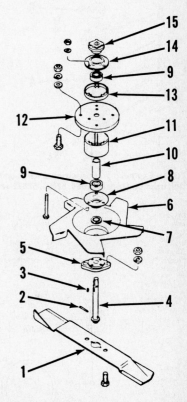

Fig. G22 — Exploded view showing mower spindle unit used on Models 52060, 52066 and 52074.

1. Blade	
2. Pin	
3. Key	9. Bearing
4. Shaft	10. Spacer
5. Adapter	11. Housing
6. Fan	12. Flange
7. Spacer	13. Housing
8. Bearing cup	14. Bearing cup
	15. Pulley hub

equipped with dual blade rotary mowers. Models 52014 and 52015 are equipped with three blade rotary mowers.

CAUTION: Always disconnect spark plug wire before performing any inspection, adjustment or other service on the mower.

On models so equipped, make certain that safety starting switches are connected and are in good working condition before returning mower to service.

The following paragraphs contain procedures for removing blades, spindles and renewing spindle bearings.

Models 52013-52031-52038

REMOVE AND REINSTALL. To remove the blade and blade spindle, first remove mower assembly as follows: Place blade clutch control lever in disengaged position. Unbolt and remove engine pulley belt guide and remove mower belt from engine pulley. Unpin front of mower from hanger arm assembly. Remove mower from right side of chassis. Unbolt and remove belt cover (13 – Fig. G12) and belt guide (11). Remove belt (12) from pulley (14) and idler (7). Remove nut and mower pulley (14), then unbolt blade (28) from spindle (27). Unbolt and remove spindle housing assembly (18 thru 27) from mower housing (15). Remove spindle (27), then remove seals (18 & 26), bearing cones (19 & 25), bearing cups (20 & 24) and retaining rings (21 & 23).

Clean and inspect all parts and renew any showing excessive wear or other damage. Reassembly is the reverse of disassembly procedure. Reinstall mower, engage clutch and adjust engine pulley belt guide for a clearance of 1/16 to ⅛-inch between the pulley and belt guide.

Models 52014-52015

REMOVE AND REINSTALL. To remove the blades and spindles, first remove mower assembly as follows: Place blade clutch control lever in disengaged position. Unbolt and remove engine pulley belt guide and remove mower drive belt from engine pulley. Unpin front of mower from hanger arm assembly. Remove mower unit out from under right side. Unbolt and remove belt cover (11 – Fig. G13), then remove mower drive belt (17) from center spindle pulley (18) and idler pulley (7). Unbolt and remove blade belt cover (13). Refer to Fig. G11 and pull idler pulley in direction shown, then remove blade belt.

Remove nut, lockwasher and two set screws, then lift off center pulley (18 – Fig. G13). Remove blade (30), then unbolt and remove center spindle housing assembly (31 thru 39) from mower housing (21). Remove spindle (39), spacer (38) and bearing (37) from bottom of spindle housing (36) and washers (31 & 33), bearings (32 & 34) and spacer (35) from top of spindle housing. Clean and inspect all parts and renew any showing excessive wear or other damage. Reassemble center spindle by reversing disassembly procedure.

Left and right spindle assemblies are identical. To remove either left or right spindle, remove nut, lockwasher and two set screws, then lift off pulley (14). Remove blade (30), then unbolt and remove spindle housing assembly (22 thru 29) from mower housing (21). Withdraw spindle (29) and spacer (28) from spindle housing (25). Remove retaining rings (22 & 28), bearings (23 & 26) and spacer (24) from spindle housing. Clean and inspect all parts and renew any showing excessive wear or other damage. Reassemble by reversing disassembly procedure.

Complete the balance of mower reassembly and reinstall the mower assembly. Engage clutch and adjust engine pulley belt guide for a clearance of 1/16 to ⅛-inch between the pulley and belt guide.

Models 52032-52033-52039-52040

REMOVE AND REINSTALL. To remove the blade and blade spindle, first remove mower assembly as follows: Place blade clutch control lever in disengaged position. Unbolt and remove engine pulley belt guide and remove mower belt from engine pulley. Unpin front of mower from hanger arm assem-

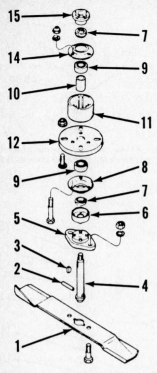

Fig. G23 — Exploded view showing mower spindle unit used on Models 52064, 52065, 52072 and 52073.

1. Blade	8. Bearing cup
2. Pin	9. Bearing
3. Key	10. Spacer
4. Shaft	11. Housing
5. Adapter	12. Flange
6. Shield	14. Bearing cup
7. Spacer	15. Pulley hub

bly. Remove mower unit out from right side. Unbolt and remove belt cover and guide (15 – Fig. G14), then remove mower belt (13) from mower pulley (16) and idler pulley (7). Remove nut, lock washer, pulley (16) and snap ring (17) from upper end of spindle. Remove blade and spindle assembly (24 thru 28) from bottom of mower housing (12). Clamp center of spindle (24) in a vise and remove nut, washers, blade (28) and dust cup (27), then remove nut (26) and bearing (25) from spindle (24). Remove retaining rings (18 & 20) and bearing (19) from mower housing.

Clean and inspect all parts and renew any showing excessive wear or other damage. Reassemble by reversing disassembly procedure. Reinstall mower unit, engage clutch and adjust engine pulley belt guide for a clearance of 1/16 to ⅛-inch between the pulley and belt guide.

Models 52051C-52061

REMOVE AND REINSTALL. To remove mower blade and blade spindle, first remove mower deck as outlined in MOWER BLADE CLUTCH AND BELT section. Remove drive belt from pulleys. Unbolt and remove spindle drive pulley. Remove spindle assembly

(1 – Fig. G21) mounting bolts, then withdraw spindle and blade as a unit. Separate blade from spindle assembly.

Clean and inspect spindle assembly for excessive wear or any other damage. Spindle assembly must be renewed as a complete unit. Reassemble by reversing disassembly procedure. Rinstall mower unit, then adjust drive belt as outlined in MOWER BLADE CLUTCH AND BELT section.

Models 52060-52066-52074

REMOVE AND REINSTALL. To remove mower blade and blade spindle, first remove mower deck as outlined in MOWER BLADE CLUTCH AND BELT section. Remove drive belt from pulleys. Remove nut, lockwasher and flat washer securing drive pulley to pulley hub (15 – Fig. G22), then withdraw drive pulley. Remove spindle assembly mounting bolts, then withdraw spindle assembly and blade as a unit. Separate blade from spindle assembly. Disassemble spindle assembly with reference to Fig. G22.

Clean and inspect all parts for excessive wear or any other damage and renew as needed. Reassemble by reversing disassembly procedure. Reinstall mower unit, then adjust drive belt as

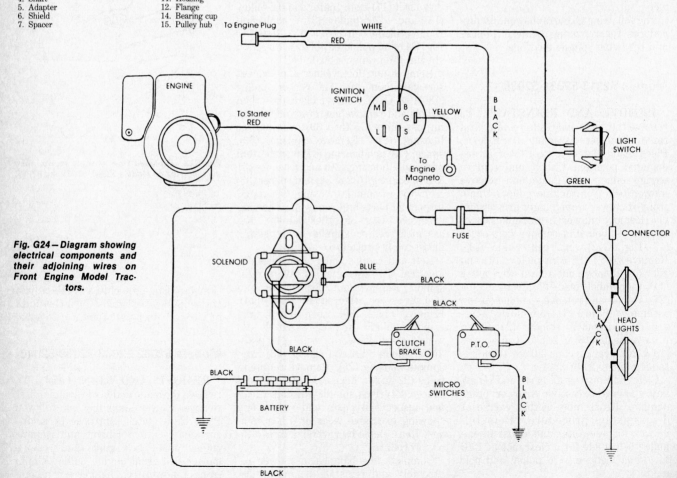

Fig. G24 — Diagram showing electrical components and their adjoining wires on Front Engine Model Tractors.

outlined in MOWER BLADE CLUTCH AND BELT section.

Models 52064-52065-52072-52073

REMOVE AND REINSTALL. To remove mower blade and blade spindle, first remove mower deck as outlined in MOWER BLADE CLUTCH AND BELT section. Remove drive belt from pulleys. Remove nut, lockwasher and flat washer securing drive pulley to pulley hub (15–Fig. G23), then withdraw drive pulley. Remove spindle assembly mounting bolts, then withdraw spindle assembly and blade as a unit. Separate blade from spindle assembly. Disassemble spindle assembly with reference to Fig. G23.

Clean and inspect all parts for excessive wear or any other damage and renew as needed. Reassemble by reversing disassembly procedure. Reinstall mower unit, then adjust drive belt as outlined in MOWER BLADE CLUTCH AND BELT section.

ELECTRICAL

Front Engine Tractors

Shown in electrical diagram Fig. G24 are electrical components and their adjoining wires.

GRAVELY

GRAVELY CORPORATION
One Gravely Lane
Clemmons, NC 27012

Model	Make	Engine Model	Horsepower	Cutting Width, In.
830	B&S	191702	8	30
830-E	B&S	191707	8	30
1130-E	B&S	252707	11	30

FRONT AXLE

All Models

REMOVE AND REINSTALL. The axle member (20–Fig. GR1) mounts within steering channel (15) and is center-pivoted on its mounting bolt (19). To remove axle, raise and securely block front of unit. Remove front wheel assemblies and unbolt outer end of tie rods (9) from spindles (21). Remove retaining pins and washers from top of spindles, then remove spindles. Remove pivot bolt (19) and spacer (16), then remove axle from steering channel. Remove nylon spindle bushings (18) if renewal is required.

Clean and inspect all parts and renew as necessary. Reassemble by reversing removal procedure. Lubricate wheel bearings and axle pivot with multi-purpose lithium base grease.

STEERING GEAR

All Models

REMOVE AND REINSTALL. To remove steering gear, remove nut (2–Fig. GR1) and pull steering wheel (1) from shaft (6). Remove cotter pin (14) and special washers (13) from bottom of steering shaft. Remove screws securing steering stand (5) to frame and disconnect pto linkage, then lift off steering stand and remove steering shaft. Remove nut (7) from below to remove steering gear (8).

CAUTION: Frequently, the performance of maintenance, adjustment or repair operations on a riding mower is more convenient if mower is standing on end and these units are provided with a stand for this purpose. This procedure can be considered a recommended practice providing the following safety recommendations are performed:

1. **Drain fuel tank or make certain that fuel level is low enough so that fuel will not drain out.**
2. **Close fuel shut-off valve is so equipped.**
3. **Remove battery on models so equipped.**
4. **Disconnect spark plug wire and tie out of way.**
5. **Although not absolutely essential, it is recommended that crankcase oil be drained to avoid flooding the combustion chamber with oil when engine is tilted.**
6. **Secure mower from tipping by lashing unit to a nearby post or overhead beam.**

Clean and inspect all parts. Renew any that are excessively worn or otherwise damaged. Lubricate bearings and steering gear with SAE 30 oil. Reassemble in reverse order of removal.

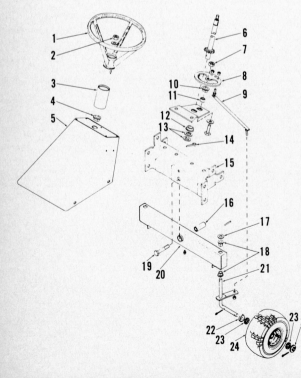

Fig. GR1—Exploded view of front axle and steering system.

1. Steering wheel
2. Nut
3. Bushing
4. Bearing
5. Steering stand
6. Steering shaft
7. Nut
8. Gear
9. Tie rod
10. Bearing
11. Spacer
12. Bearing
13. Special washers
14. Cotter pin
15. Steering channel
16. Axle pivot spacer
17. Washer
18. Nylon bushing
19. Axle pivot bolt
20. Front axle
21. Spindle
22. Special washer
23. Felt washer
24. Wheel assy.

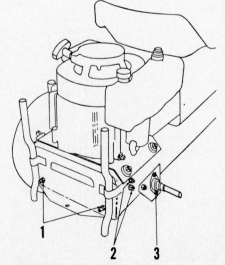

Fig. GR2—View of drive chain adjustment points and mower drive belt guide mounting nuts.

1. Axle adjustment nuts
2. Belt guide nuts
3. Axle bearing retainer

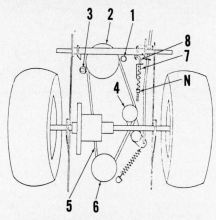

Fig. GR3-View of traction drive belt and brake adjustment rod.

1. Round belt guide
2. Transmission pulley
3. Flat belt guide
4. Idler pulley
5. Traction drive belt
6. Engine pulley
7. Spring
8. Brake adjustment rod

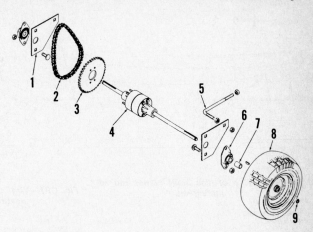

Fig. GR4—Differential and rear axle assembly.

1. Axle bearing retainer
2. Chain
3. Sprocket
4. Differential
5. Axle adjustment rod
6. Bearing
7. Spacer
8. Wheel assy.
9. Retaining ring

ENGINE

All Models

For overhaul and repair procedures on engines listed in Specification Table refer to Large Air Cooled Engines Service Manual.

REMOVE AND REINSTALL. To remove engine, disconnect spark plug wire, throttle cable and electrical wiring as needed. Move pto lever to "OFF" position. Loosen two nuts (2–Fig. GR2) on mower belt guide and slide guide to rear. Remove mower drive belt from engine pulley. Depress clutch-brake pedal and set park lock to release spring tension on traction drive belt, then remove belt from engine pulley. Remove engine mounting bolts and lift off engine.

Reinstall in reverse order of removal. Adjust mower belt as outlined in MOWER DRIVE BELT section.

TRACTION DRIVE CLUTCH AND BELT DRIVE

All Models

The main drive clutch consists of a non-adjustable, spring-loaded idler (4–Fig. GR3) when the clutch pedal is fully depressed, tension is removed from drive belt (5) and engine pulley (6) rotates freely within the belt. If drive belt slips due to excessive belt wear or stretching, renew belt as outlined under the following REMOVE AND REINSTALL paragraphs.

REMOVE AND REINSTALL. To remove the traction drive belt (5–Fig. GR3), disconnect spark plug wire and remove mower deck as outlined under

MOWER DECK paragraphs. Depress clutch-brake pedal and set park brake lock. Set unit up on its rear stand as previously outlined. Loosen nut on flat belt guide (3) and turn guide 90 degrees clockwise. Remove round belt guide (1), then remove drive belt from pulleys.

Renew drive belt by reversing removal procedure.

TRANSMISSION

All Models

A Peerless Model 706-A transmission is used on all models. For transmission overhaul procedures, refer to Peerless section in TRANSMISSION REPAIR service section. Refer to the following paragraphs for removal and installation.

REMOVE AND REINSTALL. To remove transmission, disconnect spark plug wire and remove mower unit as outlined in MOWER DECK section. Set unit up on rear stand as previously outlined. Remove traction drive belt, then remove retaining ring and pull pulley from transmission input shaft. Locate master link in drive chain and disconnect chain. Disconnect linkage at transmission brake lever. Disconnect shift linkage and necessary wiring, then unbolt and remove seat support and chain guard. Remove mounting bolts and lift transmission from frame.

Reinstall transmission by reversing removal procedure.

DRIVE CHAIN

All Models

ADJUSTMENT. To adjust drive chain, loosen nuts securing rear axle bearing retainers (3–Fig. GR2) on each side of frame. To tighten chain, move axle assembly rearward (equal distance on both sides) by turning adjusting nuts

(1) clockwise. Drive chain should deflect approximately ½-inch when about five pounds pressure is applied on chain. When chain tension is correct, tighten bearing retainer nuts.

REMOVE AND REINSTALL. To remove drive chain, rotate differential sprocket to locate master link. Disconnect master link and remove chain. Clean and inspect chain and renew if excessively worn.

Install drive chain and adjust tension as outlined in preceding paragraph. Lubricate chain with a light coat of SAE 30 oil.

DIFFERENTIAL

All Models

REMOVE AND REINSTALL. To remove, raise rear of unit and support securely. Remove retaining rings (9–Fig. GR4) and remove rear wheels (8). Remove drive chain (2). Unbolt axle bearing retainers (1), then remove differential and axle assembly. Clean rust, dirt and paint off axle ends, then slide bearings (6) from axle ends.

Reinstall differential and axle assembly by reversing the removal procedure.

OVERHAUL. A Peerless Model 100-008 differential is used on all models. Refer to Peerless section in DIFFERENTIAL REPAIR service section for overhaul procedures.

GROUND DRIVE BRAKE

All Models

The brake is applied by depressing either of the foot pedals. Brake disc is mounted on transmission output shaft at end opposite drive sprocket.

ADJUSTMENT. To adjust brake, set unit up on rear stand as previously out-

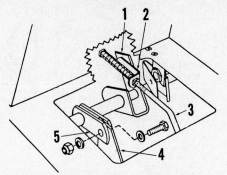

Fig. GR5—View of disc brake carrier and adjusting nut.

1. Adjusting nut
2. Carrier
3. Brake disc

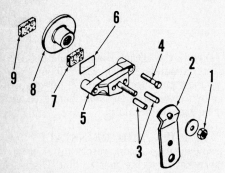

Fig. GR6—Exploded view of disc brake assembly.

1. Adjusting nut
2. Brake actuating lever
3. Actuating pins
4. Cap screw
5. Carrier
6. Back-up plate
7. Brake pad (outer)
8. Brake disc
9. Brake pad (inner)

Fig. GR7—View of mower linkage located beneath steering console.

1. Pto crank
2. Blade engagement link
3. Blade engagement rod
4. Front deck bracket
5. Cross shaft

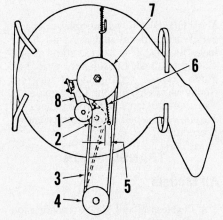

Fig. GR8—View of mower drive belt and idler.

1. Idler pulley disengaged
2. Idler pulley engaged
3. Drive belt
4. Engine pulley
5. Correct distance is one inch minimum
6. Blade brake spring
7. Blade pulley
8. Idler pulley bracket

Fig. GR9—Exploded view of mower unit.

1. Blade engagement spring
2. Blade engagement link
3. Blade engagement rod
4. Drive belt
5. Idler pulley
6. Idler bracket
7. Nut
8. Blade pulley
9. Spacer
10. Bearing
11. Spindle housing
12. Spindle shaft
13. Spinner cap
14. Blade brake spring
15. Spring bracket
16. Rear deck bracket
17. Mower deck
18. Blade
19. Support bar
20. Front deck bracket
21. Deflector chute

lined. Tighten nut (N—Fig. GR3) on brake adjustment rod (8) to compress spring (7) to length of 10 inches. Put unit back on its wheels and shift transmission to neutral. Depress clutch-brake pedal and engage pedal lock. If rear wheels turn when pushing unit, release brake and tighten adjusting nut (1—Fig. GR5) on disc brake carrier (2) ½-turn. Recheck and repeat adjustment if necessary.

REMOVE AND REINSTALL. To remove disc brake, disconnect return spring and brake rod from actuating lever (2—Fig. GR6). Remove cap screws (4) and remove pad carrier assembly. Slide brake disc (8) from shaft and remove inner brake pad (9) from holder slot in transmission case. Remove adjusting nut (1) and separate brake lever (2), actuating pins (3), back-up plate (6) and outer brake pad (7) from pad carrier (5). Renew parts as required and reassemble by reversing the removal procedure. Adjust brake as outlined in previous paragraph.

MOWER DECK

All Models

LEVEL ADJUSTMENT. To adjust

gage wheels on mower, position unit on level ground, then remove pins from gage wheel spindles letting wheels touch ground. Align one hole in left hand support with hole in the spindle and reinstall pin. Reinstall pin in corresponding holes in right hand support and spindle. If mower height is changed, readjust gage wheels.

REMOVE AND REINSTALL. To remove mower, disconnect spark plug wire, disengage pto and remove console tray. Disconnect blade engagement link (2—Fig. GR7) from pto crank (1) and cross shaft (5) from mower deck front bracket (4). Disconnect mower deck rear brackets from rear lift shaft. Loosen the two belt guide nuts (2—Fig. GR2), then slide guide rearward and remove drive belt from engine pulley. Move unit forward until blade engagement rod (3—Fig. GR7) clears cross shaft (5), then lift front wheels over mower.

Reinstall by reversing removal procedure. Be sure blade engagement rod

goes over cross shaft. Torque belt guide nuts to 25 ft.-lbs. Check belt adjustment.

MOWER DRIVE BELT

All Models

ADJUSTMENT. To adjust belt, put pto lever in "OFF" position and loosen nuts on the three mower deck brackets. Move each bracket rearward to take slack out of belt. Check belt adjustment by looking through slot in hitch plate. With pto engaged, there must be a minimum clearance of one inch (5—Fig. GR8) between sides of drive belt measured at idler pulley (2).

REMOVE AND REINSTALL. To remove belt, disconnect spark plug wire and remove mower deck. Disconnect blade brake spring (6—Fig. GR8) from deck. Remove idler pulley (1) from idler bracket (8). Remove drive belt (3) from blade pulley (7).

Reinstall in reverse of removal. Torque idler pulley nut to 25 ft.-lbs.

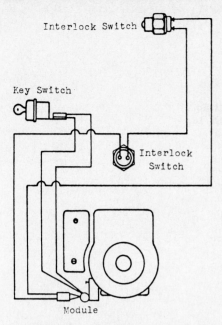

Interlock Switch

Key Switch

Interlock Switch

Module

Fig. GR10—Wiring diagram for recoil start models.

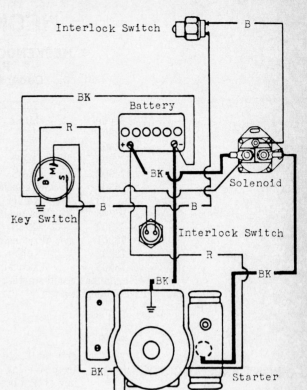

Fig. GR11—Wiring diagram for electric start models.

1. BK-Black
2. B-Brown
3. R-Red

Interlock Switch

Battery

Solenoid

Key Switch

Interlock Switch

Starter

MOWER SPINDLE

All Models

REMOVE AND REINSTALL. Remove mower deck as previously outlined. Block mower blade (18–Fig. GR9) from turning. Loosen bolts in blade support bars (19), then remove lower nut (7) from spindle shaft to remove blade. Disconnect blade brake spring (14) from mower deck. Remove idler pulley (5) and drive belt (4). Remove retaining nut and pull blade pulley (8) from spindle shaft (12). Remove idler bracket (6) and spindle housing (11) from mower deck, then remove spindle shaft (12), bearings (10) and spacers (9) from housing.

Clean and inspect all parts. Reassemble by reversing removal procedure. Spindle bearings are sealed and require no additional lubrication. When installing blade, torque spindle nuts (7) to 50 ft.-lbs., then torque blade support bar nuts to 60 ft.-lbs. Finish tightening spindle nuts to 100 ft.-lbs. torque.

ELECTRICAL

All Models

All units are equipped with safety interlock switches (Fig. GR10 and Fig. GR11) on transmission shift linkage and pto linkage. The transmission must be in neutral and pto disengaged before the unit will start.

ADJUSTMENT. To adjust the pto interlock switch (3–Fig. GR12) remove the console tray and move pto lever to "OFF" position. Pto crank (1) should compress interlock spring (2) against interlock switch (3), closing the switch. If not, loosen screws (5) holding interlock bracket (4), reposition bracket, then tighten screws (5).

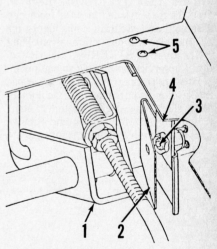

Fig. GR12—View of pto interlock switch.

1. Pto crank
2. Interlock spring
3. Interlock switch
4. Interlock bracket
5. Screws

HECKENDORN

HECKENDORN MFG. CO., INC.
P.O. Box 88
Cedar Point, KS 66843

Model	Engine		Cutting Width, In.
	Make	Horsepower	
73901*	B&S	10	36
73904	B&S	16	36
73905	B&S	10	36
75902	B&S	16	50
76901	Wisconsin	18	62
76902	Kohler	20	62
76903	Wisconsin	21**	62
77902	B&S	16	72
78901	Wisconsin	18	88
78902	Kohler	20	88
78903	Wisconsin	21**	88

*Engine has recoil starter.
**Diesel engines.

FRONT AXLE AND STEERING

All Models

R&R AND OVERHAUL. Refer to Fig. H1 for exploded view of front end assembly and steering components for all models without optional steering gear. Refer to STEERING GEAR section for service on models so equipped.

To remove steering fork shaft raise front of mower frame high enough to allow shaft to be slid out bottom of shaft frame support (6). Loosen and remove nuts retaining handlebar mounting plate and lift handlebar and plate assembly from steering shaft. Be sure not to lose key (11). Loosen and remove deck lift components as needed to remove tension from lift sleeve (1).

NOTE: Support base of steering shaft during removal. Self-injury or damage to components could occur if shaft slides out unattended.

Loosen bolt from plug (3) on collar (2), then slide steering shaft out bottom of frame support (6).

Inspect all components for excessive wear or any other damage. Renew all parts as needed, then reassemble in reverse order of disassembly. Lubricate steering shaft with No. 2 Lithium grease. Care should be used not to damage bushings (4) during installation of steering shaft.

STEERING GEAR

All Models

R&R, OVERHAUL AND ADJUST. On models so equipped refer to Fig. H2 for exploded view of steering gear assembly.

To remove steering gear assembly, first remove nut and lockwasher retaining pitman arm to output shaft (13). Withdraw arm from shaft or as needed use a suitable puller to withdraw arm. Remove nuts from U-bolt securing top of steering column. Remove steering gear mounting bolts, then lift steering gear assembly from frame. Remove all parts as needed to withdraw gear assembly clear of mower unit.

Disassemble gear with reference to Fig. H2. Remove steering wheel cover, nut and lockwasher, then lift wheel from worm shaft. Remove side cover retaining cap screws, then withdraw cover (15), gasket (14) and output shaft (13). Remove locknut (1), seal (2) and adjustment plug (3), then withdraw bearing cups (4), bearings (5) and worm shaft (6).

Inspect all parts for excessive wear, pitting, corrosion or any other damage, and renew as needed. Renew all seals and gaskets. During reassembly lubricate bearings with 40 weight gear oil.

Reassemble worm shaft unit in reverse order of disassembly. Turn adjustment plug (3) in until snug, then back off only enough to prevent worm shaft

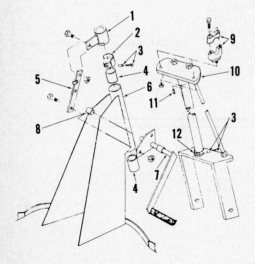

Fig. H1—Exploded view of front end assembly on all models. For models equipped with steering gear option refer to Fig. 2.

1. Lift sleeve assy.
2. Collar
3. Pipe & bolt
4. Bushing
5. Linkage bar
6. Fork frame support
7. Foot lift pedal
8. Bushing
9. Handlebar mount assy.
10. Front fork assy.
11. Key
12. Collar

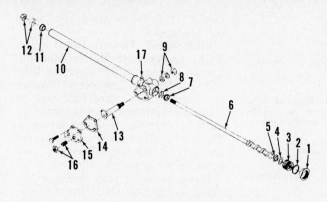

Fig. H2—Exploded view of steering gear used on all models so equipped.

1. Locknut
2. Seal
3. Adjustment plug
4. Bearing cup
5. Bearing
6. Worm shaft
7. Bearing
8. Bearing cup
9. Oil seal, lockwasher and nut
10. Housing
11. Column bearing
12. Nut and washer
13. Output shaft
14. Side cover gasket
15. Side cover
16. Adjustment screw & nut
17. Plug

and nuts, then lift engine from mounting rails and clear of mower unit. After repair, reinstall in reverse order of removal. Adjust belts as outlined in TRACTION AND MOWER DRIVE BELTS section.

TRACTION AND MOWER DRIVE BELTS

All Models

Drive belt routing and associated parts are shown in Fig. H3. Refer to chart shown below for recommended manufacturer center-distances. All distances are given in inches and are for new belt adjustment. When adjusting a used belt, obtain a good working distance with no belt slippage. If belt tension cannot be adjusted correctly, then belt must be renewed.

Refer to the following paragraphs for belt adjustment procedures. Belts must be adjusted in sequence listed below in order to maintain correct tension on each belt in relation to all others.

(Engine to Transmission). Loosen engine mounting bolts and slide engine in frame slots until correct distance is attained. Place a straightedge across faces

from binding when turned from side to side. There should be no end-play in shaft. Install locknut (1) after correct adjustment is attained. Install output shaft (13), gasket (14), cover (15) and secure with retaining bolts. To adjust output shaft, loosen locknut and turn adjustment screw (16) in until all slack is removed between worm shaft and output shaft gear mesh when output shaft is turned across center. Be sure there is no binding between gears when output shaft is rotated from side to side. If needed loosen adjustment screw in small increments until correct adjustment is attained, then tighten locknut to secure adjustment. After reassembly fill gear box with 40 weight gear oil.

Reinstall gear box assembly in reverse order of removal. Turn steering wheel to position output shaft in mid-point of travel and point front wheel straight ahead, then install pitman arm on output shaft. Secure with lockwasher and nut.

ENGINE

All Models

For overhaul and repair procedures on engines listed in Specification Table refer to Large Air Cooled Engines Service Manual for gasoline powered engines and Small Diesel Engine Service Manual for diesel powered engines.

REMOVE AND REINSTALL. Engine is mounted on frame rails and secured by bolts and nuts. To remove, all engine attachments and protective shields must be removed. When removing fuel tank caution should be used as not to allow spillage of fuel. Remove electrical wiring, throttle cable and all other components as needed.

Remove all drive belts from engine drive sheave. For assistance in removal of belts refer to Fig. H3 and TRACTION AND MOWER DRIVE BELTS section. Remove engine mounting bolts

Fig. H3—View of traction and mower drive belts with working components. Models 73901, 73904 and 73905 do not use belts 4, 5 and 6 and associated parts.

1. Engine to transmission
2. Engine to swinging spindle
3. Swinging spindle to main spindle
4. Engine to rear drive shaft
5. Rear drive shaft to right mower spindle
6. Rear drive shaft to left mower spindle

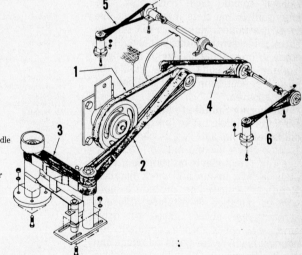

Belt Position	Model				
	73901, 904 and 905	75902	76901, 902 and 903	77902	78901, 902 and 903
Engine to transmission . . .	12	13½	13¼	13½	13¼
Engine to swinging spindle	21⅜	22⅝	28¼	22⅝	28¼
Swinging spindle to main spindle	11⅝	12½	11½	12½	11½
Engine to center power housing	. . . .	15½	18	15½	18
Sidemount power housing to right mower spindle . .	. . . .	26⅛	31	26⅛	31
Sidemount power housing to left mower spindle . . .	. . . .	26⅛	31	26⅛	31

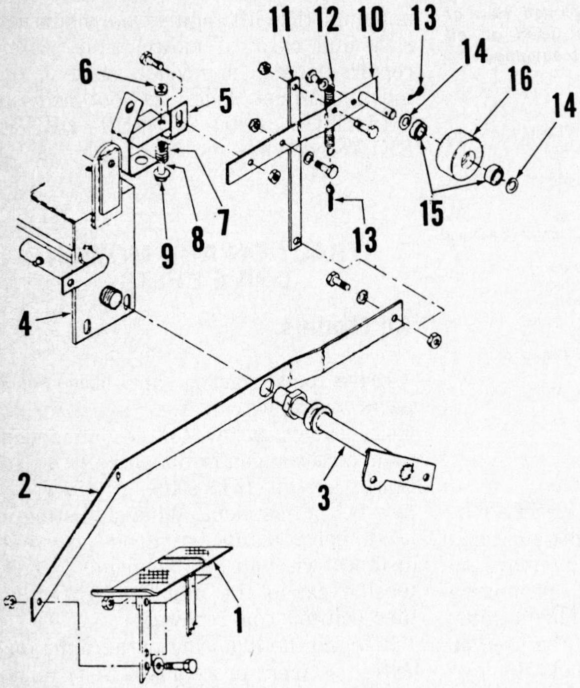

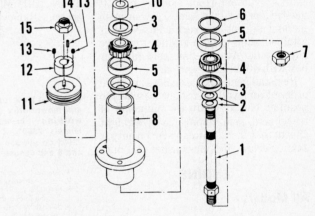

Fig. H4—View of clutch components for traction drive on all models.

1. Pedal
2. Linkage bar
3. Bell pivot tube
4. Frame assy.
5. Clutch interlock switch lever
6. Nut
7. Spring
8. Bolt
9. Flat washer
10. Clutch roller arm
11. Linkage bar
12. Spring
13. Cotter pin
14. Spacer
15. Bearing
16. Roller

¼-inch belt deflection at mid-point when belt is engaged.

(Sidemount Power Housing to Right and/or Left Mower Spindle). Loosen power housing mounting bolts and slide housing until correct distance is attained. Make sure belt runs on sheaves in a straight line, belt should not come in on spindle sheave at an angle. Tighten housing mounting bolts. There should be ¼-inch belt deflection at mid-point when belt is engaged.

To renew drive belts follow procedures as listed in adjustment sections, loosen necessary components and renew belts as needed. When adjusting belts make sure to follow sequence listed in order to maintain correct tension on each belt in relation to all others.

All drive belt guides are welded into position with the exception of one pin type, which is bolted to engine assembly. To adjust, loosen mounting bolts and rotate pin until there is ¼-inch clearance between pin and engine to transmission belt when clutch pedal is disengaged. When clutch pedal is engaged there should be slight contact between belt and pin. Drive belt guides are used to

of drive sheaves to check sheave alignment. Tighten engine mounting bolts. When clutch roller is disengaged there should be ⅛-¼ inch deflection in belt at center of drive sheaves.

(Engine to Swinging Spindle). Loosen spindle base bolts and slide engine in frame slots until correct distance is attained. Make sure belt runs on sheaves in a straight line, belt should not come in on spindle sheave at an angle. Tighten base bolts. There should be ¼-inch deflection in belt at mid-point of drive sheaves.

(Swinging Spindle to Main Spindle). Disconnect blade brake linkage rod and blade lever linkage rod at clevis ends. Loosen clevis locking nut on lever linkage rod. Turn clevis on rod, then reinstall clevis in linkage and check distance. Continue adjustment procedure until correct distance is attained. Place a straightedge across faces of drive sheaves to check sheave alignment. Reconnect blade brake clevis in linkage and check adjustment. When blade lever is in disengaged position brake shoe should contact brake drum. When blade lever is in engaged position there should be no contact between brake shoe and brake drum. Adjust clevis as needed. When blade lever is disengaged there should be ⅛-¼ inch free movement in belt.

(Engine to Center Power Housing). Place blade lever in engaged position. On center power housing loosen linkage block locknut, then turn adjusting nut until correct distance is attained. Place a straightedge across faces of drive sheaves to check sheave alignment, then tighten block locknut. There should be

Fig. H5—Exploded view of main and sidemount (on models so equipped) spindle components.

1. Spindle shaft
2. Bushings
3. Retainer
4. Bearing
5. Bearing cup
6. Backing ring
7. Locknut
8. Housing
9. Backing ring
10. Spacer
11. Drive sheave
12. Bushing
13. Set screws
14. Key
15. Locknut

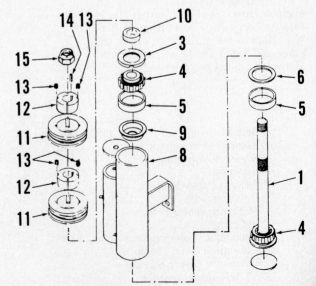

Fig. H6—Exploded view of swinging spindle components. Refer to Fig. H5 for identification of parts.

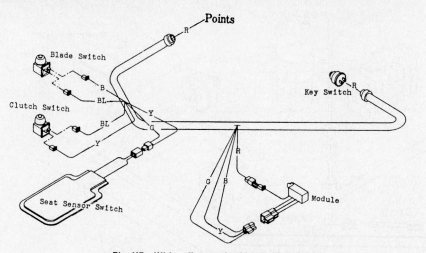

Fig. H7 — Wiring diagram for 36-inch Model 73901.

1. BK-Black	4. G-Green	8. W-White	
2. BL-Blue	5. O-Orange	9. Y-Yellow	12. G/W-Green and White
3. B-Brown	6. T-Tan	10. B/BL-Brown and Blue	13. R/BL-Red and Black
	7. V-Violet	11. G/B-Green and Brown	14. R/W-Red and White

keep belts in position and to slow down or stop belts when released from working position.

TRACTION DRIVE CLUTCH

All Models

Exploded view of clutch components for traction drive is shown in Fig. H4.

When pedal (1) is pushed down to disengage traction drive, roller (16) should be raised clear of drive belt. When pedal is released pressure from spring (12) should pull roller (16) down against drive belt to engage traction drive.

For repair, inspect all components for excessive wear, binding or any other damage. Inspect spring (12) for suitable tension and make sure spring coils are

not over-stretched. Renew all parts as needed.

MOWER SPINDLE

All Models

Spindle housings are equipped with grease fittings. Manufacturer recommends greasing spindles every two operating hours with No. 2 Lithium grease.

ADJUSTMENT. Figures H5 and H6 show exploded views of spindle assemblies used. Figure H5 pertains to main and sidemount (on models so equipped) spindles and Fig. H6 to swinging spindle.

To check, first remove all drive belts as needed. Shaft end play on all spindles should be 0.004-0.006 inch. To check end play attach a dial indicator to a suitable flat mounting surface and adjust dial needle to where it contacts end of shaft (1). Zero out dial face, then move shaft up and down in housing while observing needle movement on dial indicator. If reading is to high, tighten top locknut (15) in small increments until correct adjustment reading is attained. If reading is to low, loosen top locknut (15) ½-turn and strike shaft (1) with a hammer to free spindle components. Tighten locknut (15) in small increments until correct adjustment reading is attained.

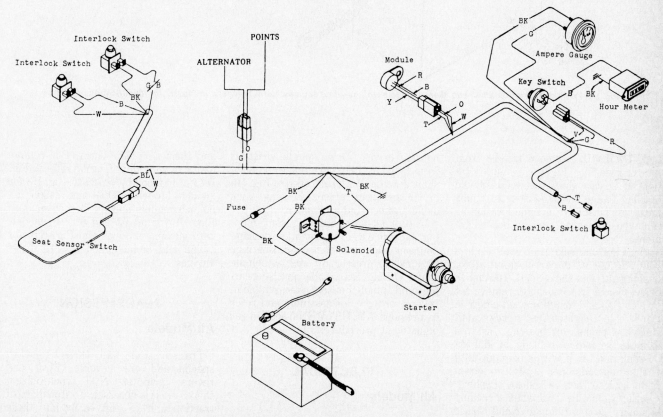

Fig. H8 — Wiring diagram for 36-inch Models 73904 and 73905, all 50-inch models and all 72-inch models. Refer to legend in Fig. H7 for identification of wires.

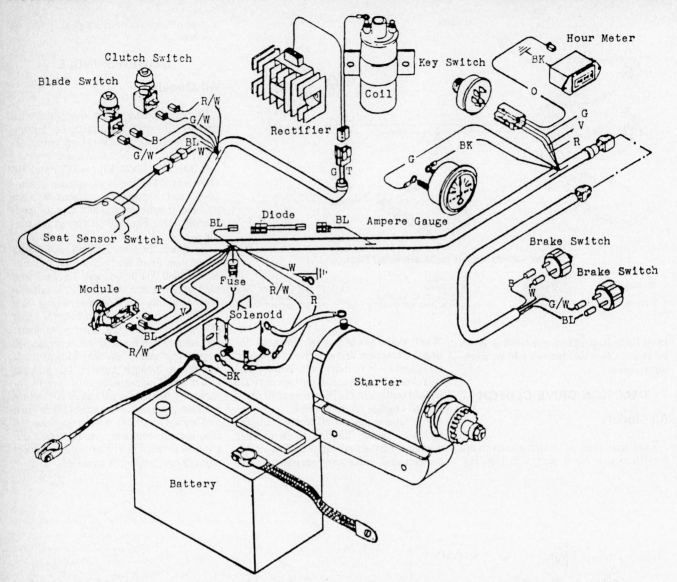

Fig. H9 — Wiring diagram for all 62 and 88-inch models with gasoline engines. Refer to legend in Fig. H7 for identification of wires.

OVERHAUL. Remove spindle from mower assembly. Be sure to mark location of spacer shims between spindle housing flange and mower deck. Shims must be reinstalled in same location as removed to insure correct belt alignment.

Place bottom locknut on shaft (1) in a vise. Remove upper locknut (15), bushing (12) and sheave (11). To remove drive sheave from bushing, remove set screws (13) and install one set screw in third hole. Tighten set screw until sheave is pushed off bushing. Install a straight screwdriver blade in slot on bushing and tap it with a hammer until bushing spreads open enough to loosen from shaft, then withdraw bushing. Caution should be used when spreading bushing, over-spreading could cause bushing to crack. Remove bottom locknut from vise and install spindle

housing in vise. Drive spindle shaft (1) from housing, take care not to damage shaft threads. With reference to Fig. H5 or H6 drive out internal housing parts using a punch and a hammer.

Inspect all parts for corrosion, binding, excessive wear or any other damage and renew all parts as needed. Bearings (4) must be pressed off and on spindle shaft. During reassembly grease bearings and spindle housing. Reassemble in reverse order of disassembly and refer to previous ADJUSTMENT section for adjustment procedures.

ELECTRICAL

All Models

Shown in electrical diagrams Figs. H7 through H10 are electrical components

and their adjoining wires. For component and wire identification refer to Fig. H7 for 36-inch Model 73901; Fig. H8 for 36-inch Models 73904 and 73905, all 50-inch models and all 72-inch models; Fig. H9 for all 62 and 88-inch models with gasoline engines and Fig. H10 for all 62 and 88-inch models with diesel engines.

TRANSMISSION

All Models

Transmission has three forward speeds and one reverse. First and reverse gears are NOT synchronized; therefore, it is necessary to stop drive of traction unit to shift easily into these gears. Second and third gears are synchronized and can be shifted before com-

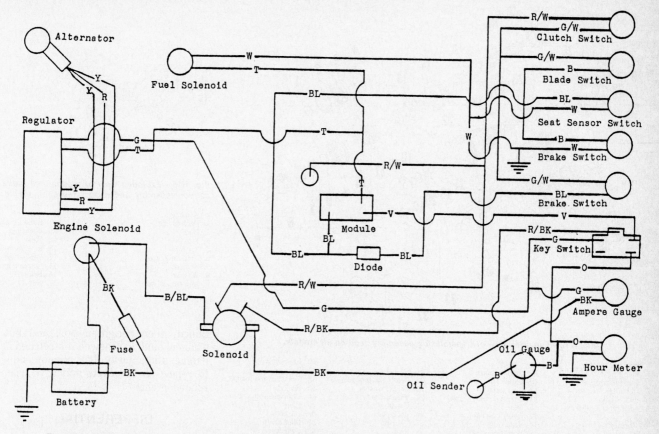

Fig. H10 — Wiring diagram for all 62 and 88-inch models with diesel engines. Refer to legend in Fig. H7 for identification of wires.

ing to a complete stop. It is not recommended using third gear while mowing.

REMOVE AND REINSTALL. To remove the transmission, first remove all protective shields and parts that will obstruct removal of transmission. Loosen and remove output shaft drive chain. Remove input shaft drive belt, then remove drive sheave retaining parts and withdraw sheave from input shaft. Remove all mounting bolts and then lift transmission unit clear of frame assembly.

To reinstall transmission unit reverse removal procedures. Refer to CHAIN DRIVE ADJUSTMENT section and DRIVE BELT ADJUSTMENT section for adjustment procedures after installation.

OVERHAUL. Unbolt and remove transmission shift tower assembly.

If disassembly of shift tower is required refer to Fig. H11. Drive shift pins (14) from shift forks and remove expansion plugs (5) from housing. Remove second and third shift rail (10) and fork (11) and then, first and reverse shift rail (16) and fork (17). A poppet spring (12) and steel ball (13) are located on top of each rail and an interlock plunger (15) is positioned between rails. Inspect all parts

Fig. H11 — Exploded view of shift tower assembly used on all models.

1. Shift tower housing
2. Pin
3. Shift lever
4. Knob
5. Plug
6. Gasket
7. Fulcrum ball
8. Pin
9. Spring
10. Shift rail
11. Shift fork
12. Spring
13. Steel ball
14. Pin
15. Interlock plunger
16. Shift rail
17. Shift fork

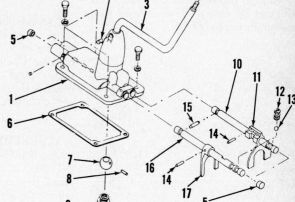

for damage and renew as needed. Reassemble shift tower in reverse order of disassembly.

To disassemble transmission, first remove cap screws from transmission mount (32 – Fig. H12) and withdraw input shaft (5) and bearing assembly. Remove snap ring (17) retaining main shaft bearing (16) in case and tap shaft out of case. Main shaft assembly can be removed as a complete unit. Remove rear housing (39) retaining cap screws and then remove housing and gasket.

Remove locking plate (26), then push idler shaft (37) and countershaft (25) from transmission housing. Lift reverse gear (38) and cluster gear (22) and associated parts from housing. Complete disassembly with reference to Fig. H12.

Inspect all components for excessive wear, broken teeth, binding or any other damage. Renew all gaskets, seals and damaged parts as needed. Reassemble transmission in reverse order of disassembly.

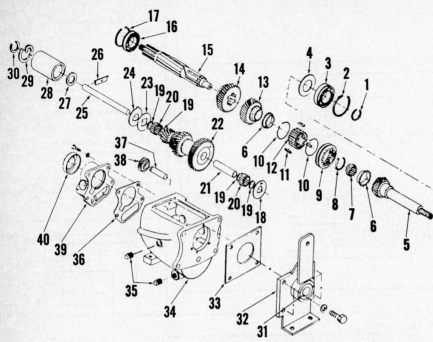

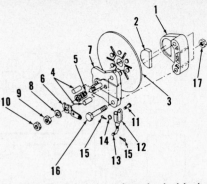

Fig. H15—Exploded view of mechanical brake caliper assembly and associated components.

1. Housing
2. Puck (2 used)
3. Disc
4. Pins
5. Spring
6. Actuating lever
7. Housing
8. Washer
9. Adjustment nut
10. Locknut
11. Clevis pin
12. Clevis
13. Linkage rod
14. Washer
15. Cotter pin
16. Bolt
17. Nut

Fig. H12—Exploded view of transmission assembly used on all models.

1. Snap ring
2. Snap ring
3. Roller bearing
4. Oil baffle
5. Input shaft
6. Synchronizer blocking ring
7. Bearing
8. Snap ring
9. Clutch sleeve
10. Synchronizer spring
11. Shifting plate
12. Clutch hub
13. Gear (second)
14. Gear (low and reverse)
15. Main shaft
16. Roller bearing
17. Snap ring
18. Thrust washer
19. Spacer
20. Bearing
21. Spacer
22. Gear (countershaft)
23. Thrust washer
24. Thrust washer
25. Countershaft
26. Lock plate
27. "O" ring
28. Spacer
29. Spacer retaining spring
30. Snap ring
31. Bearing
32. Transmission mount
33. Gasket
34. Transmission housing
35. Plug
36. Gasket
37. Idler shaft (reverse)
38. Gear (reverse)
39. Rear housing
40. Oil seal

Reassemble shift tower assembly to transmission assembly. After reassembly fill transmission unit with SAE 40 Gear Oil until oil level is even with bottom of top plug hole.

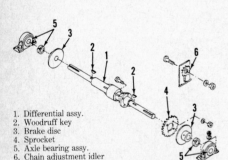

Fig. H13—Exploded view of rear end assembly used on all models. Brake disc (3) are only used on models with hydraulic disc brakes.

1. Differential assy.
2. Woodruff key
3. Brake disc
4. Sprocket
5. Axle bearing assy.
6. Chain adjustment idler

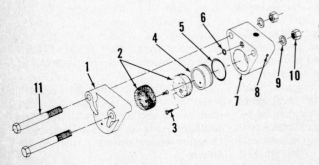

Fig. H14—Exploded view of a hydraulic brake caliper assembly.

1. Housing
2. Puck
3. Screw
4. Piston
5. "O" ring
6. "O" ring
7. Housing
8. Bleeder screw
9. Washer (2 used)
10. Locknut (2 used)
11. Bolt (2 used)

DRIVE CHAIN

All Models

Transmission output power is transfered to differential assembly through the use of a No. 50 roller chain on all models. Chain tension adjustment is very critical to the life expectancy of chain and sprockets. Chain should be kept clean and lubricated with 30 weight oil.

ADJUSTMENT. There should not be more than ½-inch or less than ¼-inch deflection at center of chain. To adjust chain, first remove all necessary protective shields. Loosen adjustment sprocket retaining bolt and nut (6—Fig. H13). Slide sprocket in adjustment bracket until correct chain tension is at-

tained, then retighten bolt and nut. Chain should be renewed if stretched or damaged in anyway. It is recommended to renew all sprockets and keys when renewing chain.

DIFFERENTIAL

All Models

Differential assembly and associated parts are shown in Fig. H13. Differential assembly (1) must be renewed as a complete unit as no individual repair parts are available. Grease assembly with No. 2 Lithium grease. Caution should be used when greasing assembly as over greasing can push out housing end seals.

BRAKE

All Models

A self adjusting hydraulic disc brake assembly as shown in Fig. H14 is used on 62 and 88-inch models. A mechanical operated disc brake assembly as shown in Fig. H15 is used on 36, 50 and 72-inch models.

PUCK RENEWAL (Hydraulic models). Renew brake pucks (2—Fig. H14) when thickness is 3/16-inch or less. To renew pucks remove bolts (12), then pull puck housings away from brake disc. Remove old pucks and renew. Reinstall housings and secure with bolts (11). To bleed air out of brake system, first depress brake pedal. Open bleeder screw (8), then close after fluid pressure is released. Continue procedure until all air is removed from system. Keep master cylinder full at all times.

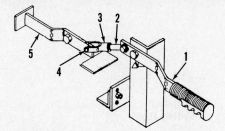

Fig. H16—View showing parking brake assembly used on models with mechanical disc brake.

1. Actuating lever
2. Linkage rod
3. Clevis
4. Clevis pin
5. Parking brake linkage

NOTE: Do not release brake pedal pressure with bleeder screw open as air will be drawn back into system.

(Mechanical models). Renew brake pucks (2–Fig. H15) when thickness is 3/16-inch or less. To renew pucks remove bolts (16), then pull puck housing away from brake disc. Puck in actuating side is of floating type and can be renewed easily. Puck in stationary side is fixed in place and must be chiseled out. Clean up caliper housing and remove all rough spots left on housing surface. Using a suitable contact glue install new puck in housing. Reinstall and secure housings, then adjust brake as outlined in following section.

ADJUSTMENT. Hydraulic models are self-adjusting and do not require periodic adjustment.

(Mechanical models). Manufacturer recommends adjusting brake at least every 40 operating hours. To adjust, raise rear of mower off the ground until wheels do not touch. Loosen locknut (10–Fig. H15) from adjustment nut (9). Fully tighten adjustment nut (9), then back-off ¼-turn. Turn drive wheels and feel for drag, if drag is felt loosen adjustment nut in small increments until wheel turns freely. Completely depress clutch/brake pedal, brake should fully lock. Renew brake pucks if full adjustment is used and brake does not fully apply when pedal is completely depressed.

PARKING BRAKE

All Models

There are two types of parking brakes used. Figure H16 shows a view of the type used on models with mechanical disc brake. Figure H17 shows a view of the type used on models with hydraulic disc brake. Model shown in Fig. H16 is adjusted by turning clevis (3). Model shown in Fig. H17 has no adjustments. Raise or lower locking lever (1) in steps on foot pedal (2) to change brake pressure.

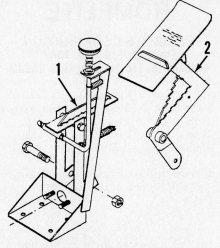

Fig. H17—View showing parking brake assembly used on models with hydraulic disc brake. Locking bar (1) is positioned in steps on foot pedal (2) to engage brake.

BLADE BRAKE

All Models

If brake lining is worn down to metal, lining should be renewed as brake will lose its effectiveness in slowing down the blades.

ADJUSTMENT. Disconnect spark plug wires and remove front belt guard. Loosen clevis (4–Fig. H18) jam nut, then remove cotter key (6), clevis pin (5) and withdraw clevis (4) clear of actuating lever. Turn clevis on linkage rod to lengthen or shorten rod. When blade lever is engaged there should be no contact between brake drum and lining. When blade lever is disengaged, brake lining should contact brake drum. After adjusting clevis on rod reconnect clevis in lever and check operation. Repeat adjustment procedure until correct adjustment is attained. Reassemble in reverse order of disassembly.

SAFETY SWITCHES

All Models

All models are equipped with six safety interlock switches. They are: brake, drive clutch, blade brake, ignition switch, seat and parking brake.

For engine to be started all switches except seat switch must be engaged. Setting parking brake and disengaging blades by-passes seat switch. If operator leaves seat while engine is running without setting parking brake and disengaging blades the engine will die.

If engine fails to turnover, check to make sure safety switches are properly engaged. If all switches are properly en-

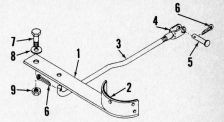

Fig. H18—View showing blade brake assembly used on all models.

1. Brake lever
2. Brake lining
3. Linkage rod
4. Clevis
5. Clevis pin
6. Cotter key
7. Bolt
8. Flat washer
9. Nut

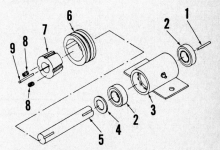

Fig. H19—Exploded view of a sidemount power housing used on models so equipped.

1. Key
2. Bearing
3. Housing
4. Spacer
5. Shaft
6. Sheave
7. Bushing
8. Set screws
9. Key

gaged refer to appropriate wiring diagram for assistance in diagnosing and repairing.

SIDEMOUNT POWER HOUSING

Models So Equipped

Figure H19 shows an exploded view of a typical power housing used on models so equipped. For repair, remove drive belt, housing mounting bolts and U-joint to shaft (5) retaining components. Be sure not to lose key (1). To remove drive sheave (6) from bushing (7), remove set screws (8) and install one set screw in third hole. Tighten set screw until sheave is pushed off bushing. Install a straight screwdriver blade in slot on bushing and tap it with a hammer until bushing spreads open enough to loosen from shaft, then withdraw bushing, key and sheave. Caution should be used when spreading bushing, over-spreading could cause bushing to crack.

Complete disassembly with reference to Fig. H19. Inspect all parts for excessive wear, binding or any other damage and renew as needed. Reassemble in reverse order of disassembly, then reinstall. Grease power housing with No. 2 Lithium grease. Readjust drive belt as outlined under DRIVE BELT ADJUSTMENT section.

HOMELITE

HOMELITE CORPORATION
Post Office Box 7047
Charlotte, North Carolina 28217

Model	Make	Engine Model	Horsepower	Cutting Width, In.
RE-5	B&S	130202	5	26
RE-8E	B&S	190707	8	30
RE-30	B&S	190402	8	30

For service information and procedures, use the following model cross reference and refer to the SIMPLICITY section of this manual.

Homelite Models	Simplicity Models
RE-5	3005
RE-8E	808
RE-30	3008-2

HUSTLER

EXCEL INDUSTRIES
P.O. Box 727
Hesston, KS 67062

Model	Make	Engine Model	Horsepower	Cutting Width, In.
261	B&S	420000	18	60
275	Kohler	K532*	20	54**
295	Continental	TC-56	25	54**

*An optional K-582 (23 horsepower) engine is available.
**Optional mower decks are available.

STEERING SYSTEM

All Models

These units do not have a conventional front axle or steering system; instead two steering control levers control two Sundstrand hydrostatic pumps which control oil flow to a left and right Ross hydraulic wheel motor. Increasing oil flow to motors will increase traction speed. When oil flow decreases, traction speed decreases. Stopping oil flow will cause tractor to stop. Reversing oil flow to motor will cause tractor to move in reverse. If oil flows faster to one motor than the other, tractor will turn. If oil flow goes in one direction to one motor and in the opposite direction to the other motor, tractor will rotate on its axis. Figure HU1 shows positioning of control levers for making directional maneuvers.

ENGINE

All Models

For overhaul and repair procedures on Briggs & Stratton and Kohler engines listed in Specification Table refer to Large Air Cooled Engines Service Manual. For repair specifications on Continental TC-56 refer to the following section.

Model Continental TC-56

Engine is of L-head design. All valves, valve lifters, cam and all other working components are contained in cylinder block assembly. Manufacturer does not

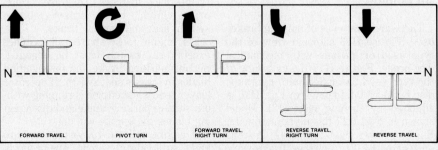

Fig. HU1 — View showing positioning of control levers for making directional maneuvers. Neutral is designated by the letter (N).

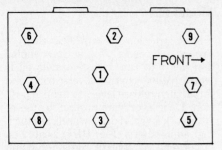

Fig. HU2 – Cylinder head tightening sequence for engine Model Continental TC-56.

recommend using fuel with an octane rating below 85.

To check ignition timing, connect timing light as instructed by manufacturer to No. 1 spark plug. Start engine and allow to warm up, then set idle speed at 400-500 rpm. Point timing light at marks located on either crankshaft dampener or flywheel ring. Timing is correct if timing pointer aligns with TDC mark on crankshaft dampener or flywheel ring. To adjust, loosen distributor securing cap screw, then turn distributor assembly until correct timing is attained. Retighten securing cap screw.

For engine specifications refer to the following chart.

Model TC-56
No. of cylinders .2
Bore . . . . .3-3/16 inches
Stroke . . . . .3½ inches
Displacement (Cu. In.) 56
Compression ratio 8:1
Maximum oil pressure 30-40
Minimum oil pressure (Idling) 7
Firing order .1-2
Breaker point gap 0.020-inch
Spark plug gap 0.025-inch
Ignition timing at
  400-500 rpm . . . . .TDC
Maximum cylinder taper 0.008-inch
Valve clearance
  Intake 0.012-inch
  Exhaust 0.020-inch
Valve seat angle (degrees)
  Intake 30
  Exhaust 45
Oil capacity (Quarts)
  Crankcase 2
  Filter . . . . .½
Water capacity (Quarts)
  Engine 2½

For torque values refer to the following chart. Cylinder head tightening sequence is shown in Fig. HU2.
Cylinder head . . . . .45-48 ft.-lbs.
Main bearing cap screws
  ⅜-inch . . . . .35-40 ft.-lbs.
  7/16-inch . . . . .70-75 ft.-lbs.
Connecting rods
  5/16-inch . . . . .25-30 ft.-lbs.
  ⅜-inch . . . . .45-50 ft.-lbs.
Flywheel
  5/16-inch . . . . .20-25 ft.-lbs.
  ⅜-inch . . . . .35-40 ft.-lbs.
Manifolds (Intake and Exhaust)
  5/16-inch . . . . .15-20 ft.-lbs.
  ⅜-inch . . . . .35-40 ft.-lbs.

CONTROL LEVER ADJUSTMENT

All Models

Tractor should not creep in either direction when control levers are placed

in neutral position. Rear stop lever as shown in Fig. HU3 is used to stop control levers from going past neutral position into reverse position when released from forward drive position. Rear stop lever must be raised up in order to pull control levers back into reverse drive position. Rear stop lever should return to the down position when control levers are released from reverse position. For engine to start, front lock lever must be in the forward position as shown in Fig. HU3 to engage neutral start switch. Raise front lock lever after engine starts to allow movement of control levers.

Should tractor creep when control levers are placed in neutral position adjust as follows:

CAUTION: If adjustment is made with engine running, use extreme caution to keep hands and feet clear of moving parts as severe personal injury could occur.

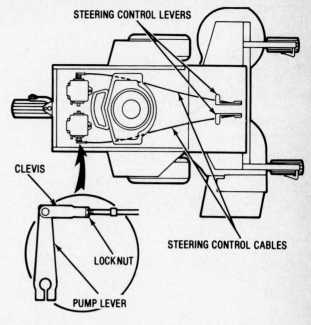

Fig. HU4 – View showing hydrostatic pump control rods and adjustment components on Model 261.

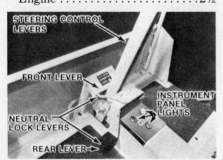

Fig. HU3 – View showing front and rear control lever locks on all models.

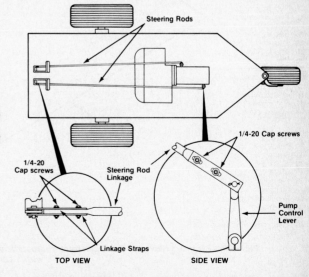

Fig. HU5 – View showing hydrostatic pump control rods and adjustment components on Models 275 and 295.

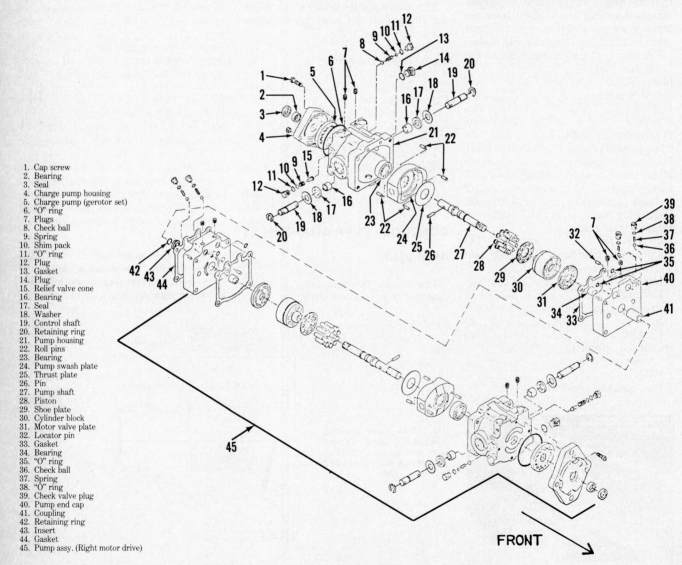

Fig. HU6 — Exploded view of split Sundstrand hydrostatic pump used on Model 261. Pumps are individually driven.

1. Cap screw
2. Bearing
3. Seal
4. Charge pump housing
5. Charge pump (gerotor set)
6. "O" ring
9. Spring
10. Shim pack
11. "O" ring
12. Plug
15. Relief valve cone
17. Seal
18. Washer
19. Control shaft
20. Retaining ring
21. Pump housing
22. Roll pins
23. Bearing
24. Pump swash plate
25. Thrust plate
26. Pin
27. Pump shaft
28. Piston
29. Shoe plate
30. Cylinder block
31. Motor valve plate
40. Pump end cap

Model 261

Raise traction drive wheels clear of ground and securely block. Start engine and allow to run until hydraulic oil is warm. Move control levers back and forth, then release levers to neutral position. Lock levers into position by use of front lock lever. If drive wheels are turning, adjust clevis (Fig. HU4) located on either end of control rod until pump lever is adjusted to neutral position. To adjust clevis, remove clevis pin, then loosen locknut on control rod and turn clevis on rod. Retighten locknut and re-install clevis on pump rod. Recheck for traction wheel creep, repeat adjustment procedure until correct adjustment is attained.

Models 275 and 295

Raise traction drive wheels clear of ground and securely block. Start engine and allow to run until hydraulic oil is

1. Cap screw
2. Bearing
3. Seal
4. Charge pump housing
5. Charge pump (gerotor set)
6. "O" ring
7. Plugs
8. Check ball
9. Spring
10. Shim pack
11. "O" ring
12. Plug
13. Gasket
14. Plug
15. Relief valve cone
16. Bearing
17. Seal
18. Washer
19. Control shaft
20. Retaining ring
21. Pump housing
22. Roll pins
23. Bearing
24. Pump swash plate
25. Thrust plate
26. Pin
27. Pump shaft
28. Piston
29. Shoe plate
30. Cylinder block
31. Motor valve plate
32. Locator pin
33. Gasket
34. Bearing
35. "O" ring
36. Check ball
37. Spring
38. "O" ring
39. Check valve plug
40. Pump end cap
41. Coupling
42. Retaining ring
43. Insert
44. Gasket
45. Pump assy. (Right motor drive)

Fig. HU7 — Exploded view of in-line Sundstrand hydrostatic pump used on Models 275 and 295. Pumps are mounted back-to-back with a common drive shaft.

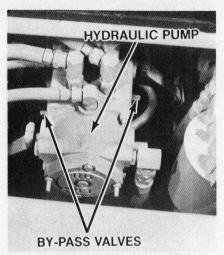

Fig. HU8—View showing location of by-pass studs on hydrostatic pumps for Models 275 and 295.

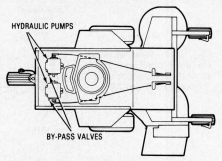

Fig. HU9—View showing location of by-pass studs on hydrostatic pumps for Model 261.

warm. Move control levers back and forth, then release levers to neutral position. Lock levers into position by use of front lock lever. If drive wheels are turning, loosen cap screws securing linkage straps to steering rod as shown in Fig. HU5, then slide steering rod until pump lever is adjusted to neutral position. Retighten cap screws.

BRAKE

All Models

If control levers are adjusted correctly, tractor should stop when levers are returned to neutral position as hydrostatic pump oil flow should be stopped to drive wheel motors.

To adjust parking brake, turn knob on top of brake lever clockwise to increase brake application. Parking brake should retain tractor from moving when brake is applied and control levers are moved forward approximately one inch.

HYDROSTATIC PUMP

All Models

Model 261 uses two split Sundstrand hydrostatic pumps as shown in Fig. HU6, each pump is individually driven. Models 275 and 295 use two inline Sundstrand hydrostatic pumps (Fig. HU7) mounted back-to-back with a common drive shaft. Each pump operates independently with its own complete circuit and is connected in a closed loop to a traction drive motor.

R&R AND OVERHAUL. To remove hydrostatic drive pump, first unlatch and lift engine hood, then tilt seat assembly forward against steering control levers. Disconnect battery ground cable from battery. Thoroughly clean exterior of

pump assembly. Disconnect control arm on control shaft (19–Fig. HU6 and HU7), then withdraw arm from shaft.

CAUTION: Use safety precautions when working with hydraulic fluid as personal injuries can occur.

Place a suitable pan under tractor, then remove reservoir drain plug and drain oil. Disconnect hydraulic hoses from pump assembly. Plug or cap all openings to prevent dirt or other foreign material from entering system. On Model 261 remove as needed all components used to drive hydrostatic pumps. On all models remove as needed all components and shields that will obstruct pump removal. Remove pump mounting cap screws, then withdraw pump assembly.

Hydrostatic pump on Models 275 and 295 can be separated into two individual pump sections with the removal of pump end caps (40–Fig. HU7) securing nuts to mounting studs. If pumps do not separate with removal of nuts use a soft mallet to tap housing. For assistance during overhaul refer Fig. HU6 for exploded view of pump used on Model 261 and Fig. HU7 for exploded view of pump used on Models 275 and 295.

To disassemble the hydrostatic drive unit, first place a scribe mark across charge pump housing (4) and pump housing (21), pump housing and pump end cap (40) for aid in reassembly. Remove all rust, paint or burrs from end of pump shaft (27). Unbolt and remove charge pump housing (4), then remove charge pump (5) and withdraw drive pin (26) from pump shaft. Remove oil seal (3) from housing. Remove relief valve assemblies (9 through 12 and 15) and (8 through 12).

CAUTION: Keep valve assemblies separated. Do not mix the components.

Separate pump end cap (40) from pump housing (21). Valve plate (31) may stick to cylinder block (30). Use care when handling plates. Remove pump cylinder block and piston assembly (28 through 30) and lay aside for later disas-

sembly. Remove thrust plate (25), drive out roll pins (22), then withdraw shafts (19) with snap ring (20) and washer (18). Lift out swashplate (24). Bearing (23) and input shaft (27) may be pressed from housing (21) as an assembly. Press bearing from input shaft. Oil seals (17) and needle bearings (16) can now be removed from housing (21). Remove check valve assemblies (36 through 39). Needle bearing (34) can be removed from end cap (40).

Carefully remove shoe plate (29) and pistons (28) from cylinder block (30). Inspect pistons and bores in cylinder blocks for excessive wear or scoring. Light scratches on piston shoes can be removed by lapping. Inspect valve plate (31) and valve plate contacting surfaces of cylinder block (30) for excessive wear or scoring and renew as needed. Check thrust plate (25) for excessive wear or any other damage. Inspect charge pump (5) for wear, pitting or scoring.

Renew all oil seals, "O" rings and gaskets, lubricate all internal parts with new engine motor oil as specified and reassemble in reverse order of disassembly. Refill hydraulic system reservoir using only SAE 10W-40, 10W-30 or 5W-30 SB/SE engine motor oil.

NOTE: Should tractor need to be moved when engine is inoperative, use a small screwdriver or punch thorough hole in by-pass valve stud on hydrostatic pump as

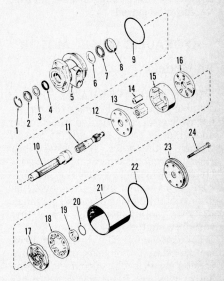

Fig. HU10—Exploded view of Ross hydraulic drive motor.

1. Snap ring	13. Rotor
2. Spacer	14. Roller (6)
3. Shim washer (0.010)	15. Stator
4. Oil seal	16. Manifold plate
5. Body	17. Manifold
6. Thrust washer	18. Commutator ring
7. Thrust bearing	19. Commutator
8. Needle bearing	20. Seal ring
9. Seal ring	21. Sleeve
10. Output shaft	22. Seal ring
11. Drive link	23. End cover
12. Wear plate	24. Capscrew (7)

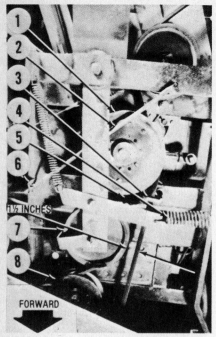

Fig. HU11 — View showing mower deck drive belt components for Model 261.

1. Engine pulley	5. Brake post
2. Removable belt guide	6. Hair pin
3. Belt guide	7. Clutch idler pulley
4. Clutch spring	8. Adjustment idler pulley

shown in Fig. HU8 for Models 275 and 295 and Fig. HU9 for Model 261. Turn stud one complete turn counterclockwise on Models 275 and 295 and one complete turn clockwise on Model 261. Oil will be allowed to flow through system enabling tractor to be moved.

HYDRAULIC WHEEL MOTOR

All Models

R&R AND OVERHAUL. Using a suitable jack, place it under the torsion bar which is located behind wheel motor and extends from one side of tractor frame to the other. Lift drive wheel clear of ground surface and remove lug bolts, then withdraw drive wheel. Disconnect hydraulic lines from hydraulic drive motor and plug or cap all openings. Remove motor mounting bolts, then withdraw motor.

To disassemble the hydraulic drive motor, clamp motor body port boss in a padded jaw vise with output shaft pointing downward. Remove the seven cap screws (24 – Fig. HU10) and remove end cover (23), seal ring (22), commutator (19) and commutator ring (18). Remove sleeve (21), manifold (17) and manifold plate (16). Lift drive link (11), wear plate (12), rotor (13), rollers (14) and stator (15) off the body (5). Remove output shaft (10), then remove snap ring (1), spacer (2), shim (3) and oil seal (4). Remove seal ring (9). Do not remove needle

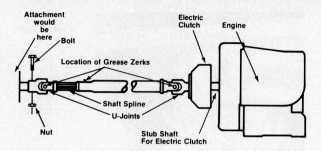

Fig. HU12 — View showing pto drive shaft and operational components for Models 275 and 295.

bearing (8), thrust bearing (7) or thrust washer (6) from body (5) as these parts are not serviced separately.

Clean and inspect all parts for excessive wear or other damage and renew as needed. A seal ring and seal kit (items 2, 3, 4, 9, 20 and 22) is available for resealing the motor. To reassemble the motor, clamp body port boss in a padded vise with the seven tapped holes upward. Insert shaft (10) and drive link (11). Install new seal ring (9) in groove on body (5). Place stator (15) on wear plate (12) and install rotor and rollers (13 and 14) with counterbore in rotor facing upward. Place wear plate and rotor assembly over the drive link and on the body.

NOTE: Two cap screws, 3/8-24x4½" with heads removed, can be used to align bolt holes in body (5) with holes in wear plate (12), stator (15), manifold plate (16), manifold (17), commutator plate (18) and end cover (23).

Install manifold plate (16) with slots toward the rotor. Install manifold (17) with swirl grooves toward the rotor and the diamond shaped holes upward. Place commutator ring (18) and commutator (19) on the manifold with the bronze ring groove facing upward. Place bronze seal ring (20) into the groove with the rubber side downward. Lubricate seal ring (9) and install sleeve (21) over the assembled components. Install new seal ring (22) on end cover (23), lubricate seal ring and install end cover. Remove line up bolts and install the seven cap screws (24). Tighten the cap screws evenly to a torque of 50 ft.-lbs.

Remove motor from vise and place it on bench with output shaft pointing upward. Lubricate and install new oil seal (4), shim (3), spacer (2) and snap ring (1). Lubricate motor by pouring new engine oil as specified in one port and rotating output shaft until oil is expelled from other port. Reinstall motor in reverse order of removal. Tighten lug bolts in a criss-cross pattern to 45-55 ft.-lbs.

ELECTRICAL

All Models

Shown in electrical diagrams Figs. HU13 through HU15 are electrical com-

ponents and their adjoining wires. For component and wire identification refer to Fig. HU13 for Model 261, Fig. HU14 for Model 275 and Fig. HU15 for Model 295.

MOWER DECK DRIVE

Model 261

Install tractor tool bar arms through openings in mower deck until arms are fully inserted in deck openings. Install clevis pins through holes in tool bar arms and mower deck, then lock clevis pins into place using hair pin clips. Deck weight is balanced by hooking deck balance spring on left side of deck, then attaching spring to appropriate chain link on chain under drivers seat until deck weight is evenly distributed.

To install mower deck drive belt proceed as follows. Disengage pto clutch lever, then remove hair pin clip (6 – Fig.

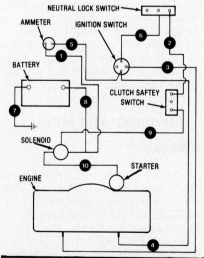

INDEX	COLOR	GA.	DESCRIPTION
1	Brn	14	•Amp. Mtr.-Solenoid
2	Blu	14	Neutral Lk. Sw.-Clutch. Sw.
3	Blk	16	Mag-Ign. Sw. "M" Post
4	Red	16	Alternator-Ign. Sw. "B" Post
5	Red	14	-Amp.-Ign. Sw."B" Post
6	Grn	14	Neutral Lk. Sw.-Ign. Sw. "S" Post
7	Blu	4	Battery Cable
8	Blu	4	Battery Cable
9	Blu	14	Clutch Sw.-Solenoid Wire Assy
10	Blu	4	Switch Cable

Fig. HU13 — Wiring diagram for Model 261.

HU11) from frame slot and slide belt guide (2) out of slot. Position drive belt over clutch cable and route belt to left side of adjustment idler (8), between clutch idler (7) and brake post (5). Slip belt onto engine drive pulley (1). Reinstall belt guide and hair pin clip. Engage pto clutch lever and measure distance between belt as shown in Fig. HU11. Distance should be 1½ inches. To adjust, loosen bolt on adjustment idler (8) and slide idler until correct distance is attained, then retighten bolt.

Drive belt tension can also be changed by repositioning clutch cable in holes on pto lever.

Models 275 and 295

Tractors use a telescoping pto shaft to power mower deck gear box as shown in Fig. 12. Yoke on shaft slides over splines on gear box input shaft and is secured by a bolt and nut. Pto shaft is engaged and disengaged by use of an electric clutch.

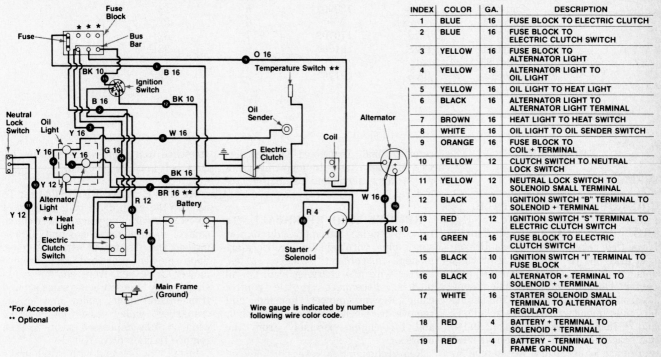

INDEX	COLOR	GA.	DESCRIPTION
1	BLUE	16	FUSE BLOCK TO ELECTRIC CLUTCH
2	BLUE	16	FUSE BLOCK TO ELECTRIC CLUTCH SWITCH
3	YELLOW	16	FUSE BLOCK TO ALTERNATOR LIGHT
4	YELLOW	16	ALTERNATOR LIGHT TO OIL LIGHT
5	YELLOW	16	OIL LIGHT TO HEAT LIGHT
6	BLACK	16	ALTERNATOR LIGHT TO ALTERNATOR LIGHT TERMINAL
7	BROWN	16	HEAT LIGHT TO HEAT SWITCH
8	WHITE	16	OIL LIGHT TO OIL SENDER SWITCH
9	ORANGE	16	FUSE BLOCK TO COIL + TERMINAL
10	YELLOW	12	CLUTCH SWITCH TO NEUTRAL LOCK SWITCH
11	YELLOW	16	NEUTRAL LOCK SWITCH TO SOLENOID SMALL TERMINAL
12	BLACK	10	IGNITION SWITCH "B" TERMINAL TO SOLENOID + TERMINAL
13	RED	12	IGNITION SWITCH "S" TERMINAL TO ELECTRIC CLUTCH SWITCH
14	GREEN	16	FUSE BLOCK TO ELECTRIC CLUTCH SWITCH
15	BLACK	10	IGNITION SWITCH "I" TERMINAL TO FUSE BLOCK
16	BLACK	10	ALTERNATOR + TERMINAL TO SOLENOID + TERMINAL
17	WHITE	16	STARTER SOLENOID SMALL TERMINAL TO ALTERNATOR REGULATOR
18	RED	4	BATTERY + TERMINAL TO SOLENOID + TERMINAL
19	RED	4	BATTERY - TERMINAL TO FRAME GROUND

*For Accessories
** Optional

Wire gauge is indicated by number following wire color code.

Fig. HU14 — Wiring diagram for Model 275.

1. B-Blue   3. BR-Brown   5. O-Orange   7. W-White
2. BK-Black   4. G-Green   6. R-Red   8. Y-Yellow

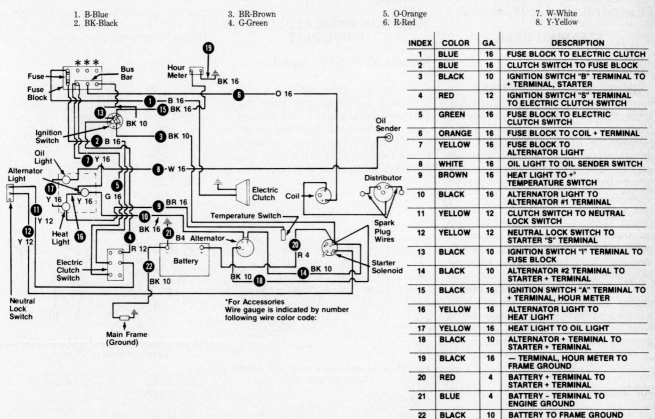

INDEX	COLOR	GA.	DESCRIPTION
1	BLUE	16	FUSE BLOCK TO ELECTRIC CLUTCH
2	BLUE	16	CLUTCH SWITCH TO FUSE BLOCK
3	BLACK	10	IGNITION SWITCH "B" TERMINAL TO + TERMINAL, STARTER
4	RED	12	IGNITION SWITCH "S" TERMINAL TO ELECTRIC CLUTCH SWITCH
5	GREEN	16	FUSE BLOCK TO ELECTRIC CLUTCH SWITCH
6	ORANGE	16	FUSE BLOCK TO COIL + TERMINAL
7	YELLOW	16	FUSE BLOCK TO ALTERNATOR LIGHT
8	WHITE	16	OIL LIGHT TO OIL SENDER SWITCH
9	BROWN	16	HEAT LIGHT TO +° TEMPERATURE SWITCH
10	BLACK	16	ALTERNATOR LIGHT TO ALTERNATOR #1 TERMINAL
11	YELLOW	12	CLUTCH SWITCH TO NEUTRAL LOCK SWITCH
12	YELLOW	12	NEUTRAL LOCK SWITCH TO STARTER "S" TERMINAL
13	BLACK	10	IGNITION SWITCH "I" TERMINAL TO FUSE BLOCK
14	BLACK	16	ALTERNATOR #2 TERMINAL TO STARTER + TERMINAL
15	BLACK	16	IGNITION SWITCH "A" TERMINAL TO + TERMINAL, HOUR METER
16	YELLOW	16	ALTERNATOR LIGHT TO HEAT LIGHT
17	YELLOW	16	HEAT LIGHT TO OIL LIGHT
18	BLACK	10	ALTERNATOR + TERMINAL TO STARTER + TERMINAL
19	BLACK	16	— TERMINAL, HOUR METER TO FRAME GROUND
20	RED	4	BATTERY + TERMINAL TO STARTER + TERMINAL
21	BLUE	4	BATTERY - TERMINAL TO ENGINE GROUND
22	BLACK	10	BATTERY TO FRAME GROUND

*For Accessories
Wire gauge is indicated by number following wire color code:

Fig. HU15 — Wiring diagram for Model 295. Refer to legend in Fig. HU14 for identification of wires.

INTERNATIONAL HARVESTER

INTERNATIONAL HARVESTER CO.
401 North Michigan Avenue
Chicago, IL 60611

Model	Make	Engine Model	Horsepower	Cutting Width, In.
Cadet 55	B&S	130902	5	28
Cadet 60	Tecumseh	V60	6	32
Cadet 75	B&S	170702	7	32
Cadet 85	B&S	191707	8	32
Cadet 85 Special	B&S	191707	8	28

FRONT AXLE

All Models

REMOVE AND REINSTALL. To remove axle main member (5–Fig. IH1), raise front of unit and securely block, then remove front wheels. Disconnect drag link (8–Fig. IH2) from right spindle (7–Fig. IH1). Remove axle center pivot bolt and lower axle assembly from chassis. Remove tie rod assembly, then unbolt and remove spindle (1 and 7). Inspect nylon bushings (2 and 4) and spacers (3 and 6) for excessive wear and renew as necessary. Reassemble by reversing disassembly procedure and adjust tie rod for 0 to 1/16-inch toe-in.

STEERING GEAR

All Models

REMOVE AND REINSTALL. To remove the steering gear assembly (Fig. IH2), remove cotter pin and washer from lower end of steering shaft (12). Remove the three cap screws and withdraw covers (9 and 10) and quadrant gear (6). Drive out roll pin and remove upper shaft (14) and steering wheel (1) from steering shaft (12). Raise front of unit and withdraw steering shaft and pinion. Remove drag link end (7) from quadrant gear. Clean and inspect all parts and renew any showing excessive wear or other damage. When reassembling, lubricate steering gears and bushings with No. 2 multi-purpose lithium grease. With steering wheel and front wheels in straight ahead position, center teeth of quadrant gear (6) should be engaged with pinion on shaft (12).

ENGINE

All Models

For overhaul and repair procedures on engines listed in Specification Table, refer to Small Air Cooled Engines Service Manual. Refer to following paragraphs for removal and installation procedures.

REMOVE AND REINSTALL. To remove engine assembly, disconnect spark plug wire and on electric start models, disconnect battery cables, starter wire and charging lead. On all models, disconnect throttle control cable. Lower mower housing and remove drive belts from engine. Unbolt and lift engine assembly from the chassis.

Reinstall engine by reversing removal procedure.

TRACTION DRIVE CLUTCH AND DRIVE BELT

Models 55-75-85-85 Special

The traction drive clutch is a spring-tensioned, belt idler type operated by the clutch-brake pedal on left side. When pedal is depressed, belt tension is removed, allowing engine drive pulley to rotate freely within drive belt. No adjustment is required; if belt is worn or stretched to a point where slippage occurs, belt must be renewed.

REMOVE AND REINSTALL. To remove the main drive belt, disconnect spark plug wire and place mower clutch control in disengaged position. Lower mower and disconnect attaching pins. Slide mower rearward and remove mower drive belt from engine crankshaft pulley. Loosen belt guides around transmission input pulley and engine crankshaft pulley. Lock clutch-brake pedal in fully depressed position, then remove traction drive belt.

Install new belt and with clutch-brake pedal in up (clutch engaged) position, adjust belt guides to clearances shown in Fig. IH3. Reinstall mower by reversing the removal procedure. Check and ad-

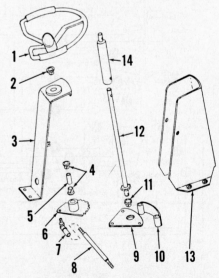

Fig. IH2 — Exploded view of typical steering gear assembly used on all models.

1. Steering wheel	8. Drag link
2. Nylon bushing	9. Bottom cover
3. Steering support	10. Front cover
4. Nylon bushing	11. Nylon bushing
5. Spacer	12. Steering shaft &
6. Quadrant gear	pinion
7. Ball joint end (2	13. Shroud
used)	14. Upper shaft

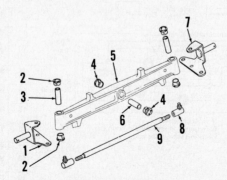

Fig. IH1 — Exploded view of front axle assembly used on all models.

1. Spindle L.H.	
2. Nylon bushing	6. Spacer
3. Spacer	7. Spindle R.H.
4. Nylon bushing	8. Ball joint end (2 used)
5. Axle main member	9. Tie rod

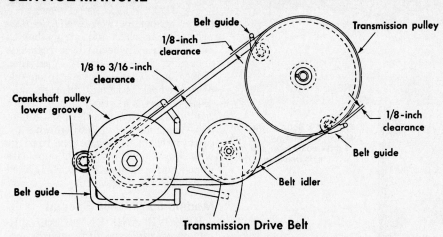

Fig. IH3—Bottom view of typical main drive belt installation on Models 55, 75, 85 and 85 Special. With clutch engaged, adjust belt guides to clearances shown.

just brake as outlined in BRAKE paragraph.

Model 60

A forward-reverse type drive is used on the Model 60. Two drive belts and a single idler pulley are used on this drive. One belt is driven by the engine crankshaft pulley (forward drive) and the other belt is driven by the engine camshaft pulley (reverse drive). The idler is located between the two drive belts. See Fig. IH4. The idler applies tension to only one belt at a time, depending on which foot pedal is depressed. Left foot pedal is for reverse drive and right foot pedal is for forward drive. When neither pedal is depressed, tension is removed from both drive belts and the transmission input pulley stops rotating allowing transmission to be shifted to any of the four speeds. No adjustment is required on the forward-reverse drive belts. If belt slippage occurs due to excessive wear or stretching, renew the belts.

REMOVE AND REINSTALL. To remove the forward and reverse traction drive belts, disconnect spark plug wire and place mower clutch control lever in disengaged position. Lower mower and remove attaching pins at front lift arms. Slide mower rearward and remove mower drive belt from engine crankshaft pulley. Unbolt and remove drive belt guides. Remove forward drive belt and then reverse drive belt.

Install new belts and bolt belt guides in position. Reinstall mower by reversing the removal procedure.

TRANSMISSION

All Models

The transmissions used on all models

are manufactured by the J.B. Foote Foundry Company. The transmission used on Models 55, 75, 85 and 85 Special is equipped with three forward gears and one reverse. Model 60 is equipped with a 4-speed transmission. A reverse gear is not required on this model since the forward-reverse drive allows operation in forward or reverse in any of the four gears.

REMOVE AND REINSTALL. To remove transmission, disconnect spark plug wire and on electric start models, disconnect battery cables. Remove mower as outlined in MOWER DECK paragraphs and remove riding mower body as follows: Remove steering wheel, upper steering shaft and seat. Remove transmission shift lever knob, mower clutch lever and throttle control knob. Unbolt throttle control from body and disconnect ignition switch. Remove mower lift lever grip, then unscrew spring loaded cap screw and remove lift lever. Remove clutch-brake pedal (Models 55, 75, 85 and 85 Special) or forward

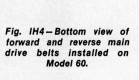

Fig. IH4—Bottom view of forward and reverse main drive belts installed on Model 60.

and reverse pedals (Model 60) and when necessary, remove exhaust muffler. Remove body mounting bolts and note location of body mounting spacers. Tilt body slightly rearward and lift body from chassis.

Unbolt and remove chain guard and transmission belt guides. On Model 60, loosen chain idler and remove chain from transmission output sprocket. On Models 55, 75, 85 and 85 Special loosen rear axle bearing mounting bolts to loosen drive chain and remove chain from transmission output sprocket. Disconnect brake rod and transmission neutral start switch. On all models, remove drive belt(s) from transmission input pulley, then remove snap ring and pulley from input shaft. Unbolt and remove transmission assembly.

Reinstall by reversing the removal procedure. Adjust drive chain, belts and belt guides as required.

NOTE: On Models 55, 75, 85 and 85 Special, make certain safety starting switches are connected and in good condition before returning mower to service.

OVERHAUL. Foote transmissions are used on all models. For overhaul procedure refer to the Foote paragraphs in the TRANSMISSION REPAIR section of this manual.

DRIVE CHAIN

Model 60

R&R AND ADJUST. To remove chain, rotate differential sprocket to locate master link. Disconnect master link and remove chain. Clean and inspect chain and renew if excessively worn. Reinstall in reverse order of removal.

Adjust chain idler (Fig. IH5) until ⅛ to ¼-inch slack exists in chain on side op-

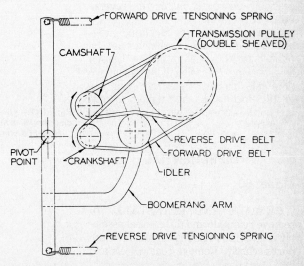

Fig. IH5—Bottom view of Model 60 showing location of drive chain idler and rear axle bearings.

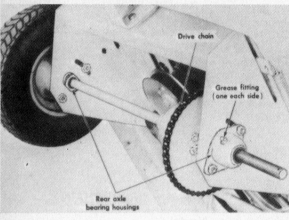

Fig. IH6—On Models 55, 75, 85 and 85 Special, rear axle assembly is pivoted rearward to tighten the drive chain.

spacer washers between each axle bearing and rear wheel and one washer on outside of each wheel. Check rear axle end play. If end play is more than ⅛-inch, add washers equally to outside of rear wheels, starting with right side, until end play is less than ⅛-inch. Secure rear wheels with new cotter pins. Adjust drive chain idler until ⅛ to ¼-inch slack exists in chain. Lubricate axle bearings with No. 2 multi-purpose lithium base grease and apply light coat of SAE 30 oil to drive chain.

Models 55-75-85-85 Special

REMOVE AND REINSTALL. To remove rear axle and differential assembly (Fig. IH6), raise rear of unit and support securely. Remove rear wheels and washers. Note location of washers. They must be reinstalled in original positions to maintain proper drive chain alignment. Unbolt and remove both rear axle bearings, then remove drive chain from differential sprocket. Remove four bolts from differential housing and slide sprocket off housing. Separate differential housing and carefully remove the two halves from main frame.

Lubricate differential with No. 2 multi-purpose lithium base grease and reinstall by reversing removal procedure. When installing wheels and washers, be sure washers are assembled in original order to maintain drive chain alignment. Check rear axle end play. If end play is more than ⅛-inch, add washers to outside of wheels, starting with right side, until end play is less than ⅛-inch. Rear axle bearing mounting holes are slotted so axle can be pivoted to adjust drive chain. Adjust both sides equally until chain has ⅛ to ¼-inch slack, then tighten mounting. Lubricate axle bearings with No. 2 multi-purpose lithium base grease and apply light coat of SAE 30 oil to chain.

OVERHAUL. Peerless Series 100 differentials are used on all models. For

posite idler. Lubricate with light coat of SAE 30 oil.

Models 55-75-85-85 Special

R&R AND ADJUST. To remove chain (Fig. IH6), rotate differential sprocket to locate master link. Disconnect master link and remove chain. Clean and inspect chain and renew if excessively worn. Reinstall in reverse order of removal.

The rear axle bearing mounting plates are slotted so axle assembly can be pivoted to adjust chain tension. Loosen axle bearing mounting bolts and move both sides equally until drive chain has ⅛ to ¼-inch slack, then tighten mounting bolts. Lubricate with light coat of SAE 30 oil.

DIFFERENTIAL

Model 60

REMOVE AND REINSTALL. To remove rear axle and differential assembly (Fig. IH5), raise rear of unit and support securely. Remove rear wheels and

keys. Note spacer washers between wheels and axle bearings. Washers must be reinstalled in original positions to align differential drive sprocket with drive chain. Loosen drive chain idler and remove chain from differential sprocket. Unbolt axle bearings and remove axle and differential assembly.

When reinstalling, tighten axle bearing retaining bolts securely. Install four

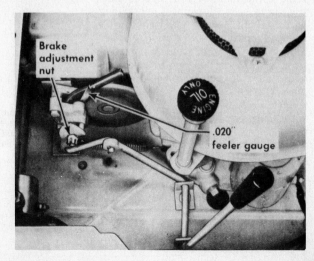

Fig. IH7—View showing disc brake adjustment on Model 60. Refer to text.

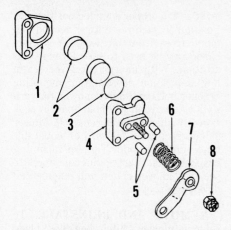

Fig. IH8—Exploded view of disc brake caliper assembly used on Model 60.

1. Brake pad carrier (inner)
2. Brake pads
3. Back-up plates
4. Brake pad carrier (outer)
5. Actuating pins
6. Spring
7. Cam lever
8. Adjusting nut

overhaul procedures, refer to the DIFFERENTIAL REPAIR section of this manual.

GROUND DRIVE BRAKE

Model 60

The hand lever applied caliper type disc brake is used primarily as a parking brake. It can also be used as an emergency type brake. Normal braking is accomplished by lightly depressing opposite direction (forward or reverse) pedal. The brake disc is pinned to transmission output shaft.

ADJUSTMENT. To adjust disc brake, place brake hand lever in disengaged position. Remove cotter pin from brake adjustment nut and insert a 0.020-inch feeler gage between outer brake pad and brake disc. See Fig. IH7. Turn

adjustment nut clockwise until a slight drag is felt on feeler gage. Install cotter pin to secure nut.

R&R AND OVERHAUL. To remove disc brake, disconnect brake return spring. Remove cap screws securing brake lever bracket and brake caliper bracket to the chassis, then lift out caliper assembly. To disassemble brake caliper, remove the two through-bolts and adjusting nut (8–Fig. IH8), then separate the parts. Clean and inspect all parts and renew any showing excessive wear or other damage. Check to see that brake disc is free to float axially on transmission shaft. Reinstall brake and adjust as outlined in the preceding paragraph.

Models 55-75-85-85 Special

The brake used is a caliper type disc brake located on output shaft of transmission. Brake is operated by clutch-brake pedal located on left side.

ADJUSTMENT. To adjust brake, lock clutch-brake pedal in fully depressed position. On underside (near differential), adjust jam nuts on brake rod until spring length is 1½ inches measured as shown in Fig. IH9.

R&R AND OVERHAUL. To remove disc brake, disconnect brake rod from cam actuating lever (3–Fig. IH10). Remove shoulder bolt (2) and remove brake parts (1 through 6–Fig. IH11). Slide disc (7) from transmission shaft and remove inner brake pad (8) from holder slot in transmission housing. Clean and inspect all parts for excessive wear or other damage. Renew parts as required and reassemble by reversing the removal procedure. Adjust brake as outlined in the preceding paragraph.

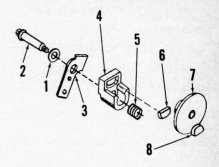

Fig. IH11—Exploded view of disc brake assembly used on Models 55, 75, 85 and 85 Special.

1. Washer
2. Shoulder bolt
3. Cam lever
4. Caliper
5. Spring
6. Brake pad (outer)
7. Brake disc
8. Brake pad (inner)

MOWER DRIVE BELTS

All Models

Models 55 and 85 Special are equipped with single blade rotary mowers while the mowers used on Models 60, 75 and 85 are equipped with twin blades.

CAUTION: Always disconnect spark plug wire before performing any inspection, adjustments or other service on the mower.

Models 55-85 Special

ADJUSTMENT. The mower blade clutch is a belt idler type. When clutch is disengaged, a spring loaded brake arm comes into contact with drive belt at mower spindle pulley to stop blade rotation. A cable connected to brake arm and clutch control linkage pulls brake arm free from belt when clutch is engaged. No adjustment is required on blade brake. When mower drive belt has worn or stretched to a point where slippage occurs, a new belt should be installed as no adjustment can be made on clutch idler.

REMOVE AND REINSTALL. To remove mower drive belt, place mower

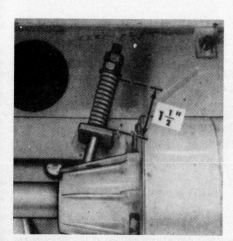

Fig. IH9—Bottom view showing brake adjustment on Models 55, 75, 85 and 85 Special.

Fig. IH10—Disc brake assembly on transmission of Models 55, 75, 85 or 88 Special. Brake disc (1) should be free to move axially on transmission shaft.

1. Brake disc
2. Shoulder bolt
3. Cam lever

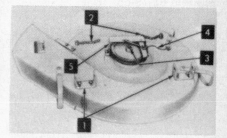

Fig. IH12—Mower unit used on Models 55 and 85 Special.

1. Front mounting brackets
2. Rear hanger chains
3. Blade spindle
4. Blade brake cable
5. Belt guide & blade brake bracket

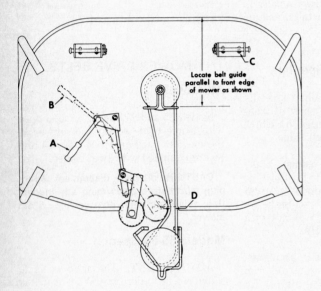

Crankshaft pulley lower groove

1/8 to 3/16-inch clearance

Mower spindle pulley

Belt guide

Belt idler

Mower Drive Belt

Fig. IH13— Mower drive belt installation on Models 55 and 85 Special. Note belt guide clearance.

Locate belt guide parallel to front edge of mower as shown

B

A

C

D

Fig. IH14 — Mower drive clutch belt adjustment on Models 75 and 85.

A. Clutch disengaged
B. Clutch engaged
C. Front mounting bracket bolts
D. Correct distance is 1½ inches

lever to the proper engaged position (B). Position (B) is directly over second diamond from front on decal. A wooden block may be used to hold lever in this position. With mower lift lever set in lowest position, loosen front mounting bracket bolts and move mower forward until belt tension will hold lever in position (B) when block is removed. Mower should be moved forward an equal distance on each side. Tighten mounting bracket bolts. Start engine and check to see that mower clutch will engage and disengage properly.

REMOVE AND REINSTALL. To remove mower clutch belt and blade spindle belt on Models 60, 75 or 85, place mower clutch control lever in disengaged position and set mower lift lever in lowest position. Unpin front of mower from lift arms and slide mower rearward so mower drive belt can be removed from engine crankshaft pulley. Remove belt from mower drive pulley. To remove blade spindle belt, unhook idler tension spring and move idler out of the way. Remove belt from spindle pulleys.

Install new blade spindle belt and attach idler tension spring. See Fig. IH16 or IH17. Install new mower drive clutch belt and reinstall mower on lift arms. Make certain belt is correctly installed on belt guides as shown in Fig. IH18. Adjust mower clutch belt as outlined in preceding ADJUSTMENT paragraph.

clutch control lever in disengaged position. Set mower lift lever in lowest position and disconnect blade brake cable (4–Fig. IH12). Unpin front of mower and unhook chains at rear of mower. Move mower rearward and remove belt from crankshaft pulley. Place mower lift lever in highest position and slide mower out right side. Unbolt blade brake bracket (5) and remove belt.

Renew belt and reinstall mower by reversing removal procedure. Make certain that belt is correctly installed in belt guides as shown in Fig. IH13.

Models 60-75-85

ADJUSTMENT. The mower clutch is belt idler type. On Models 75 and 85, if mower clutch belt slips under normal operation, adjust belt as follows: Place mower lift lever in lowest position and engage mower clutch. Mark a pencil line on mower housing at front of each mounting bracket. Loosen front mounting bracket bolts (C–Fig. IH14) and move mower forward until a distance (D) of 1½ inches exists between inner sides of belt at idler pulley. Mower should be moved forward an equal distance on each side. Tighten mounting

bracket bolts. Start engine and with mower lift lever in third position from bottom, check to see that mower clutch will engage and disengage properly.

On Model 60, when mower clutch control lever moves to position (C–Fig. IH15) when clutch is engaged, mower clutch belt should be adjusted. Position (C) is over front diamond on decal. To adjust clutch belt, move clutch control

MOWER SPINDLE

All Models

LUBRICATION. Blade and center spindle bearings should be lubricated with No. 2 multi-purpose lithium grease when mower is disassembled and after each 10 hours of operation.

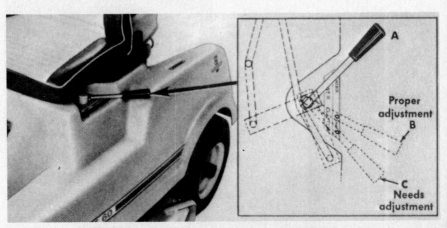

Proper adjustment
B

A

C
Needs adjustment

Fig. IH15 — On Model 60, mower clutch control lever is in disengaged position at "A". When mower clutch belt is properly adjusted position "B" is engaged position. Belt adjustment is required when lever moves to position "C" when clutch is engaged.

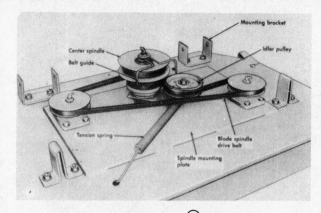

Fig. IH16—View of rear discharge, twin blade mower assembly used on all Model 60 and some Models 75 and 85. Note installation of blade spindle belt and idler tension spring.

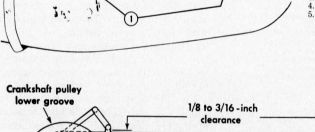

Fig. IH17—View of side discharge, twin blade mower assembly used on some 75 and 85 models.

1. Spindle cap
2. Pulley nut
3. Blade spindle drive belt
4. Belt guide
5. Idler tension spring

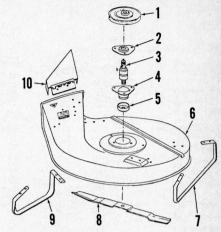

Fig. IH20—Exploded view of mower unit used on Models 55 and 85 Special.

1. Mower pulley
2. Spindle cap
3. Spindle shaft & bearing assy.
4. Spindle housing
5. Spindle cup
6. Mower housing
7. Mower runner L.H.
8. Blade 28"
9. Mower runner R.H.
10. Deflector

renew any showing excessive wear or other damage.

Reassemble mower by reversing disassembly. Tighten blade spindle nut to a torque of 55-60 ft.-lbs. Reinstall mower and make certain mower drive belt is correctly installed in belt guides as shown in Fig. IH13.

Models 60-75-85

R&R AND OVERHAUL. To remove the blades and spindles, first remove mower assembly as follows: Place mower blade clutch control lever in disengaged position and move mower lift lever to lowest position. Unpin front of mower from lift arms and slide mower rearward until lift arms are free of rear lift brackets. Remove mower clutch belt from mower drive pulley. Raise mower lift lever to highest position and slide mower out right side.

On rear discharge mowers, unhook idler tension spring (18–Fig. IH21) and remove belt (15). Refer to Fig. IH23 and remove shaft bolt from center of each

Fig. IH18—Mower drive clutch belt installation on Models 60, 75 or 85. Note belt guide clearance.

Crankshaft pulley lower groove — 1/8 to 3/16-inch clearance — Mower drive pulley upper groove — Belt guide — Drive belt idler — Belt guide

Models 55-85 Special

R&R AND OVERHAUL. To remove the single blade and blade spindle, first remove mower assembly as follows: Place mower blade clutch control lever in disengaged position. Set mower lift lever in lowest position and disconnect blade brake cable. Unpin front of mower and unhook chains at rear of mower. Move mower rearward and remove mower drive belt from crankshaft pulley. Place mower lift lever in highest position and slide mower out right side. Unbolt and remove blade brake bracket and belt. Place a wood block between blade and housing, then remove pulley retaining nut and mower pulley. Invert mower housing assembly, block the blade, then remove blade retaining nut, blade and spindle cup. See Fig. IH19. Unbolt and remove spindle assembly.

Remove spindle housing (4–Fig. IH20) and spindle cap (2). Spindle shaft and bearing are available only as an assembly (3). Clean and inspect all parts and

Fig. IH19—On Models 55 and 85 Special, place a wood block between blade and housing and remove blade retaining nut.

1. Blade retaining nut
2. Blade
3. Spindle cup
4. Spindle retaining nuts

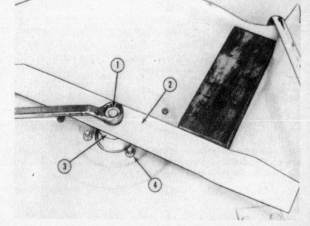

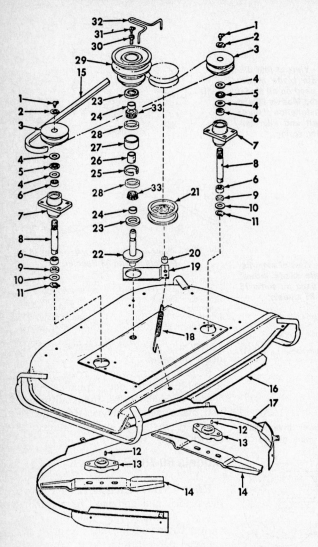

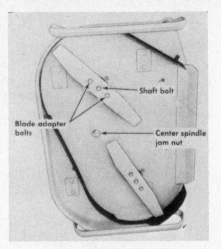

Fig. IH21 — Exploded view of rear discharge mower assembly used on Model 60. Rear discharge mower used on Models 75 and 85 is similar. Refer to Fig. IH22.

1. Lube fitting
2. Snap ring
3. Blade pulley
4. Thrust washer
5. Thrust bearing
6. Needle bearing
7. Spindle housing
8. Blade spindle shaft
9. Oil seal
10. Thrust washer
11. Snap ring
12. Key
13. Blade adapter
14. Blade
15. Blade spindle belt
16. Mower housing
17. Shroud
18. Idler tension spring
19. Idler arm
20. Spacer
21. Idler pulley
22. Center spindle shaft
23. Oil seal
24. Spacer
25. Split outer spacer
26. Bearing spacer (inner)
27. Bearing spacer (outer)
28. Bearing cup
29. Center spindle pulley
30. End bolt
31. Lube fitting
32. Belt guide
33. Bearing cone

Fig. IH23 — Bottom view of rear discharge mower used on Models 60, 75 and 85.

spacers (24), oil seals (23), bearing cones (33), cups (28) and spacers (25, 26 and 27) from pulley. Renew parts as necessary and reassemble by reversing disassembly procedure. Install oil seals (23) with lips toward outside of pulley. Tighten center spindle jam nut to a torque of 185 ft.-lbs. Reinstall blade spindle belt (15) and hook idler tension spring (18) in hole in mower housing.

On side discharge mowers used on Models 75 and 85, unhook idler tension spring (14 – Fig. IH24) and remove belt. Place a wood block between blades and mower housing, then remove pulley retaining nuts and pulleys (1). Invert mower housing assembly, block the blades (Fig. IH25) and remove blade retaining nuts (1), blades (19 – Fig. IH24) and spindle cups (18). Unbolt and remove the blade spindle assemblies. Remove spindle caps (2) and spindle housings (4). Spindle shafts and bearings are available only as assemblies (3). Clean and inspect all parts and renew any showing excessive wear or other damage. When reassembling, tighten blade spindle nuts to a torque of 55-60 ft.-lbs.

To remove the center spindle assembly, refer to Fig. IH25 and remove center spindle nut (4). Lift center spindle assembly from mower. Refer to Fig. IH24 and remove lube fitting, end bolt (6) and belt guide (5). Remove the two snap rings (7) and press shaft and bearing assembly (8) from pulley (9). Press only on outer race of bearing to prevent damage to bearing. Shaft and bearing are available only as an assembly (8). When reassembling, tighten center spindle jam nut (4 – Fig. IH25) to a torque of 185 ft.-lbs. Reinstall blade spindle belt and hook idler tension spring (14 – Fig. IH24) in hole in mower housing.

On all models, reinstall mower and make certain mower drive clutch belt is correctly installed in belt guides as shown in Fig. IH18.

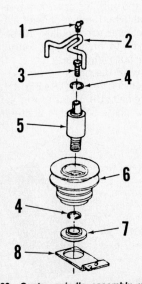

Fig. IH22 — Center spindle assembly used on Models 75 and 85 rear discharge mower. Refer to Fig. IH21 for the balance of mower components.

1. Lube fitting
2. Belt guide
3. End bolt
4. Retaining ring
5. Shaft & bearing assy.
6. Center spindle pulley
7. Spacer
8. Idler arm

blade. Remove blades (14 – Fig. IH21) and blade adapters (13). Unbolt and remove blade spindle assemblies from mower housing. Remove snap ring (11), washer (10) and withdraw pulley (3) and shaft (8). Remove lube fitting (1) and snap ring (2), then press shaft out of pulley. Remove thrust washers (4) and thrust bearing (5). Needle bearings (6) and oil seal (9) can now be pressed from spindle housing (7). When reassembling blade spindles, install oil seal (9) with lip towards washer (10). Torque blade shaft bolts to 57 ft.-lbs.

To remove the center spindle assembly, refer to Fig. IH23 and remove center spindle jam nut. Lift center spindle assembly from mower. On Models 75 and 85, refer to Fig. IH22 and remove lube fitting (1), end bolt (3) and belt guide (2). Remove snap rings (4) and press shaft and bearing assembly (5) from pulley (6). If bearing is to be reused, press only on outer race. Shaft and bearing is available only as an assembly. On Model 60, refer to Fig. IH21 and remove lube fitting (31), end bolt (30) and belt guide (32). Remove pulley (29) with bearings from shaft (22). Remove

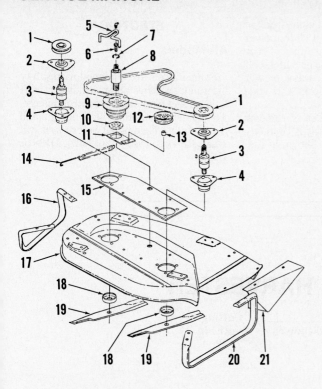

Fig. IH24—Exploded view of side discharge mower used on Models 75 and 85.

1. Blade pulley
2. Spindle cap
3. Blade spindle shaft & bearing assy.
4. Spindle housing
5. Belt guide
6. End bolt
7. Retaining ring (2)
8. Center spindle shaft & bearing assy.
9. Center spindle pulley
10. Spacer
11. Idler arm
12. Idler pulley
13. Spacer
14. Idler tension spring
15. Spindle mounting plate
16. Mower runner L.H.
17. Mower housing
18. Spindle cup
19. Blades
20. Mower runner R.H.
21. Deflector

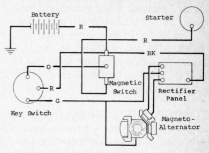

Fig. IH26—Wiring schematic for electric start Model 60. Refer to legend in Fig. IH29 for identification of wires.

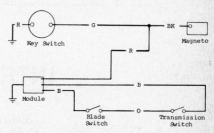

Fig. IH27—Typical wiring schematic for models with recoil start only. Refer to legend in Fig. IH29 for identification of wires.

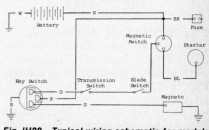

Fig. IH28—Typical wiring schematic for models with electric start only. Refer to legend in Fig. IH29 for identification of wires.

MOWER DECK

Models 55-85 Special

LEVEL ADJUSTMENT. Position unit on level surface, then disconnect spark plug wire. Measure distance from ground to blade from side to side and front to rear. Blade should be approximately 1/8-inch lower in front. Adjust front to rear by hooking rear hanger chains (2–Fig. IH12) in different links or add 11/32-inch flat washers between mower deck and hanger chains. To adjust side to side, add washers between mower deck and front mounting bracket (1) and rear hanger chain (2) on side that measures high.

REMOVE AND REINSTALL. Place mower blade clutch control lever in disengaged position. Set mower lift lever in lowest position and disconnect blade brake cable. Unpin front of mower and unhook chains at rear of mower. Move mower rearward and remove mower drive belt from crankshaft pulley. Place mower lift lever in highest position and slide mower out right side.

Reinstall in reverse order of removal and make certain drive belt is correctly installed in belt guides as shown in Fig. IH13.

Models 60-75-85

LEVEL ADJUSTMENT. Position unit on level surface, then disconnect spark plug wire. Measure distance from ground to blade from side to side and front to rear. Blade should be approximately 1/8-inch lower in front. To adjust front to rear, add 11/32-inch washers between mower deck and front mounting brackets (Fig. IH16) or rear mounting brackets depending on which measures

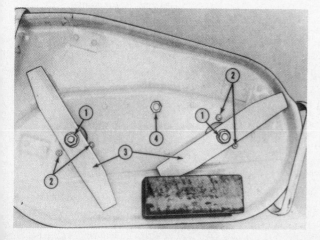

Fig. IH25—Bottom view of side discharge mower used on Models 75 and 85. Place a wood block between blade and mower housing when removing blade nut.

1. Blade retaining nut
2. Spindle retaining nuts
3. Blades
4. Center spindle nut

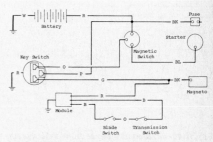

Fig. IH29—Typical wiring schematic for models with electric and recoil start.

1. G-Green	5. W-White
2. R-Red	6. B-Brown
3. O-Orange	7. BL-Blue
4. P-Purple	8. BK-Black

high. Note that front mounting brackets must be reinstalled in original positions to maintain proper drive belt adjustment. To adjust side to side, add washers between mower deck and front and rear mounting bracket on side that measures high.

REMOVE AND REINSTALL. Place mower blade clutch control lever in disengaged position and move mower lift lever to lowest position. Unpin front of mower from lift arms and slide mower rearward until lift arms are free of rear lift brackets. Remove mower clutch belt from mower drive pulley. Raise mower lift lever to highest position and slide mower out right side.

Reinstall in reverse order of removal and make certain drive belt is correctly installed in belt guides as shown in Fig. IH18.

ELECTRICAL

All Models

Some models are equipped with safety interlock switches on transmission shift linkage and blade clutch control linkage. The transmission must be in neutral and blade clutch disengaged before unit will start. Refer to Fig. IH26, IH27, IH28 or IH29.

INTERNATIONAL HARVESTER

Model	Make	Engine		Cutting Width, In.
		Make	Horsepower	
Cadet 95	Electric Motor	1		32

FRONT AXLE

REMOVE AND REINSTALL. To remove axle main member (5–Fig. IH40), support front of unit and remove front wheels. Disconnect drag link (8–Fig. IH41) from right spindle (7–Fig. IH40). Remove axle center pivot bolt and lower axle assembly from chassis. Remove tie rod assembly, then unbolt and remove spindle (1 and 7). Inspect nylon bushings (2 and 4) and spacers (3 and 6) for excessive wear and renew as necessary. Reassemble by reversing disassembly procedure and adjust tie rod for 0 to 1/16-inch toe-in.

STEERING GEAR

REMOVE AND REINSTALL. To remove the steering gear assembly (Fig. IH41), remove cotter pin and washer from lower end of steering shaft (12). Remove the three cap screws and withdraw covers (9 and 10) and quadrant gear (6). Drive out roll pin and remove upper shaft (14) and steering wheel (1) from steering shaft (12). Raise front of unit and withdraw steering shaft and pinion (12). Remove drag link end (7) from quadrant gear. Clean and inspect all parts and renew any showing excessive wear or other damage. When reassembling, lubricate steering gears and bushings with No. 2 multi-purpose lithium grease. With steering wheel and front wheels in straight ahead position, center teeth of quadrant gear (6) should be engaged with pinion gear.

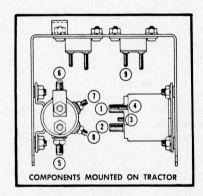

COMPONENTS MOUNTED ON TRACTOR

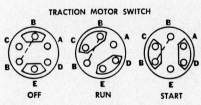

TRACTION MOTOR SWITCH

OFF   RUN   START

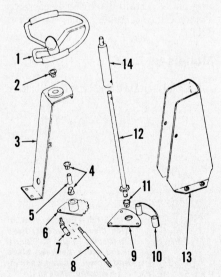

Fig. IH41—Exploded view of typical steering gear assembly.

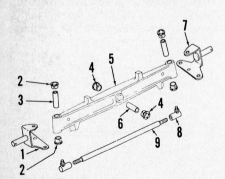

Fig. IH40—Exploded view of front axle assembly.

1. Spindle L.H.
2. Nylon bushing
3. Spacer
4. Nylon bushing
5. Axle main member
6. Spacer
7. Spindle R.H.
8. Ball joint end (2 used)
9. Tie rod

1. Steering wheel
2. Nylon bushing
3. Steering support
4. Nylon bushing
5. Spacer
6. Quadrant gear
7. Ball joint end (2 used)
8. Drag link
9. Bottom cover
10. Front cover
11. Nylon bushing
12. Steering shaft & pinion
13. Shroud
14. Upper shaft

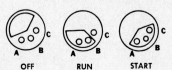

BLADE MOTORS SWITCH

OFF   RUN   START

Fig. IH42—View of switches showing operating positions.

Fig. IH43—Schematic of electrical system for traction motor and blade motors.

OVERHAUL. Clamp flat portion of armature in a vise. Remove through bolts and pry end frame off armature end bearing. Pull motor housing off armature with a quick upward thrust. Press armature and bearing from drive end frame. Clean and inspect all parts and renew any showing excessive wear or damage. Check armature for open or short circuits.

NOTE: When installing motor housing, hold housing by sides only. The magnets will pull motor housing down sharply against end plate as it goes over the armature.

Reassemble in reverse order of disassembly. Push brushes back into holder and rest spring against brush to hold in position. Install commutator end plate part way, then use a small screwdriver to push the brushes in against the commutator.

TRACTION DRIVE CLUTCH AND DRIVE BELT

The traction drive clutch is spring-tensioned, belt idler type operated by clutch-brake pedal on left side. When pedal is depressed, belt tension is removed, allowing drive pulley to rotate freely within drive belt. No adjustment is required; if belt is worn or stretched to a point where slippage occurs, belt must be renewed.

REMOVE AND REINSTALL. Turn traction motor key switch and mower control switch off. Remove mower deck as outlined in MOWER DECK paragraphs. Remove traction motor pulley belt guide (3–Fig. IH45) and loosen transmission pulley belt guides (1). Remove idler pulley (5) and remove drive belt (6).

Renew belt by reversing removal procedure and adjust belt guides to clearance "B" shown in Fig. IH45.

TRANSMISSION

REMOVE AND REINSTALL. Remove the mower as outlined in MOWER

TRACTION DRIVE ELECTRIC MOTOR

Three 12 volt batteries (Fig. IH43) connected in series provide 36 volts D.C. for operation of traction and blade motors. The traction motor is protected from overload by a thermal switch in brush holder assembly and a 100 amp circuit fuse. Refer to Fig. IH42 and IH43.

REMOVE AND REINSTALL. Remove mower as outlined in MOWER DECK paragraphs and remove mower body as follows: Lift off battery cover shroud and disconnect and remove batteries and mounting plate. Drive out roll pin and remove steering wheel and upper steering shaft. Disconnect and remove clutch-brake pedal. Remove gear shift knob. Remove body mounting bolts and note location of body spacers. Raise body and disconnect wiring harness from traction motor and blade motor switches. Move mower lift handle to highest position, then raise body clear of handle and remove body.

Loosen drive pulley belt guide and slip drive belt off pulley. Remove snap ring and drive pulley from motor shaft. Remove motor end cover and disconnect wires from terminals. Unbolt and remove motor.

Reinstall by reversing removal procedure. Refer to Fig. IH44 for location of wires.

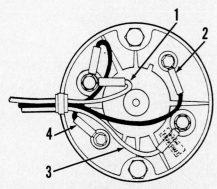

Fig. IH44 — View of traction motor electrical connections. Note red wire is attached to post marked +CCW.

1. Brown wire to 100 amp fuse
2. Gray wire to magnetic switch
3. Red wire to post marked +CCW
4. Pink wire to brown wire on post No. 1

Fig. IH45 — Top view of traction drive belt installation. Note belt guide clearance "B" of ⅛ to 3/16-inch between guide and outside surface of belt or pulley.

1. Belt guides
2. Traction motor drive pulley
3. Belt guide
4. Belt guide
5. Idler pulley
6. Drive belt
7. Transmission pulley

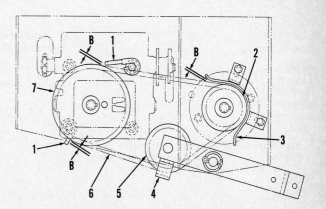

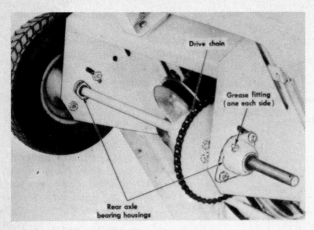

Fig. IH46 — Rear axle assembly is pivoted rearward to tighten drive chain.

DECK paragraphs and remove mower body as follows: Lift off battery cover shroud and disconnect and remove batteries and mounting plate. Drive out roll pin and remove steering wheel and upper steering shaft. Disconnect and remove clutch-brake pedal. Remove gear shift knob. Remove body mounting bolts and note location of body spacers. Raise body and disconnect wiring harness from traction motor and blade motor switches. Move mower lift handle to highest position, then raise body clear of handle and remove body.

Remove the traction drive belt as previously outlined and remove snap ring and drive pulley from transmission shaft. Loosen rear axle bearing mounting bolts to loosen drive chain and remove chain from transmission sprocket. Disconnect transmission neutral start switch wiring. Disconnect brake rod and return spring from brake actuating lever. Unbolt and remove transmission.

OVERHAUL. A Foote model transmission is used. For overhaul procedure refer to Foote paragraphs in TRANSMISSION REPAIR section of this manual.

DRIVE CHAIN

R&R AND ADJUST. To remove chain (Fig. IH46), rotate differential sprocket to locate master link. Disconnect master link and remove chain. Clean and inspect chain and renew if excessively worn. Reinstall in reverse order of removal.

The rear axle bearing mounting plates are slotted so axle assembly can be pivoted to adjust chain tension. Loosen axle bearing mounting bolts and move both sides equally until chain has 1/8 to 1/4-inch slack, then tighten mounting bolts. Lubricate with light coat of SAE 30 oil.

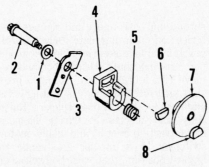

Fig. IH48 — Exploded view of disc brake assembly.

1. Washer	5. Spring
2. Shoulder bolt	6. Brake pad (outer)
3. Cam lever	7. Brake disc
4. Caliper	8. Brake pad (inner)

DIFFERENTIAL

REMOVE AND REINSTALL. To remove rear axle and differential assembly (Fig. IH46), raise rear of unit and support securely. Remove rear wheels and washers. Note location of washers. They must be reinstalled in original positions to maintain proper drive chain alignment. Unbolt and remove axle bearings. Remove drive chain from differential sprocket and remove differential and axle from frame.

Reinstall by reversing removal procedure. When installing wheels and washers, be sure washers are assembled in original order to maintain drive chain alignment. Check rear axle end play. If end play is more than 1/8-inch, add washers to outside of wheels, starting with right side, until end play is less than 1/8-inch. Rear axle bearing mounting holes are slotted so axle can be pivoted to adjust drive chain. Adjust both sides equally until chain has 1/8 to 1/4-inch slack, then tighten mounting bolts. Lubricate axle bearings with No. 2 multi-purpose lithium base grease and apply light coat of SAE 30 oil to chain.

GROUND DRIVE BRAKE

The brake is a caliper type disc brake located on output shaft of transmission. The brake is operated by clutch-brake pedal located on left side.

ADJUSTMENT. To adjust brake, lock clutch-brake pedal in fully depressed position. Loosen jam nut on clevis (4 – Fig. IH47), located near drive belt idler pulley, and adjust brake adjusting bolt (1) until spring length "A" is 1½ inches. Tighten jam nut on clevis and recheck brake operation.

R&R AND OVERHAUL. To remove disc brake, disconnect brake rod from

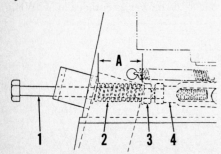

Fig. IH47 — Brake adjusting bolt located near traction drive idler pulley. Adjust bolt until spring length "A" is 1½ inches.

1. Adjusting bolt	3. Jam nut
2. Spring	4. Clevis

Fig. IH49 — Twin blade mower and lift linkage.

1. Front mower bracket
2. Lift arm
3. Rear hanger bracket
4. Blade motors
5. Lift chains
6. Eyebolt
7. Lift linkage

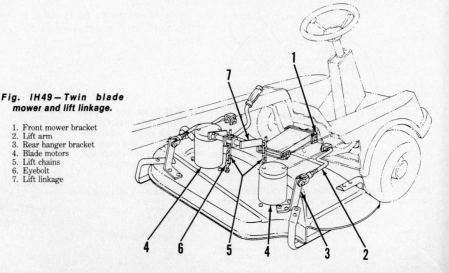

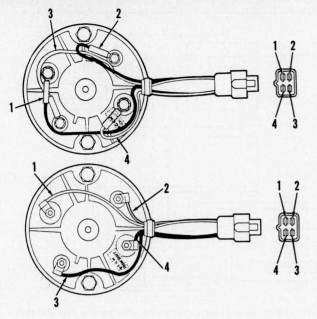

Fig. IH50 — Blade motor electrical connections. Note blue wire connects to terminal marked + CCW.

1. Green wire
2. Black w/white tracer wire
3. Black wire
4. Blue wire

DECK paragraphs. Remove mower blades. Remove mounting bolts and slide motor out bottom of mower deck.

Reinstall motor by reversing removal procedure. Be sure motor is wired correctly. Refer to Fig. IH50. Torque blade nut to 45-55 ft.-lbs.

ELECTRICAL

TESTING ELECTRICAL SYSTEM. Refer to Fig. IH42 and IH43. If traction motor fails to run, check the following: Check batteries and cable connections. Check continuity of 100 amp fuse. Traction motor magnetic switch should "click" when key switch is turned to start position. If click is not heard, check continuity of transmission neutral switch, key switch, magnetic switch and thermal switch. If click is heard, magnetic switch is defective or traction motor has internal damage. If motor starts but will not continue to run when switch is released to on position, the key switch is defective.

If blade motors fail to run, check the following: Traction motor circuit must be functional before blade motors will operate. Blade motor magnetic switch should "click" when blade motor switch is turned to start position. If click is not heard, check continuity of blade motor switch, blade magnetic switch, motor thermal switches and circuit breakers. If click is heard, magnetic switch is defective or blade motors have internal damage. If motors start but do not continue to run when switch is released to run position, the diode or blade motor switch is defective.

cam actuating lever (3 – Fig. IH48). Remove the shoulder bolt (2) and remove brake parts (1 through 6). Slide disc (7) from transmission shaft and remove inner brake pad (8) from holder slot in transmission housing. Clean and inspect all parts for excessive wear or other damage. Renew parts as required and reassemble by reversing removal procedure. Adjust brake as outlined in the preceding paragraph.

MOWER DECK

LEVEL ADJUSTMENT. Position unit on level surface and turn traction and blade motor switches off. Measure distance from ground to blade from side to side and front to rear. Blade should be approximately 1/8-inch lower in front. To adjust front to rear, loosen clevis jam nuts on leveling arms (2 – Fig. IH49). Remove pin from one clevis at a time and adjust as needed. To adjust side to side, loosen nut securing lift chain eyebolt (6) to mower deck, depending on which side measures high, and adjust bolt as necessary.

REMOVE AND REINSTALL. Turn traction and blade motor off and lower mower to lowest position. Disconnect electrical connectors at each motor (4 – Fig. IH49). Disconnect lift chains (5) and remove pin from front mower bracket (1). Disconnect lift arms (2) from rear hanger brackets (3), and slide mower out.

Reinstall in reverse order of removal. Place height control lever in lowest position. Be sure electrical connectors are properly installed.

BLADE DRIVE ELECTRIC MOTORS

Two electric motors (Fig. IH43) are used on the twin blade mower with the blades bolted directly to motor shaft. The blade motors are protected by circuit breakers and thermal switches. If either switch opens, the blade motor magnetic switch contact points open by spring action and short circuit blade motors providing a braking action.

REMOVE AND REINSTALL. Remove mower as outlined in MOWER

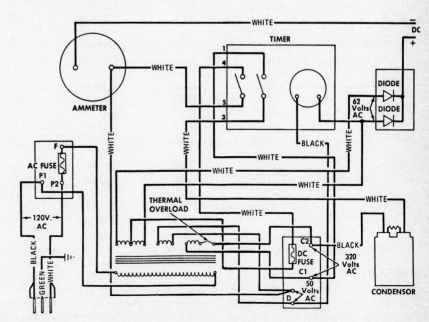

Fig. IH51 — Wiring schematic for battery charger used on units with serial number 8921 and below.

R&R BATTERY CHARGER. Remove the mower body as follows: Lift off battery cover shroud and disconnect and remove batteries and mounting plate. Drive out roll pin and remove steering wheel and upper steering shaft. Disconnect and remove clutch-brake pedal. Remove gear shift knob. Remove body mounting bolts and note location of body spacers. Raise body and disconnect wiring harness from traction motor and blade motor switches. Move mower lift handle to highest position, then raise body clear of handle and remove body.

Disconnect the D.C. leads to the circuit and remove the charger. Reinstall in reverse of removal.

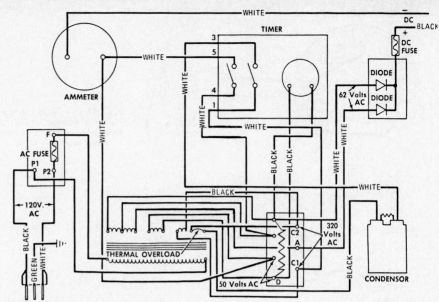

Fig. IH52—Wiring schematic for battery charger used on units with serial number 8922 and above.

JACOBSEN

HOMELITE CORP.
P.O. Box 7047
Charlotte, NC 28217

Model	Make	Engine Model	Horsepower	Cutting Width, In.
42635*	B&S	130000	5	25
42636(526)	B&S	130000	5	26
42638	B&S	130000	5	26
42641	B&S	130000	5	26
42642	B&S	130000	5	26
Mark II				
43010	B&S	140000	6	30
43015	B&S	170000	7	30
43020	B&S	170000	7	30
Mark III				
43025	B&S	190000	8	30
43026	B&S	190000	8	30
43027	B&S	190000	8	30
43030	B&S	190000	8	30
43031	B&S	190000	8	30
43032	B&S	190000	8	30
43036	B&S	190000	8	30
43037	B&S	190000	8	30
43042	B&S	190000	8	30
43043	B&S	190000	8	30
RMX8	B&S	190000	8	30
RMX11	B&S	250000	11	30

*Gear drive model—all others, friction drive.

STEERING SYSTEM

All Models

For all models except RMX models steering is conventional and non-adjustable, with a center-mounted, lever-type steering shaft. Most models are fitted with a steering wheel (1–Fig. J1); however, Models 42636 and 42638 are equipped with a steering handle (2). Model 43635 (not shown) has a center-pivoted front axle. All others have their steering spindles (14) mounted in front support (10) with steering shaft (12). Pivot action of front end to allow for side-to-side twist is by rotation of frame tube in bushings (37). Single steering link (13) serves as a tie rod between spindles (14).

Lubrication of front wheel bushings, shoulder spindle bushings, steering shaft and wear points in linkage calls for use of SAE 30 engine oil. Wipe off excess to prevent dirt accumulation.

For RMX models Fig. J1A shows complete view of front steering components. By turning steering wheel, pinion gear (1–Fig. J1B) will rotate steering sector outer ring (2) on center bolt from side-to-side. Steering bracket (3) will move causing steering spindles (44–Fig. J1A) to pivot front wheels.

Periodically lubricate front wheel assemblies through grease fittings on inside of each wheel hub with a general purpose grease.

CAUTION: On some models the performance of maintenance, adjustment or repair operations is more convenient if mower is standing upright. This procedure can be considered a recommended practice providing the following safety recommendations are performed:

1. **Drain fuel tank or make certain that fuel level is low enough so that fuel will not drain out.**
2. **Close fuel shut-off valve if so equipped.**
3. **Remove battery on models so equipped.**
4. **Disconnect spark plug wire and tie out of way.**
5. **Although not absolutely essential, it is recommended that crankcase oil be drained to avoid flooding the combustion chamber with oil when engine is tilted.**
6. **Secure mower from tipping by lashing unit to a nearby post or overhead beam.**

For repair procedures refer to the following paragraphs.

R&R AND OVERHAUL(All Models Except RMX Models). With mower on its rear stand, carefully inspect for wear or damage to parts of steering gear, then disassemble as follows:

Remove hub caps (17–Fig. J1), cotter pins and wheels. Set wheel washers (16) and dust shields (15) aside with wheels. Clean thoroughly and determine need for renewal of any parts, especially wheel bearings (bushing type) or dust shields. Steering link assembly (13) can be removed after cotter pins are pulled at spindle ends. Use a small pin punch to drive out roll pins at top of steering spindles (14). Lower spindles out of place and carefully check condition of shouldered bushings (19 and 21) for possible renewal.

On models with steering wheel, drive out roll pin for removal. Unbolt steering handle of models so equipped. Lower steering shaft (12) out of front platform and check bushings (6).

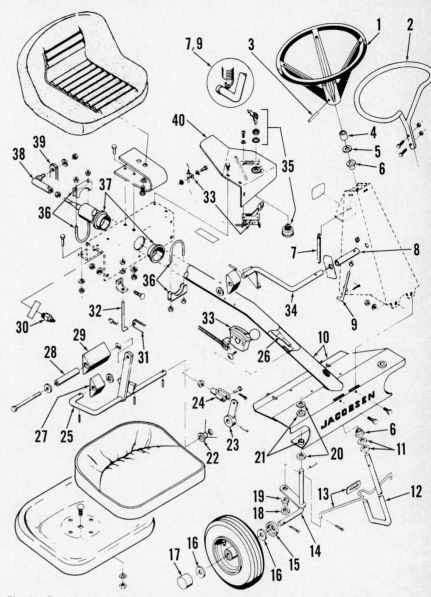

Fig. J1 — Exploded view of steering system with drive, brake and operating controls, typical of all models except 42635.

1. Steering wheel		32. Support rod
2. Steering handle		33. Throttle control
3. Roll pin (¼X2)	12. Steering shaft	22. Treadle bushing (2)	34. Neutral brake pedal
4. Spacer	13. Steering link assy.	23. Drive treadle lever	35. Ignition switch**
5. Washer	14. Spindle RH	24. Push rod clevis	36. Tube clamp (2)
6. Shaft bushing (2)	15. Dust shield (2)	25. Treadle assy.	37. Tube bushing (2)
7. Brake return spring	16. Wheel washer (4)	26. Push rod	38. Control ball joint
8. Brake tube	17. Hub cap	27. Reverse/brake pedal	39. Control lever
9. Brake rod	18. Washer, ⅜ (2)	28. Spacer bushing	40. Control panel
10. Frame tube/	19. Steering link	29. Forward control	
front support	bushing (2)	pedal	
11. Washers (2)	20. Spindle washer (2)	30. Ignition switch*	*Manual start
	21. Spindle bushings (2)	31. Deck lift arm	**Electric start

NOTE: On Model 42635, disassembly is similar except that an axle is used, and the unit must be raised and blocked for service, not set up on a rear stand.

Renew all damaged or defective parts and reassemble by reversal of disassembly procedure. Lubricate, using SAE 30 engine oil.

(RMX Models). Remove eight lock nuts retaining front axle pivot shaft and bearing halves (45–Fig. J1A). Remove ball bushing retaining bolt securing steering sector outer ring (2–Fig. J1B) to steering bracket (3). Remove axle assembly (46–Fig. J1A), spindles (44) and steering bracket (42) as a complete unit.

Complete disassembly and inspect all parts for excessive wear or any other damage. Renew all parts as needed and reassemble in reverse order of disassembly. Reinstall front steering unit and lubricate all pivoting joints with a suitable lubricant.

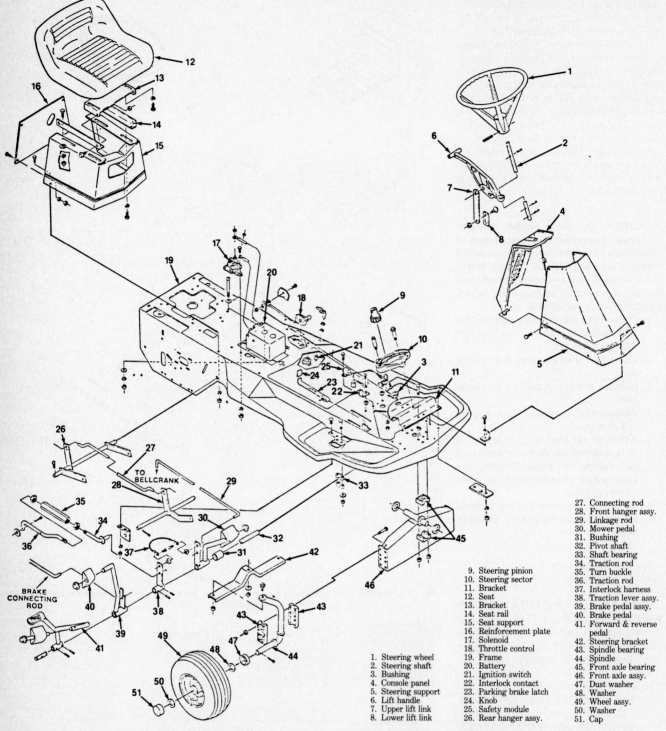

9. Steering pinion
10. Steering sector
11. Bracket
12. Seat
13. Bracket
14. Seat rail
15. Seat support
16. Reinforcement plate
17. Solenoid
18. Throttle control
19. Frame
20. Battery
21. Ignition switch
22. Interlock contact
23. Parking brake latch
24. Knob
25. Safety module
26. Rear hanger assy.

27. Connecting rod
28. Front hanger assy.
29. Linkage rod
30. Mower pedal
31. Bushing
32. Pivot shaft
33. Shaft bearing
34. Traction rod
35. Turn buckle
36. Traction rod
37. Interlock harness
38. Traction lever assy.
39. Brake pedal assy.
40. Brake pedal
41. Forward & reverse pedal
42. Steering bracket
43. Spindle bearing
44. Spindle
45. Front axle bearing
46. Front axle assy.
47. Dust washer
48. Washer
49. Wheel assy.
50. Washer
51. Cap

1. Steering wheel
2. Steering shaft
3. Bushing
4. Console panel
5. Steering support
6. Lift handle
7. Upper lift link
8. Lower lift link

Fig. J1A — Exploded view of chassis, steering and linkage components for RMX models. Parts 17 and 20 are used only on electric start models.

ENGINE

All Models

For overhaul and repair procedures on engines listed in Specification Table refer to Small or Large Air Cooled Engines Service Manual.

Refer to the following paragraphs for removal and installation procedures.

All Models Except 42635 and RMX

REMOVE AND REINSTALL. To remove engine, disconnect spark plug lead, and disconnect battery cables on electric start models. Remove cover plate (6–Fig. J10) after mower engagement lever (4) is set in OFF position. Remove idler arm belt guide bolt (15)

from idler arm (17) and release belt from idler pulley (13). Manually release cutter brake (7 and 8) so that brake pad (10) will clear sheave and remove belt from cutter spindle pulley. Remove belt by pulling to the rear.

Now, remove rear panel (19–Fig. J2). It will be noted that during removal of this panel that one of the bolts also serves to anchor top end of chain case

Fig. J1B—View of front steering sector components for RMX models.

1. Pinion gear
2. Outer ring   3. Steering bracket

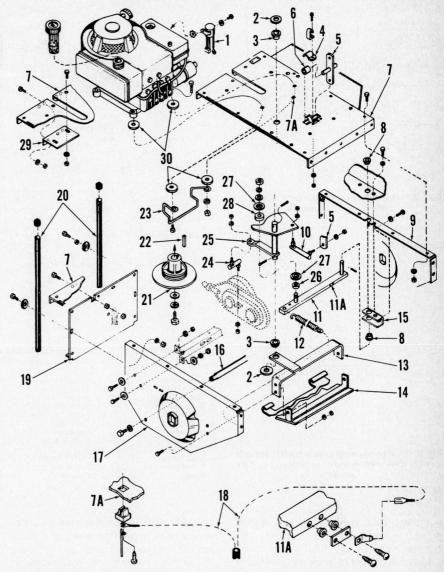

Fig. J2—Exploded view of engine platform/rear frame assemblies typical for all friction drive models.

1. Safety interlock module	8. Support bushing (2)	15. Pivot bracket
2. Washer	9. Side plate LH	16. Tie rod
3. Control cam bushing	10. Ball joint link (90°)	17. Side plate RH
4. Bushing clamp	11. Roller support assy.	18. Interlock leads
5. Control lever	12. Drive cam spring	19. Rear panel
6. Control lever bushing (2)	13. Cross brace	20. Rear stand tubes
7. Rear platform	14. Drive shield	21. Pulley/drive disc
		22. Crankshaft key
23. Belt guide		
24. Ball joint link (180°)		
25. Control cam assy.		
26. Spacer		
27. Cam follower guide		
28. Roller bearing		
29. Stiffener		
30. Mounting washer (8)		

support spring. Prior to unbolting, it is advisable to release this spring from chain case bearing shaft. Take care not to lose the damper tube fitted over this spring on most models. Also, remove lock nut (68–Fig. J4) from rear end of brake shaft (71) which is likewise fitted through rear panel. After rear panel has been removed, slip mower drive belt rearward and clear of the drive disc on engine crankshaft. Remove stiffener (29–Fig. J2) which reinforces and blocks rear of crankshaft slot in rear platform (7).

Carefully check all other items connected to engine. Disconnect safety interlock module (1) at engine and the black cable from starter solenoid to engine starting motor on models so equipped. Disconnect throttle control cable (33–Fig. J1) at carburetor and be sure that ignition and interlock grounding wires are disconnected at engine end.

Unbolt engine from rear platform (7–Fig. J2) then slide rearward and out of crankshaft slot in platform. Note assembly order of eight mounting washers (30) and placement of belt guide (23).

Reinstall engines by reversing removal procedure. Check control adjustments before returning mower to operation.

Model 42635

REMOVE AND REINSTALL. To remove engine, unbolt and remove operator's seat for easier access to engine compartment. Pull starter cord out and, while extended, remove T-handle. Then, carefully allow cord to retract under cover and tie a knot to prevent wind-up of cord into manual starter. Remove rear cover from engine housing. Disconnect spark plug lead. Unbolt engine brace (1–Fig. J3) and disconnect throttle cable at carburetor. Remove safety interlock grounding wires from engine.

Remove all belts, as follows: Set mower lift lever in ENGAGED position, and lock parking brake; then, remove belt keeper from right side of engine belt guard (30) and remove nut which retains mower clutch idler support (42) on left side. Remove belt keepers and shoulder bolt from mower deck, disengage lift lever and remove belt from spindle pulley (39). Unbolt and remove belt guard (30). Back off nut from variable speed pulley shaft, then nut from

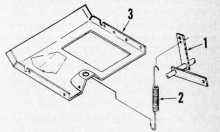

Fig. J2A—View showing rear mower deck hanger assembly (1), counterbalance spring (2) and bottom plate (3) on RMX models.

transmission input shaft. Take care not to lose key or washers. Remove cap screw from end of engine crankshaft, then carefully remove belts and pulleys

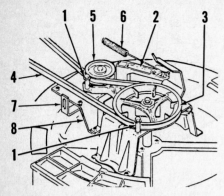

Fig. J2B — View showing mower deck drive components for RMX models.

1. Belt guide
2. Clutch arm
3. Spindle pulley
4. Drive belt
5. Idler pulley
6. Spring
7. Rear lift bracket
8. Support bracket

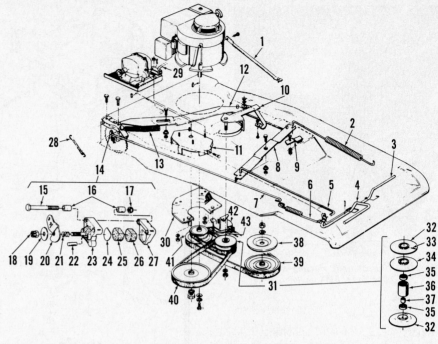

Fig. J3 — Exploded view of chassis parts, traction control linkages, and belt arrangement for Model 42635.

1. Engine brace	12. Variable drive arm	23. Casting cam side
2. Spring (¾X11)	13. Variable drive spring	24. Back-up disc
3. Clutch pedal rod	14. Disc brake assy.	25. Brake pad (0.450)
4. Brake pedal rod	15. Bridge bolt - 2½ (2)	26. Brake pad (0.250)
5. Link rod	16. Spacer (2)	27. Casting - inner
6. Brake spring	17. Locknut 5/16- 18	28. Brake return spring
7. Brake rod	18. Locknut ⅜-24	29. Shift lever
8. Clutch bar	19. Washer	30. Belt guard assy.
9. Blade brake	20. Brake cam lever	31. Variable speed pulley
10. Guide bracket	21. Cam spring	32. Sheave half (2)
11. Transmission belt guard	22. Push pin	33. Spirol pin 5/32

34. Movable sheave
35. Ball bearing (2)
36. Steel tube
37. Spacer
38. Mower brake disc
39. Blade spindle pulley
40. Transmission pulley
41. Two-step engine pulley
42. Idler support
43. Idler

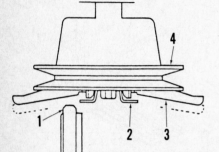

Fig. J2C — View showing drive wheel (1), retainer disc (2), drive disc (3) and drive pulley (4) on RMX models.

together. Take care not to lose engine pulley-to-crankshaft key. Unbolt and remove engine.

Reverse details of removal procedure to reinstall engine in mower chassis.

Model RMX

REMOVE AND REINSTALL. Stand unit upright and remove mower drive belt as follows: Disconnect counterbalance spring (2–Fig. J2A) from bottom cover (3). Move lift handle selector to lowest position. Remove front hanger bracket shoulder bolt and lower mower deck until rear portion rest on rear wheels. Loosen two spindle pulley belt guides (1–Fig. J2B), then remove belt from idler pulley. Rotate clutch arm (2) away from spindle pulley, then lift belt off pulley. Depress forward pedal until drive wheel (1–Fig. J2C) clears edge of retainer (2) on drive disc. Remove belt from engine drive pulley and slip belt between drive wheel and disc.

Remove crankshaft bolt and washers securing engine drive disc (3) and pulley (4) assembly. Withdraw drive assembly from engine crankshaft. Lower mower unit, then remove all engine attaching cables, wires and any other component

that will obstruct engine removal. Remove engine mounting bolts and nuts, then lift engine clear of unit.

After repair of engine, reassemble in reverse order of disassembly.

FRICTION-DRIVE CLUTCH

Function of the friction-drive clutch is to essentially place friction drive parts in a neutral or no-drive attitude when not in drive-forward or drive-reverse.

All Models Except RMX

ADJUSTMENT. If slippage occurs and it becomes evident that forward or reverse drive is ineffective, or if there is noise chatter or drag in neutral, place unit on rear stand as previously outlined and proceed as follows:

Block or lash forward-reverse pedal in full FORWARD position which will shift and hold chain case and driven roller at right. Check for ⅜-inch clearance between chain case roller (59 – Fig. J4) and angle-iron rail which supports neutral cam (60). Also see Fig. J5. If clearance is not correct, loosen rail mounting screws in right hand side plate to adjust, then

retighten. Now, depress reverse-drive pedal and secure by blocking at full REVERSE, then check for ⅜-inch clearance between chain case roller and rail at left hand side and adjust at left side mounting bolts if required. Release pedal to allow chain case to return to neutral position with case bearing centered on neutral cam. When in neutral, there must be 0.030-inch clearance between friction face of drive disc (21 – Fig. J2) and rubber tread of driven disc (49 – Fig. J4). Use feeler gage to measure. Rotate driven disc while checking to determine clearance at several points around disc circumference. To set required 0.030-inch clearance, loosen mounting bolts which attach neutral cam (60) to support rail, then shift cam plate for correct setting.

NOTE: Support rail is slotted at right angles to slots in cam plate. This will permit proper neutral centering (side-to-side) at the same time that vertical clearance of driven disc is being adjusted.

IMPORTANT: When making traction drive adjustment, carefully evaluate friction drive parts for condition. Drive disc on engine crankshaft must be smooth, clean

and dry. Be especially watchful for oily film or a slick, burnished surface. Clean thoroughly with a non-contaminating solvent, mineral spirits or denatured alcohol. Check driven disc on chain case for excessive wear, damaged rubber tread or flat spots. If driven disc is defective, unbolt drive wheel (49) from its hub (50) and renew; then, follow adjustment procedure outlined previously. Be sure to observe chain case and its neutral cam roller for wear, misalignment or lubricant leakage. Note if operating linkage functions without binding or excessive looseness. Do not return mower to service if there is an apparent problem in drive system.

R&R AND OVERHAUL (Sliding Chain Case). To remove chain case proceed as follows:

For convenience and to eliminate weight from up-ended mower, unbolt housing cover from cutting deck, remove mower drive belt, pull mower deck hanger shaft and suspension pins and remove mower from under chassis. Partially loosen bolts for right hand stand tube as this tube will be removed when right hand side plate is unbolted. Set unit up on stands as previously outlined, use some extra blocking (scrap 2 X 4 lumber will do) under rear panel to provide clearance above floor and lash upper end so as to hold unit securely with most weight on left hand stand tube.

With free-wheeling hub engaged (driving position) press inward on spring keeper (2 – Fig. J4) to compress free-wheeling lock spring (3) and remove snap ring (1) from axle end. Unbolt and remove wheel. When wheel has been removed so as to expose end of axle (13), carefully drive out roll pin (7) and remove hub (6), hub sleeve (8) and dust cap (9).

Disengage drive cam spring (12 – Fig. J2) at its top end by reaching through frame and lifting upper hooked end from the hole which it engages in right hand side plate (17).

Fig. J4 — Exploded view of rear axle and JACOBSEN friction drive. View A shows differential, reduction gearing, axles and hub assemblies. View B shows differential case assembled and in place on left hand side plate with exploded view of variable drive chain case and parts.

1. Snap ring	21. Differential case	39. Pinion shaft/snap
2. Spring keeper	22. Gasket	rings (6)
3. Compression spring	23. Thrust washer	40. Thrust washer
4. Free wheeling hub	24. Sprocket thrust	41. Axle bearing LH
5. Hub washer	washer	42. Felt ring gasket
6. Rear hub, RH	25. Bushing (3)	43. Dust cap LH
7. Roll pin, 3/16X1¼	26. Hex. sleeve &	44. Rear hub LH
8. Hub sleeve, RH	sprocket	45. Hub cap LH
9. Dust cap	27. Thrust washer	46. Boot clamp (4)
10. Axle hub w/grease	28. Differential gear	47. Axle bellows (boot)
fitting	29. Woodruff key	48. Locknut ⅝-18
11. Hub seal spring	30. Differential sleeve &	49. Drive wheel
12. Hub seal	gear	50. Hub
13. Axle shaft	31. Sun gear	51. Chain case, left half
14. Spacer bushing	32. Allen set screw	52. Bushing
15. Bushing (2)	33. Shim washer	53. Thrust washer
16. Gear and sprocket	34. Bushing (2)	54. Drive hex &
17. Thrust washer	35. Pinion spacer (6)	sprocket
18. Bushing	36. Pinion gear (6)	55. Thrust washer
19. Double sprocket	37. Drive gear key	56. Bushing
20. Case seal	38. Differential plate	57. Chain case gasket

58. Chain case, right	68. Nuts ¼-20 (3)	
half	69. Brake rod spring	
59. Chain case roller	70. Washer	
60. Neutral cam	71. Brake rod	
61. Brake shaft	72. Drive wheel bearing	
62. Brake	73. Spacer/slinger	
63. Roll pin ⅛X¾	74. Woodruff key	
64. Snap rings	⅛X½	
65. Brake pivot shaft	75. Shaft & sprocket	
66. Damper	76. Slinger	
67. Spring	77. Drive wheel bearing	

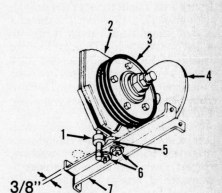

Fig. J5 — View of adjustment points for friction drive. When in neutral, as shown, clearance between driven disc (3) and drive disc (4) on crankshaft must be 0.030-inch. Refer to text for adjustment procedure.

1. Chain case roller	
2. Chain case	5. Neutral cam
3. Driven disc	6. Mounting bolts
4. Drive disc	7. Cam support rail

Now unbolt side plate (17) from rear platform (7) and from cross brace (13) which supports drive cam and roller support assembly (11), then unbolt right hand end of neutral cam support rail (7–Fig. J5). Back out screw from end of tie rod (16–Fig. J2), then loosen boot clamps (46–Fig. J4) on each side of chain case to release bellows (47). Disconnect spring (67) from chain case. Remove nut from ball joint link (24–Fig. J2) and swing ball joint away from chain case. Carefully lift off right hand side plate, exposing chain case for removal.

NOTE: When side plate is being lifted away, hub seal spring (11–Fig. J4) and seal (12) with hex tube thrust washer (23) will usually remain inside bellows (47); however, they may fall out. Do not lose.

Chain case can now be slipped off right hand end of axle and hex sleeve (26). A light pull outward on axle will help chain case to clear linkage parts during removal.

Set chain case up in a well-padded vise with case roller bearing shaft secured in vise jaws. Use a strap wrench to hold rubber tread of driven disc against turning and back off flexloc nut from shaft. If no strap wrench (Jacobsen No. 545286) is on hand, a scrap of fabric or leather belt can be clamped in heavy pliers or vice grips to serve the purpose. After driven disc (49) and hub (50) are removed, pry Woodruff key (74) from recess in sprocket shaft (75). Unbolt case halves (51 and 58), noting that heads of all twelve assembly bolts are on driven disc side of chain case. Keep this in mind for reassembly. Because No. 3 Permatex is used on chain case gasket (57), case halves will not separate readily. Case halves may be forced apart by careful driving with a light hammer and drift on input shaft (75). However, the use of a strong, broad-bladed putty knife inserted into case joint will be less likely to damage parts. Carefully support case halves when separated and press or bump input shaft (75) lightly and remove chain, drive hex and sprocket (54), along with shaft.

Use clean solvent to thoroughly wash parts and case halves and carefully examine all bearings, spacers, washers, slingers, chain and flanged bushings to determine if renewal is called for. If breakage, wear or dry galling of parts is evident, or if wear appears inside chain case, consider renewal of chain or sprockets. A new case gasket (57) must always be used.

Note parts arrangement in Fig. J4 and reassemble chain case as follows:

Apply gasket sealer to gasket flange of each chain case half, and position new gasket (57) on left side of chain case (51). Coat inside of each bearing boss with Loctite so that outer race of each drive wheel bearing (72 and 77) will be secure. Inner race of each bearing must also be coated to adhere to input shaft (75). Be sure that spacer is located on left side (keyway end) of shaft.

NOTE: Some earlier production models may not be equipped with grease slingers (73 and 76). In such cases, obtain JACOBSEN part number 351566 from parts outlet for installation.

Set shouldered bushing (52) in place, and assemble continuous chain on sprockets (54 and 75). Do not overlook thrust washer (53). Install complete assembly on left hand case half (51). Be sure that slinger (76) and thrust washer (55) are in place, then fit right hand case half over shaft ends, pressing carefully by hand until case flanges meet. DO NOT FORCE. Test by hand for freedom of movement of sprockets in bearings and bushings. Install case assembly bolts with heads on drive wheel side and tighten nuts alternately in a criss-cross pattern to 55-85 inch-pounds (5-7 ft.-lbs.) Again, check freedom of shaft rotation.

Lubricate with **no more** than 1½ ounces of Fibrex grease through filler hole and install plug.

Reinstall driven disc and its bolt-on hub on input shaft, noting that Woodruff key must be angled in its keyway to correspond with taper (slant) of keyway in hub. Torque retainer nut to 18-22 ft.-lbs.

Chain case assembly can now be reinstalled in reverse of removal order.

(Differential Gear Case). If checkout of drive train indicates that drive problem is located in differential/final drive assembly, follow this order for gear case removal:

Remove mowing deck, and loosen left hand stand tube screws in preparation for removal. Raise mower up on rear stand as previously outlined, place extra blocking under rear panel and lash upper end so as to hold unit securely with most weight on right hand stand tube.

With free-wheeling hub in drive position, press inward on spring keeper (2–Fig. J4), remove snap ring (1) and balance of free-wheeling hub parts. Remove right rear wheel and when hub (6) is exposed, check condition of roll pin (7). If this roll pin has been sheared in axle (13), a no-drive condition will result. Renewal of pin may be a solution to a breakdown problem.

Now, remove left rear wheel from hub (44) and the bolt which attaches hub to differential sleeve (30) and remove hub.

Loosen outer boot clamps (46) so that hex shaft cover bellows (47) will be loose at each outboard end. Unbolt left hand

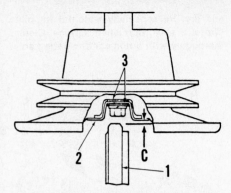

Fig. J5A – View showing drive clutch components on RMX models. Clearance (C) distance between drive wheel (1) and retainer disc (2) should be between 0.02-0.04 inch, adjust by washers (3).

side plate from rear platform (7–Fig. J2).

IMPORTANT: Loosen but DO NOT REMOVE bolt which holds pivot bracket (15) beneath engine platform, so that bracket will remain attached to platform when side plate (9) is unbolted and removed. Unbolt left side plate (9) from rear panel (19) and stand tube (20) at left side.

Unbolt left side plate (9) from cross brace (13) and back out cap screw from left end of tie rod (16), then remove two bolts at left end of neutral cam support rail (7–Fig. J5). Differential case (21–Fig. J4) and left hand side plate with hex sleeve (26) and axle (13) can now be withdrawn outward and separated from bellows and chain case. Check inside right side bellows for loose parts from axle when withdrawn.

Disassemble chain case from left side plate as follows:

Secure hex sleeve in a well-padded vise so that side plate is horizontal, then unbolt from differential case (21). Gasket (22) will not be reusable. Remove through-bolt for bushing (14) and lift off side plate. Axle bearing (41) will remain with left side plate. Differential gears and reduction sprockets and chains will now be open for further service.

Lift thrust washer (40) from differential sleeve (30), then lift complete axle and differential assembly out of gear case and hex drive tube. Release hex tube from vise, and remove reduction sprocket (26) all together from differential gear case. Thoroughly wash all parts in solvent and clean fragments of gasket (22) from inner surface of side plate and from flange of case (21), then carefully check all parts for wear or damage.

If axle-differential assembly appears to be in good operating condition, further disassembly of its parts may be unnecessary; however, if there is any roughness in mesh of pinions or if there is a possibility that set screw (32) is loose in hub of sun gear (31) or that key (29 or

37) is damaged, further disassembly must be performed.

Set long axle (13) up in padded vise jaws with differential sleeve (30) upward as in Fig. J6. Remove retainers from each of six pinion shafts, then lift off differential sleeve (30–Fig. J4), followed by six pinions (36) and six spacers (35). Push pinion shafts (39) out and remove from under differential ring gear (28). Remove gear (28) from hub of sun gear (31) and separate from axle (13).

When reassembling, check bushings (34) for wear and do not re-use if excessively loose on axle (13). Check condition of shim washer (33). If pinion shafts must be renewed, **press** new shafts into their six bores in ring gear (28). Carefully locate sun gear (31) on axle shaft (13), being sure that Woodruff key (29) is in place. Use Loctite on threads, and torque set screw (32) to 9-14 ft.-lbs. Be sure that this (Allen) set screw does not extend above outer circumference of hub of sun gear (31). If set screw (32) is renewed, do not substitute a longer screw. When installing pinion gears (36) on pinion shafts (39), be sure that spacers (35) are arranged as shown in Fig. J6, in alternating order, with a pinion over spacer on every other shaft.

VERY IMPORTANT: When shim washer (33 – Fig. J4) is reinstalled between sun gear (31) on axle (13) and sun gear of differential sleeve (30), ONE shim washer is usually sufficient. However, if both sun gears can contact teeth of the same pinion, differential will be locked up and not function. In such a case, use two shim washers during assembly to insure that each sun gear contacts only three of the six pinion gears, alternate and opposite. Use Molygrease on all differential gears, washers, spacers and shims during reassembly. After differential plate (38) is reinstalled over pinion shafts, differential sleeve and its sun gear (30), reinstall all six retainers on pinion shafts and set differential-axle assembly aside.

Reassemble differential gear case and reduction sprockets as follows:

Renew gear case seal (20) and use Loctite on its outer circumference to secure seal in shaft boss. Set differential sprocket thrust washer (24) in place inside case. Remove grease screw from hex sleeve and use a hand grease gun to preload hex sleeve with grease, then assemble drive chains on hex sleeve sprocket and reduction sprockets in this order: Longer of two continuous chains connects hex sleeve sprocket (26) to larger side of double sprocket (19). Shorter continuous chain runs between smaller side of double sprocket (19) and the sprocket portion of gear and sprocket (16). Install complete chain and sprocket assemblies with hex sleeve into gear case. Do not lose three bushings (25) and be sure that thrust washer (23) is in place at bottom of hex sleeve after fitting into case. Bushings (15) should be in place inside gear and sprocket (16) and bushing (14) and thrust washer (17) must be in place. Insert gear and sprocket mounting screw up through bushing (14) from under gear case. Be sure that double sprocket (19) has its shaft portion squarely into bushing (18) in wall of gear case. Check for free movement of chains and sprockets, then carefully insert long axle (13) down through hex sleeve (26) and engage teeth of ring gear (28) with gear portion of gear and sprocket assembly (16). Again, check rotation by turning shaft slightly. Pour one pint of JACOBSEN gear lube, Part No. 500650, into open side of gear case. Install thrust washer (40) over shaft of differential sleeve and gear (30).

Now, apply No. 3 Permatex to gasket flange of gear case (21), install gasket (22), then coat face of gasket with Permatex.

Carefully align left hand side plate with differential case, and lower side plate into place over shaft of differential sleeve (30). Thread gear and sprocket

mounting screw upwards through well in side plate and reinstall jam nut. Be sure that shaft of double sprocket is properly fitted into side plate. Reinstall all assembly bolts, check rotation of all exposed shafts for freedom of movement and tighten all case to side plate assembly bolts to 7-9 ft.-lbs. torque in an alternating criss-cross pattern.

Before reinstalling side plate/gear case assembly on chassis, coat outer surface of hex sleeve with Molygrease along its entire length. This will insure lubrication for easy side-to-side movement of sliding chain case.

Complete installation of side plate and differential case in reverse of removal order, then check adjustments as outlined in appropriate sections.

(Control Push Rod and Linkage). Drive controls are not adjustable, except as previously covered under CLUTCH. If it should occur that clutch adjustment cannot be made, even with a new rubber-faced driven disc and other parts apparently in order, a careful check should be made of all linkage parts as shown in Figs. J1 and J2. Pay particular attention to ball joints (10 and 24–Fig. J2), to bushings (3, 6 and 8) and to cam roller bearing (28). If worn, loose or damaged, these items can cause trouble. Renew any which are defective.

If it is determined that control push rod (26–Fig. J1) is at fault, proceed as follows:

Remove cutting deck from under mower. At rear end of push rod, under operator's seat, remove nut from stud of ball joint (38) and separate rod from top end of control lever (39). Remove roll pin from hub of drive treadle lever (23) to release shaft of treadle assembly (25).

NOTE: Later Mark III models may have a set screw at this point. If so, loosen.

Slip pedal shaft out of treadle lever (23). Place mower on rear stand, as previously outlined. To gain access to push rod, front chassis tube will have to be removed from rear platform. On most models, if rearmost tube clamp (36) is removed, front of mower can be lifted clear of rear platform for removal. Some Mark III models require that seat bracket be unbolted and that a front end mounting plate be unbolted from rear platform. These same models must also have control panel unbolted from front console at steering column so that throttle cable and ignition switch wiring can be disconnected.

After front chassis tube is separated from rear platform, pull push rod from rear end of tube, then carefully measure working length of push rod from center of clevis yoke pin bore to center of ball joint stud as shown in Fig. J7. This length should be 32½ inches. If a new

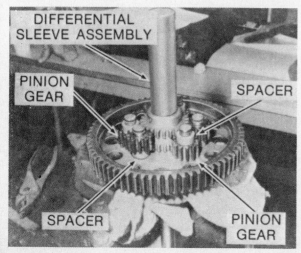

Fig. J6 – View of axle-differential assembly removed from differential case. Note alternating arrangement of pinions and spacers. Refer to text for details.

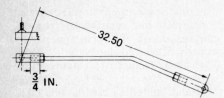

Fig. J7 — Measure preset length of control push rod for friction drive models as shown. About ¾-inch of thread should be engaged at each end of rod. See text.

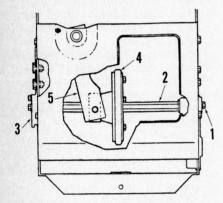

Fig. J7A — View showing clutch drive wheel and associated parts for RMX models.

1. Cap screw
2. Hex shaft
3. Traction arm
4. Drive wheel
5. Pivot shaft

push rod is installed, it is particularly important that this length be preset before installing in chassis tube.

Reverse disassembly order to reinstall forward section of mower chassis, then check drive adjustments before returning unit to service.

Model RMX

ADJUSTMENT. Place forward-reverse pedal in "Neutral" position. Stand mower unit in upright position, observe safety precautions as outlined in earlier section. With a feeler gage measure clearance between top of rubber drive wheel (1 – Fig. J5A) and bottom of retainer disc (2). Recommended clearance is between 0.02-0.04 inch.

To change clearance distance, remove cap screw in crankshaft end. Reposition or add washers (3) above or below retainer (2) until correct clearance distance is attained. Retighten cap screw and lower unit.

R&R AND OVERHAUL (Drive Wheel and Hex Shaft). While observing safety precautions as outlined in earlier section stand unit upright. Remove right rear wheel assembly, then remove cap screw (1 – Fig. J7A) securing hex shaft assembly (2) on left side. Remove cap screws, lockwashers and nuts retaining traction arm (3) on right side. Withdraw traction arm, hex sleeve and shaft from

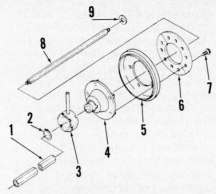

Fig. J7B — Exploded view of drive wheel assembly used on RMX models.

1. Hex slide
2. Snap ring
3. Sleeve & bearing assy.
4. Driven hub
5. Drive wheel
6. Plate
7. Cap screw
8. Hex shaft
9. Thrust washer

right side of frame as an assembly. Be sure to note thrust washer located between hex sleeve and gear reducer housing. Disconnect and remove drive wheel assembly (4) from pivot shaft (5). Withdraw drive wheel assembly through access hole located in bottom plate.

With reference to Fig. J7B inspect and renew all parts that are excessively worn or damaged in any way. Inspect left frame bearing and traction arm bearing for binding or excessive wear, renew as needed. Reassemble in reverse order of disassembly. Refer to ADJUSTMENT section for adjustment procedures after reassembly.

(Gear Reducer). While observing safety precautions as outlined in earlier section stand unit upright. Remove drive wheel and hex shaft assembly as outlined in previous section. Locate and remove master link in drive chain (9 – Fig. J7C), then remove chain from drive sprockets. Noting their location remove five cap screws securing gear housing to frame. Withdraw gear reduction assembly from mower unit.

With reference to Fig. J7C inspect parts 10 through 16 for excessive wear, roughness, broken teeth or any other damage. Renew all parts as needed and reassemble in reverse order of disassembly. Recommended gear box lubricant is two ounces of Kendall Kenlube L424 or a suitable Moly E.P. No. 2 Lithium grease.

(Differential). While observing safety precautions as outlined in earlier section stand unit upright, then remove both rear wheels. Disconnect rear hanger (1 – Fig. J2A) counterbalance spring (2) and remove bottom cover (3). Remove two cap screws, washers and nuts securing right tube assembly and withdraw tube from axle shaft. Loosen cap screws and nuts retaining right bearing strap (18 – Fig. J7C), then remove right differential bearing assembly (3) from frame. For reassembly, be sure to note location of spacer washers (2). Locate and remove master link in drive chain (9), then remove chain from drive sprockets. Remove cap screws and nuts retaining left bearing strap (18). Remove bolts securing left differential bearing to frame. Remove cap screws securing frame support (25), then withdraw support from mower frame. Lift differential assembly out left frame slot and slide shaft out right side.

Complete differential disassembly with reference to Fig. J7C. Remove four bolts, washers and nuts securing differential hub to sprocket. Note alignment marks on differential hub and sprocket, marks must be aligned during reassembly to insure proper alignment.

Separate hub and sprocket assembly. For reassembly, be sure to note location of shim washers installed between pinion gears and differential sprocket.

Inspect sprocket and gears for worn, chipped or missing teeth. Inspect all bushings, bearings and shafts for excessive wear or any other damage. Renew

Fig. J7C — Exploded view of differential and gear reduction unit on RMX models.

1. Right tube assy.
2. Spacer washers
3. Bearing assy.
4. Differential sprocket
5. Spacer
6. Pinion gears
7. Gear & shaft assy.
8. Left tube, bearing, hub & gear assy.
9. Drive chain
10. Sprocket shaft assy.
11. Gear reduction housing
12. Pinion bushing
13. Bearing
14. Primary drive pinion
15. Primary drive gear
16. Pin
17. Spacer
18. Strap
19. Bushing
20. Spacer
21. Pin
22. Bearing
23. Bearing
24. Housing cover
25. Frame support

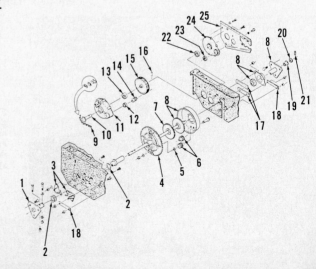

all parts as needed and reassemble in reverse order of disassembly. Lubricate drive chain with a good grade SAE 30 weight oil. Caution should be used not to get oil on traction drive wheel or disc. Refer to DRIVE CHAIN ADJUSTMENT AND ALIGNMENT section after reassembly.

(Drive Disc and Pulley). While observing safety precautions as outlined in earlier section stand unit upright and remove drive wheel and hex shaft as outlined in earlier section.

Disconnect traction drive brake rod (1–Fig. J7D), then remove retainer plate (2) cap screw and withdraw traction brake assembly. Remove cap screw (15–Fig. J7E) securing disc and pulley assembly to crankshaft. Be sure to note spacer washers (12) installed above and below disc retainer (13). Move pivot rod clear of disc, then lift disc and pulley assembly off crankshaft. Account for all shims installed between pulley and crankshaft end.

Complete disassembly of drive disc and pulley assembly with reference to Fig. J7E. Inspect all bushings for excessive wear or any other damage. Inspect disc and pulley for roughness, scoring, excessive wear or any other damage.

Reassemble in reverse order of disassembly. It is recommended to lubricate bushings (8) and (9) with camshaft oil or a suitable equivalent to prevent binding during operation.

TRACTION DRIVE CLUTCH

Model 42635

ADJUSTMENT. There is no actual adjustment of the clutch mechanism on this gear drive model. Clutch pedal (3–Fig. J3) operates through rod (5) and pivot bar (8) against tension of springs (2 and 13) to shift variable speed pulley

Fig. J7D — View showing traction drive brake rod (1) and retainer plate (2) used on RMX models.

(31), breaking drive contact between engine pulley (41) and transmission pulley (40). If clutch operation becomes unsatisfactory, check belts for glazing, wear or stretching, check for worn or damaged rods, examine springs for loss of tension or breakage and make sure there is no damage or binding in linkage. Worn or damaged parts should be renewed.

TRANSMISSION

Model 42635

R&R AND OVERHAUL. To remove transmission, proceed as follows:

Relieve drive chain adjustment by backing off chain tensioner bolt, then separate and remove drive chain. Remove rear cover from engine compartment and unbolt and remove operator's seat from its mounting bracket.

Block clutch pedal or linkage in disengaged position to release belt tension, then back off nut and remove pulley (40–Fig. J3) from transmission input shaft. When pulley is lowered from shaft end, take care not to lose Woodruff key. Disconnect control lever (29) from transmission shift arm, then remove four nuts at transmission mounting flanges on platform which are threaded on through-bolts from transmission belt guard (11). Lift transmission from chassis.

NOTE: Procedure above is for removal of transmission only. If extensive work is planned for mower chassis, mower deck should also be removed along with all belts and pulleys as covered under R&R – ENGINE.

Reverse removal steps to reinstall transmission.

For complete disassembly and detailed repair service to this transmission refer to TRANSMISSION REPAIR section under FOOTE.

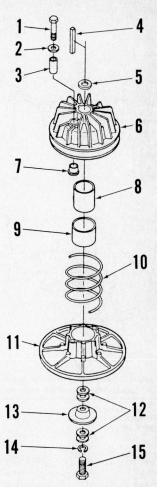

Fig. J7E — Exploded view of drive disc and pulley assembly used on RMX models.

1. Cap screw	9. Bushing
2. Washer	10. Spring
3. Spacer	11. Disc
4. Square key	12. Spacer washer
5. Spacer washer	13. Disc retainer
6. Pulley	14. Lockwasher
7. Bushing	15. Cap screw
8. Bushing	

DIFFERENTIAL AND REAR AXLE

Model 42635

R&R AND OVERHAUL. To remove differential and rear axle, first release drive chain tensioner, pull out master link and remove chain. Disconnect brake rod from disc brake cam lever, then unbolt and remove disc brake mounting bracket.

Raise and block up rear frame and remove each rear wheel, then remove hub flanges from axle shaft ends by driving out 5/16 X 1⅜-inch spirol pins. Unbolt rear axle support (8–Fig. J8) under chassis, then slip axle assembly to left to disengage short axle (2) from right side flange bearing (3) and remove axle assembly from frame. Check axles and bearings (3, 6 and 10) for wear or damage. Renew as necessary. Reinstall axle/differential assembly in reverse of removal order.

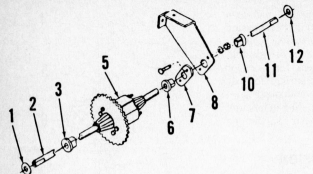

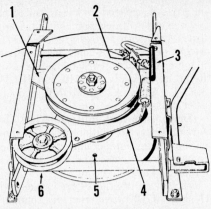

Fig. J8 — View of differential and rear axle with bearings as used in Model 42635.

1. Wheel washer
2. Axle RH
3. Axle bearing RH
5. Differential assy.
6. Axle bearing, center
7. Bearing bracket
8. Axle support
10. Axle bearing LH
11. Axle LH
12. Wheel washer

Fig. J9 — View of mowing deck pulley housing with top cover removed. Note that blade brake (2) contacts spindle pulley rim when control handle (3) is in "OFF" position.

1. Stiffener plate
2. Blade brake
3. Control lever
4. Idler arm
5. Belt guide mounting hole
6. Idler pulley

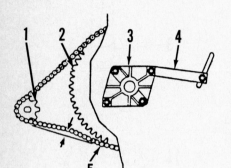

Fig. J8A — View showing drive chain adjustment components for RMX models.

1. Gear reducer sprocket
2. Differential sprocket
3. Bearing assy.
4. Strap
5. Chain

For disassembly and complete overhaul of differential, see DIFFERENTIAL REPAIRS section of this manual. This differential unit is covered under MTD Products.

DRIVE CHAIN

RMX Models

ALIGNMENT. A misaligned drive chain may cause premature chain and sprocket wear or cause chain to jump off sprockets. To align sprockets proceed as follows: While observing safety precautions as outlined in earlier section, stand unit upright. Remove right rear wheel, then remove cap screws, flat washers and nuts securing right tube assembly (1 – Fig. J7C) to axle shaft. Loosen bolt retaining right adjustment strap (18), then remove right bearing assembly (3) and spacer washers (2). Realign sprockets and reposition or add washers as needed to maintain correct alignment. Reassemble in reverse order of disassembly. Refer to ADJUSTMENT section for chain tension adjustment.

ADJUSTMENT. While observing safety precautions as outlined in earlier section, stand unit upright. Check chain deflection at center of reducer sprocket (1 – Fig. J8A) and differential sprocket (2). Recommended deflection is ⅛-inch as shown by arrows.

To adjust proceed as follows: Remove both rear wheels, then loosen left and right differential bearing (3) and strap (4) retaining bolts. Slide differential in adjustment slots until ⅛-inch deflection is attained, then retighten retaining bolts. Lubricate drive chain with a good grade SAE 30 weight oil. Caution should be used as not to get oil on traction drive wheel or disc.

GROUND DRIVE BRAKES

Friction-Drive Models Except RMX

The single brake system which operates both as service brake and parking brake is applied automatically whenever drive is in neutral.

CAUTION: Never park unit with free wheeling hub of right rear wheel disengaged. Brake is designed to lock the drive by preventing rotation of chain case driven disc and with it, the entire rear axle. If a rear wheel is unlocked from its axle, then opposite wheel is also free to rotate due to differential action, and parking brake will be ineffective.

When adjustment to brake is apparently needed, simple procedure is to be sure that brake rod spring (69 – Fig. J4) is under compression when drive is in neutral and brake is applied. As a quick check, inner face of outside nut (68) on brake rod (71) should have a clearance of 1/32 to 1/16-inch from outer surface of rear panel. If there is no clearance, set mower up on rear stand as previously outlined and with drive in neutral, adjust inner nuts on brake rod (71) so that brake (62) is in solid contact with rubber tread of driven disc (49). Outer nut must be clear of rear panel so as not to restrict movement of brake rod. Adjust outer nut, if necessary, for minimum 1/32-inch clearance, then operate mower to test brake.

RMX Models

There is no adjustment provided on ground drive brake. If brake slips when applied, check condition of drive wheel and check brake linkage for binding or insufficient application. Make sure brake rod (1 – Fig. J7D) is not bent or excessively worn. Renew all parts as needed in order to restore good working operation.

To apply parking brake, first push brake pedal down. Pull parking brake latch (23 – Fig. J1A) rearward, then release foot from brake pedal.

Model 42635

This gear-drive model is equipped with a disc brake which engages axle sprocket inboard of right rear wheel. If operation is not satisfactory, adjust as follows:

Tighten locknut (18 – Fig. J3) against cam lever (20) one-half turn at a time until brake action is suitable to operator.

If brake parts are worn to such extent that adjustment is no longer possible, uncouple brake rod (7) from brake spring (6), remove brake rod from brake cam lever (20), then unbolt disc brake bracket from frame.

When two bridge bolts (15) are removed, followed by locknut (18), all brake internal parts can be checked for condition and possible renewal. Brake pads (25 and 26) cannot be interchanged. Note difference in thickness. Be sure that rounded end of push pin (22) is installed toward cam lever (20), and reassemble in order shown in exploded view of Fig. J3.

BLADE DRIVE BRAKE

Friction-Drive Models

Non-adjustable blade brake is applied whenever cutting control lever is in

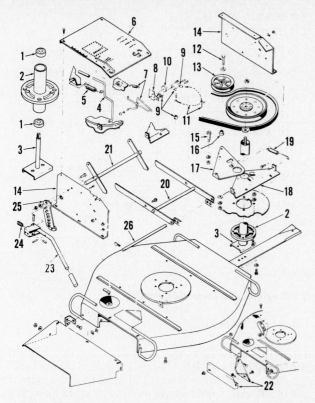

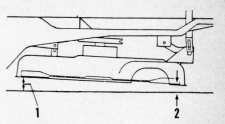

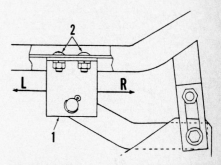

Fig. J10 — Exploded view of friction drive mower cutting deck assembly to show parts arrangement.

1. Spindle bearing (2)
2. Spindle housing
3. Spindle shaft/blade saddle
4. Off-On control lever
5. Idler tension spring
6. Cover plate
7. Brake rod & spring
8. Brake bracket
9. Blade safety switch parts
10. Blade brake pad
11. Interlock switch wires
12. Idler bolt 3/8-16X1 3/4
13. Idler pulley
14. Side bracket (2)
15. Idler arm belt guide
16. Idler pulley spacer
17. Idler arm
18. Support plate
19. Idler return spring
20. Front lift arm
21. Rear lift arm
22. Rear discharge conversion parts
23. Lift handle
24. Compression spring
25. Height setting quadrant
26. Hanger shaft

Fig. J10B — View showing procedure for checking mower deck height on RMX models. Rear deck height (1) should be 1/4-3/8 inch higher than front (2).

Fig. J10C — View showing carriage bolts (2) and mower deck hanger (1) used to adjust mower deck height. Refer to text for adjustment procedures.

"OFF" position. For all models except RMX refer to Fig. J9. If brake pad (10 – Fig. J10) is badly worn it can be renewed or complete new brake bracket assembly (8) is available. For RMX models refer to Fig. J10A. If brake bracket (4) should become excessively worn or broken, bracket may be replaced separately or as an assembly with clutch arm (3).

Model 42635

On this model, blade brake (9 – Fig. J3) contacts brake disc (38) when mower is disengaged. If blade brake is worn to the point of being ineffective, renew parts as needed.

NOTE: Components of mower cutting decks are usually conventional in design. Refer to applicable illustrations for assembly order of parts should it become necessary to renew a spindle shaft, a bearing or other part which fails in service.

Fig. J10A — View showing mower drive components for RMX models.

1. Mower drive pulley
2. Idler pulley
3. Clutch arm
4. Brake bracket

LEVEL MOWING DECK

Friction-Drive Models Except RMX

Park mower on a level floor for all deck level measurements. Disconnect spark plug wire and set control lever to "OFF". Rotate blade so that tip is toward front. Adjust height control lever for **low cut,** then measure blade height at front and rear blade tips. Rear of blade should be 1/4 to 3/8-inch higher than front. If adjustment is required, use adjusting nuts on threaded portion of support rod (32 – Fig. J1) to set angle of cutting deck. When front-to-rear adjustment is correct, be sure that mower deck is level, side-to-side, by adjusting one or the other support rod (32), then recheck front-to-rear adjustment. Repair or renew all parts of mowing deck suspension or controls which are damaged or defective.

RMX Model

Place mower unit on level ground. Disconnect spark plug wire and tie out of way. Check tire pressure, recommended tire pressure is 10-12 lbs. per square inch. Measure rear mower deck height (1 – Fig. J10B) and front mower deck height (2). Rear deck height (1) should be 1/4-3/8 inch higher than front (2).

To adjust mower deck height proceed as follows: Loosen carriage bolts (2 – Fig. J10C) on left and right front mower deck hanger (1), then slide brackets forward (R) to raise front and backward (L) to lower front. Make sure sides are adjusted evenly, then retighten carriage bolts. Remeasure front and rear deck height and readjust as needed.

If mower deck side-to-side height is uneven, then add washers between rear lift bracket (7 – Fig. J2B) and support bracket (8) until even.

Model 42635

This gear model has no provision for leveling the deck in relation to mower chassis. Only gage wheels at rear of deck are adjustable and these should be set just clear of floor when lift and disengagement lever is in low position.

MOWER DRIVE BELT

Model RMX

REMOVE AND REINSTALL. Stand unit upright and remove mower

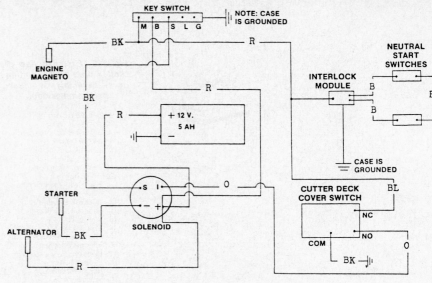

Fig. J12 — Wiring diagram for RMX models.

1. BK-Black
2. BL-Blue
3. B-Brown
4. O-Orange
5. R-Red

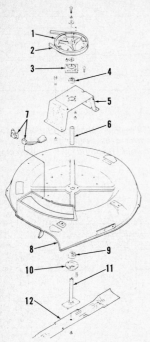

Fig. J10D — Exploded view showing mower deck parts on RMX models.

1. Pulley
2. Drive belt
3. Retainer
4. Bearing
5. Bracket assy.
6. Spindle spacer
7. Switch & bracket assy.
8. Mower deck
9. Bearing
10. Retainer
11. Spindle
12. Blade

Fig. J11 — View to show placement of safety switches in Model 42635 interlock system.

1. Starter safety switch
2. Ignition switch key
3. Ignition switch
4. Safety switches NC (2)
5. Wiring harness
6. Engine magneto connection

drive belt as follows: Disconnect counterbalance spring (2 – Fig. J2A) from bottom cover (3). Move lift handle selector to lowest position. Remove front hanger bracket shoulder bolt and

lower mower deck until rear portion rests on rear wheels. Loosen two spindle pulley belt guides (1 – Fig. J2B), then remove belt from idler pulley. Rotate clutch arm (2) away from spindle pulley, then lift belt off pulley. Depress forward pedal until drive wheel (1 – Fig. J2C) clears edge of retainer (2) on drive disc. Remove belt from engine drive pulley and slip belt between drive wheel and disc.

Inspect all pulleys and renew if damaged or excessively worn. Renew drive belt and reassemble in reverse order of disassembly.

MOWER DECK

Model RMX

R&R AND OVERHAUL. Raise front of mower unit high enough to allow access to underneath side. First remove mower drive belt, proceed as follows: Disconnect counterbalance spring (2 – Fig. J2A) from bottom cover (3). Move lift handle selector to lowest position. Remove front hanger bracket shoulder bolt and lower mower deck. Loosen two spindle pulley belt guides (1 – Fig. J2B), then remove belt from idler pulley. Rotate clutch arm (2) away

from spindle pulley, then lift belt off pulley. Depress forward pedal until drive wheel (1 – Fig. J2C) clears edge of retainer (2) on drive disc. Remove belt from engine drive pulley and slip belt between drive wheel and disc. Disconnect clutch arm spring (6), then remove shoulder bolts retaining rear lift bracket to hanger assemblies and lower deck unit. Withdraw deck assembly from mower unit.

Complete disassembly of mower deck with reference to Fig. J10D. Inspect and renew pulleys if damaged or excessively worn. Inspect bearings for excessive wear, binding or any other damage. Check spindle shaft for straightness, excessive wear or any other damage. Renew all parts as needed and reassemble in reverse order of disassembly.

ELECTRICAL

Model RMX

Shown in electrical diagram Fig. J12 are electrical components and their adjoining wires. Shown is a diagram for an electric start model, recoil starter models are not equipped with a battery, alternator, starter or adjoining wiring harnesses.

MASSEY-FERGUSON

MASSEY-FERGUSON, INC.
1901 Bell Avenue
Des Moines, Iowa 50315

Model	Make	Engine Model	Horsepower	Cutting Width, In.
MF5	Tecumseh	V50	5	26
MF6*	Tecumseh	V60	6	32
MF626	Tecumseh	V60	6	26
MF8*	Tecumseh	V80	8	32
MF832*	Tecumseh	V80	8	32

*Electric starting optional. All others are equipped for manual recoil start.

STEERING SYSTEM

Rigid front axle is center-pivoted under frame and steering linkage includes a steering arm at base of steering shaft which is connected by threaded ball joints at drag link ends to arm portion of left wheel spindle. Steering spindles are connected by a single tie rod with counter-threaded ball joints at each end.

All Models

ADJUSTMENT. Steering adjustment is limited to setting of front wheel toe-in, and calls for adjusting lengths of drag link (15–Fig. MF1) and tie rod (13), measuring center-to-center on ball joint studs (9 and 12). To do so, set front wheels straight ahead, then loosen jam nut at right hand end of each rod and rotate rod (13 or 15) to set length. On Models MF5, MF6 and MF626, drag link (15) should measure 9-1/32 inch and tie rod (13) should measure 16 inches. On Models MF8 and MF832, drag link (15) should measure 10-15/16 inches and tie rod (13) should measure 19-15/16 inches. Be sure threads are engaged the same depth at each end.

R&R AND OVERHAUL. To remove front axle, suspension and steering parts, raise and securely block forward portion of chassis.

Remove "X" lockwasher (17–Fig. MF1) and washer (19), then remove each front wheel (23). Remove nuts from ball joint studs (9 and 12) and separate tie rod (13) and drag link (15) from steering arm of spindles (20 and 22) and from arm (11). Loosen set screw (14), remove steering arm (11), pull steering shaft (8) out of tube (6) and support shaft. Need for further disassembly is based on determination of condition of parts at wear points, particularly bushing sets (7, 18 and 24), ball joints (9 and 12), steering arm (11) and shaft (8). If looseness or heavy wear is apparent, parts should be

renewed. To reinstall, reverse order of disassembly. Steering wheel retainer nut (3) should be torqued to 30-35 ft.-lbs. Lubricate reassembled front system with Multi-Purpose lithium base grease (MF Spec. M-1105).

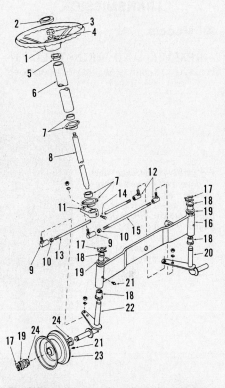

Fig. MF1 – Exploded view of steering and front suspension parts typical of all models. On Models MF5, MF6 and MF626, two washers (5) are used, one at top and one at bottom of tube (6). Models MF8 and MF832 are assembled as shown.

1. Steering wheel
2. Cap/logo
3. Nut (⅝-18)
4. Washer, flat
5. Washer, special
6. Tube
7. Bushing set (2)
8. Shaft
9. Ball joint (RH thread)
10. Nut (⅜-24)
11. Arm
12. Ball joint (LH thread)
13. Tie rod
14. Set screw
15. Drag link
16. Axle assy.
17. "X" lockwasher
18. Spindle bushing (4)
19. Washer (4)
20. Spindle, LH
21. Grease fitting
22. Spindle, RH
23. Wheel & rim
24. Wheel bushing (4)

ENGINE

All Models

For overhaul and repair procedures on engines listed in Specification Table refer to Small Air Cooled Engines Service Manual.

Refer to the following paragraphs for removal and installation procedures.

REMOVE AND REINSTALL. Set mower engagement lever in "OFF" position and lower mower by height control handle to low position. Disconnect blade brake cable from brake band. Remove front pivot bolts and pins. Remove rear lift pin, shift cutting deck to rear for maximum slack in mower drive belt, then disengage belt from engine pulley and mower clutch idler. Slide mower out at right side of chassis. Depress and lock mower drive clutch-brake pedal to allow slack in main drive belt. For easier access to underside of chassis, it may be desirable to place mower up on its rear stand.

CAUTION: Frequently, the performance of maintenance, adjustment or repair operations on a riding mower is more convenient if mower is standing on end. This procedure can be considered a recommended practice providing the following safety recommendations are performed:

1. Drain fuel tank or make certain that fuel level is low enough so that fuel will not drain out.
2. Close fuel shut-off valve if so equipped.
3. Remove battery on models so equipped.
4. Disconnect spark plug wire and tie out of way.
5. Although not absolutely essential, it is recommended that crankcase oil be drained to avoid flooding the combustion chamber with oil when engine is tilted.
6. Secure mower from tipping by lashing unit to a nearby post or overhead beam.

7. Adjustment nuts (5/16)
8. Brake rod
9. Brake spring
10. Spring retainer
11. Return spring
12. Brake link
13. Fastener
14. Adjustment nuts (5/16)
15. Pedal latch
16. Latch spring
17. Belt keepers
18. Clutch spring
19. Spring anchor
20. Support
21. Spacer
22. Crankshaft pulley
23. Transmission pulley
24. Snap ring
25. Mower clutch idler
26. Mower clutch arm
27. Control springs
28. Drive clutch arm
29. Front idler
30. Rear idler
CS. Crankshaft
TS. Transmission shaft

1. Clutch-brake pedal
2. Bushing
3. Spirol pin (3/16)
4. Pedal arm
5. Clutch rod
6. Clutch clevis

Loosen belt keeper bolts to allow greater clearance and work belt off transmission and engine pulleys. Loosen engine mounting bolts slightly and remove any belt keepers or supports which may be in the way. Lower mower to rest on all four wheels, then disconnect throttle control bowden cable at carburetor, magneto lead wire from ignition switch connector, and if so equipped, remove red cable from electric starter. Remove four bolts which attach rear stand to frame, then complete unbolting of engine base and lift engine from mower for further service. Be sure to check for possible interference from wiring for safety interlock systems, battery charger or other supplemental equipment. Observe circuitry and disconnect with care.

To reinstall engine, reverse order used for removal, and torque engine mounting bolts as follows: ¼-inch, 12-14 ft.-lbs.; 5/16-inch, 22-24 ft.-lbs.

TRACTION DRIVE CLUTCH

All Models

ADJUSTMENT. Clutch-brake pedal (1 – Fig. MF2) should be depressed to its limit and locked down by parking latch (15). Tip unit up on rear stand as previously outlined.

Make traction clutch adjustment by turning nuts (7), at rear of clutch rod (5), against clevis (6) to set distance between circumferences of front idler pulley (29) and smaller (2¼ inch diameter) sheave of crankshaft pulley (22). Correct clearance between pulleys should be from 15/16 to 1-1/16 inches. If adjustment cannot be made satisfactorily, carefully check condition of all parts of clutch system. Bent or binding control rod, stripped fasteners, defective or damaged springs, worn, damaged pulleys or bushings or frayed, worn belts must be

renewed or repaired for satisfactory operation of unit.

Test before returning to service.

TRANSMISSION

All Models

REMOVE AND REINSTALL. To remove transmission, raise operator's

seat, and if unit is equipped for electric starting, disconnect and remove battery. Detach safety start switch connector plug. If unit is so equipped, remove chain guard. Disconnect brake rod from disc brake assembly on transmission.

NOTE: As a matter of convenience, operator's seat support (1 and 2 – Fig. MF6) and shroud (4) can be removed. Remove master link from drive chain, separate and remove chain. On Models 8 and 832 only, drain and remove fuel tank.

Release clutch and remove drive belt. Then extract snap ring from bottom end of transmission input shaft and remove transmission pulley. Take care not to lose Woodruff key. Remove shift lever which on Model MF5, requires removal of cam plate cover first. When lever has been removed, take care not to lose detent springs or balls. Remove retaining cap screws and lift transmission from frame.

To reinstall transmission, observe the reverse of this procedure. Readjust drive chain tension as covered in preceding section and torque the mounting cap screws evenly to 18-20 ft.-lbs. Adjust transmission brake.

Fig. MF3 — Underside view of drive system and mower clutch assembled. As shown, drive clutch is engaged, mower idler is disengaged. Note mounting of axle shaft bearings. See text for cautions regarding use of rear stand.

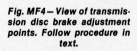

Fig. MF4 — View of transmission disc brake adjustment points. Follow procedure in text.

Fig. MF5—View of traction chain adjustment points. Refer to text for tension adjustment procedure.

1. Shaft bolt
2. Roller
3. Chain

OVERHAUL. After transmission is removed, refer to FOOTE TRANSMISSION REPAIR section of this manual for detailed service procedures.

DRIVE CHAIN

All Models

To set chain tension, loosen nut at outer end of roller bolt (1–Fig. MF5) and shift roller in its slot so there will be ½-inch slack in chain, measured at center of top strand. When adjustment is made, retighten nut to a torque of 40-45 ft.-lbs. Be sure that inner nut on shaft allows roller to turn freely.

To remove chain for renewal or cleaning or in preparation for removal of transmission or rear axle, loosen tension adjustment roller, disconnect master link and withdraw chain. Easiest position for removal of master link is next to transmission sprocket under operator's seat. If a new chain is installed, tension should be adjusted during installation and rechecked after two or three hours operation.

DIFFERENTIAL AND REAR AXLE

All Models

REMOVE AND REINSTALL. Tip unit up on rear stand as previously outlined.

Loosen drive chain tensioner bolt so chain will go slack, extract master link and remove chain. Remove through-bolt at each rear wheel hub and remove both rear wheels. Remove nuts from carriage bolts which retain flange portion of rear wheel bearings to frame and lift axle-differential assembly from its "U" notches in frame rails. Thoroughly clean ends of axles before loosening set screws and removing bearings from axles. Reverse this procedure to reinstall. Tighten nuts on carriage bolts to 8-10 ft.-lbs. torque.

OVERHAUL. Refer to DIFFERENTIAL REPAIR section of this manual under PEERLESS for overhaul procedures.

GROUND DRIVE BRAKE

All Models

ADJUSTMENT. Service brakes operate from a single clutch-brake pedal, with linkage designed to release traction clutch just before brake is applied. Disc brake is mounted on transmission output shaft at opposite end from drive sprocket.

Depress clutch-brake pedal to its limit and engage locking latch. Raise operator's seat for access, and use nuts (14–Fig. MF2) at rear end of brake rod (8) to compress brake spring (9) to a length of 3⅞ to 4 inches on Models 5, 6 and 8, or 4⅛ inches on Models 626 and 832.

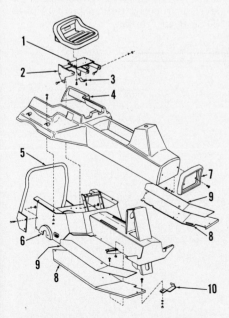

Fig. MF6—Exploded view of frame, body shroud removed, to show assembly details.

1. Seat plate & hinge	6. Frame
2. Support	7. Cap
3. Spring	8. Plate (2)
4. Shroud	9. Tread (2)
5. Rear stand	10. Clip (2)

Fig. MF7—Exploded view of typical mowing deck with lift operating controls and linkages. Two-spindle type shown is used on MF6, MF8 and MF832. See text for service.

1. Blade brake cable (2)
2. Blade brake band (2)
3. Shield
4. Lift lever assy.
5. Latch
6. Latch spring
7. Quadrant
8. Latch rod
9. Lift straps
10. Support bar
11. Link rod
12. Crank assy.
13. Spindle pulley (2)
14. Spacer
15. Guide pulley
16. Belt guide
17. Spacer
18. Stud (2)
19. Deck
20. Spindle bearing
21. Spacer
22. Spindle housing
23. Snap ring
24. Bearing
25. Snap ring
26. Spindle shaft
27. Blade
28. Spring cup washer

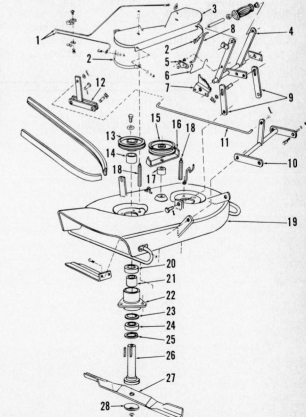

Fig. MF8—Schematic views of typical electrical circuitry to show color coding and connector arrangement. Diagram for electric start models is at left, manual start models at right.

Carefully inspect all brake components paying particular attention to condition of disc brake pads and operating linkage, especially rod (8), to note if threads are damaged or if elongated loop at forward end of rod is sprung or distorted. Repair or renew parts as necessary. Brake parts for disc assembly will be found in FOOTE portion of TRANSMISSION REPAIR section of this manual.

NOTE: Mower blade brake is covered under MOWER SERVICE.

MOWER BLADE BRAKE

Single Spindle Models

ADJUSTMENT. During routine operation, an occasional visual check should be made to determine if proper clearance exists between brake band and drive belt at left hand side of mower pulley. If clearance is insufficient (band dragging) or if braking is ineffective, adjust as follows:

Engage mower clutch control by setting handle in "ON" position. Adjust brake bracket (fixed end of brake band) for 1/32 to 1/16-inch clearance between brake band and drive belt. Measure at bracket end. Move control handle to "OFF" position. Now, manually swing idler pulley to a mid-point between "ON" and "OFF" positions, take up slack in cable with stop on brake band against stop post on mower deck and tighten cable set screw. Allow idler pulley to swing back to "OFF" position and check tension of drive belt which should be relaxed at engine drive pulley. Observe blade brake action during normal mower operation.

Double Spindle Models

ADJUSTMENT. On these models there is a separate blade brake band on each spindle pulley. If band clearance is incorrect, set control handle in "ON" position, then loosen two nuts which hold shield (3 – Fig. MF7) to mounting studs (18) on mower deck. Shift shield (3) in its slotted holes to obtain 1/32 to 1/16-inch clearance between each brake band and drive belt. Retighten nuts. Move control handle to "OFF". Swing mower clutch idler pulley to a mid-point between "ON" and "OFF" positions and adjust cable to remove slack but leaving running clearance between brake bands and belt. Tighten cable clips and set screw. Let idler pulley return to "OFF" position and check drive belt tension. Belt should be relaxed at engine drive pulley. Check function of blade brake during normal mower operation.

LEVEL MOWER

All Models

Because the mowing deck is attached to the center-pivoted front axle and not suspended from rigid frame member only, it is imperative that unit be parked on a hard, level surface such as a concrete floor before attempting to level the cutting deck. Level deck by means of pivot bolts which attach support bar (10 – Fig. MF7) to deck (19). When mower is leveled properly, tighten pivot bolts to 30-35 ft.-lbs. torque.

MTD

MTD PRODUCTS, INC.
5965 Grafton Rd.
P.O. Box 36900
Cleveland, OH 44136

		Engine		
Model	Make	Model	Horsepower	Cutting Width, In.
360, 362	B&S	130000	5	25
380, 385	B&S	130000	5	*
390, 395	B&S	190000	8	30
400, 402	B&S	130000	5	26
405	B&S	130000	5	26
406, 407	Tecumseh	VM70	7	26
410, 412	B&S	190000	8	30
420, 425	B&S	190000	8	30
430, 435	B&S	190000	8	30
440, 445	B&S	190000	8	34
460, 465	B&S	190000	8	34
470, 475	B&S	190000	8	34
480, 485	B&S	190000	8	34
495	B&S	190000	8	38
497	B&S	220000	10	38
498	B&S	252000	11	38
520, 525	B&S	190000	8	26
630, 632	B&S	190000	8	34
638, 698	B&S	252000	11	38
760	B&S	252000	11	38
780, 784, 786	B&S	326000	16	44
796, 797	B&S	190000	8	30
820	B&S	326000	16	44

The first digit of the model number indicates general equipment classification. The second digit indicates type of equipment. The third digit indicates the year. (The first three digits have been deleted from model numbers listed in Specification Table.) The numbers following the decimal are the specific model type. An "A" following the model numbers indicates a model manufactured between 1974 and 1984. Example: Model 132.410 is a 1972 and Model 132.410A is a 1982 model of the 410 type. *Early production 25"; 1974 and later 26".

FRONT AXLE AND STEERING SYSTEM

Models 360-362

R&R AND OVERHAUL. To remove axle main member (2–Fig. MT1), support front of unit and disconnect tie rods (7) from arms on spindles (1 and 5). Unbolt and remove front wheel assemblies. Remove nuts securing axle to frame channel (3) and remove axle assembly. Remove spindle pins (4) and spindles (1 and 5). When reinstalling, bolt front wheels to spindles using same holes from which they were removed.

NOTE: The four holes in spindles are used to adjust mower cutting height at front. Four corresponding holes are located in rear axle mounting brackets for setting cutting height at rear. Check toe-in and if necessary, adjust tie rods to obtain a toe-in of ⅛-inch. Lubricate spindle pins (4) with SAE 30 oil.

To remove steering gear, unbolt and remove hood. Refer to Fig. MT2 and un-

bolt tube clamp (7). Unbolt front mounting bracket and remove upper steering shaft assembly (1 through 6) with steer-

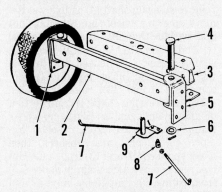

Fig. MT1—Front axle assembly used on Models 360 and 362.

1. Spindle R.H.
2. Axle main member
3. Frame channel & axle support
4. Spindle pin
5. Spindle L.H.
6. Washer
7. Tie rods
8. Tie rod end
9. Lower steering shaft

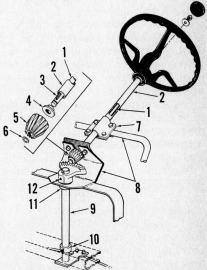

Fig. MT2—Steering gear assembly used on Models 360 and 362. Flange bushings (4) are used at both ends of steering tube (2).

1. Steering shaft (upper)
2. Steering tube
3. Key
4. Flange bushing
5. Pinion gear
6. Retaining ring
7. Tube clamp
8. Steering support
9. Steering shaft (lower)
10. Collar
11. Flange bushing
12. Quadrant gear

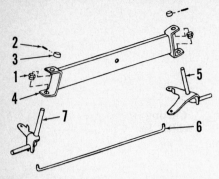

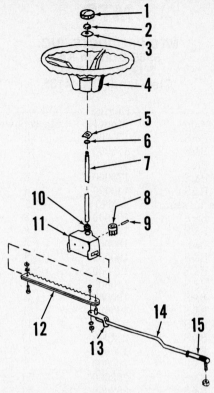

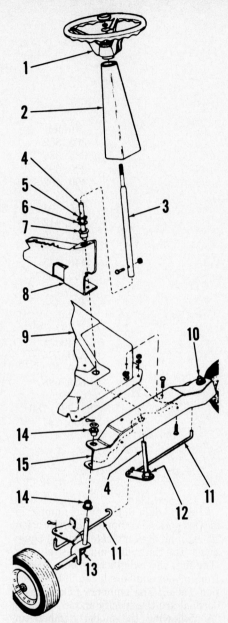

Fig. MT3 — Exploded view of front axle assembly used on Models 380, 385, 390 and 395.

1. Spindle bushings
2. Set screw
3. Collar
4. Axle main member
5. Spindle L.H.
6. Tie rod
7. Spindle R.H.

ing wheel. Remove cap, nut, washer and steering wheel, then remove retaining ring (6), pinion gear (5), key (3) and shaft (1) from steering tube (2). Loosen set screw in quadrant gear (12) and remove gear and key. Raise front of unit and disconnect tie rods from arm on lower steering shaft (9). Loosen set screw in collar (10) and withdraw steering shaft from bottom.

Clean and inspect all parts and renew any showing excessive wear or other damage. Reassemble by reversing disassembly procedure. Lubricate flange bushings with SAE 30 oil and apply multi-purpose lithium base grease to pinion and quadrant gear teeth.

Models 380-385-390-395

R&R AND OVERHAUL. To remove axle main member (4 – Fig. MT3), raise front of unit and remove front wheel assemblies. Disconnect drag link from arm on spindle (5) and remove tie rod (6). Loosen set screws (2), remove collars (3) and lower spindles (5 and 7) from axle. Remove pivot bolt, then remove axle main member.

Check spindle bushings (1) and spindles (5 and 7) for excessive wear and renew as necessary. Reinstall by reversing removal procedure. Lubricate bushings and all pivot points with SAE 30 oil.

To remove steering gear, first remove mower unit as outlined in MOWER DECK paragraphs. Raise front of chassis and block securely. Refer to Fig. MT4, unbolt bracket (13) and disconnect drag link (14) from steering rack (12). Drive roll pin (9) from pinion (8). Withdraw steering shaft (7) with steering wheel (4) from top and remove steering rack (12), housing (11) and pinion (8) from bottom. Remove pinion (8) and two-piece rack (12) from the housing. Remove cap (1), nut (2), washer (3) and steering wheel (4) from shaft (7).

Inspect all parts and renew as necessary. When reassembling, oil steering

Fig. MT4 — Exploded view of typical steering gear used on Models 380, 385, 395, 460, 465, 470, 475, 480 and 485.

1. Cap
2. Nut
3. Belleville washer
4. Steering wheel
5. Retainer
6. Washer
7. Steering shaft
8. Pinion
9. Roll pin
10. Bushing
11. Housing
12. Steering rack
13. Bracket
14. Drag link
15. Drag link end

shaft bushings and lubricate pinion (8) and rack (12) with multi-purpose lithium grease.

Models 400-402-405-406-407-410 (Late Production)-412

R&R AND OVERHAUL. To remove axle main member (15 – Fig. MT5), first remove steering shaft (4) as follows: Remove steering wheel cap, retaining nut and washer, then lift off steering wheel (1) and cover (2). Unbolt and remove shaft extension (3), speed nut (5) and washer (6). Disconnect tie rods (11) from spindles (10 and 13), raise front and withdraw steering shaft (4) from bottom of axle main member. Block up under frame and remove front wheels. Remove cotter pins and washers, then lower spindles (10 and 13) from axle. Unbolt and remove axle main member (15) from frame (9). Flange bushings (7 and 12) and spindle bushings (14) can now be removed if necessary.

Clean and inspect all parts and renew any showing excessive wear or other damage. When reassembling, reverse removal procedure and lubricate bush-

Fig. MT5 — Exploded view of front axle and steering system used on Models 400, 402, 405, 406, 407, 410 (late production) and 412.

1. Steering wheel
2. Cover
3. Shaft extension
4. Steering shaft
5. Retainer
6. Washer
7. Flange bushing
8. Support
9. Frame
10. Spindle L.H.
11. Tie rods
12. Flange bushing
13. Spindle R.H.
14. Spindle bushings
15. Axle main member

ings with SAE 30 oil. Tie rods (11) are non-adjustable.

Models 410 (Early Production)-420-425-430-435

R&R AND OVERHAUL. To remove axle main member (8 – Fig. MT6), first remove steering shaft (7) as follows: Remove steering wheel cap, retaining nut and washer, then remove steering wheel (1) and spacer tube (4). Drive out roll pin (5) and remove shaft extension (2). Raise

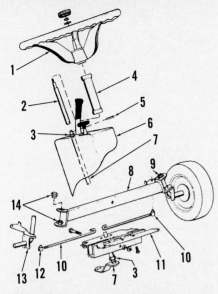

Fig. MT6 — Partially exploded view of front axle and steering system used on Models 410 (early production), 420, 425, 430 and 435.

1. Steering wheel
2. Shaft extension
3. Flange bushings
4. Spacer tube
5. Roll pin
6. Console
7. Steering shaft
8. Axle main member
9. Spindle L.H.
10. Tie rods
11. Axle support
12. Tie rod end
13. Spindle R.H.
14. Spindle bushings

front and block securely. Remove tie rods (10), then withdraw steering shaft (7) from bottom of axle support (11). Remove front wheel assemblies and spindles (9 and 13). Unbolt axle support from frame and remove front axle and support assembly. Remove axle pivot bolt and separate axle main member from support. Remove flange bushings (3) and spindle bushings (14).

Clean and inspect all parts and renew any showing excessive wear or other damage. Reassemble by reversing the disassembly procedure and lubricate bushings and all pivot points with SAE 30 oil. Check front wheel toe-in and if necessary, adjust tie rods to obtain a toe-in of ⅛-inch.

Models 440-445

R&R AND OVERHAUL. To remove axle main member (15 – Fig. MT7), support front of unit and remove front wheels. Disconnect tie rods (14) from spindles (12 and 18). Remove cotter pins and washers, then lower spindles from axle main member. Remove pivot bolt (17) and unhook mower tension springs from axle support (16). Unbolt axle sup-

port from frame (9) and separate axle main member from support.

Inspect spindle bushings (13) for excessive wear and renew as necessary. When reinstalling, reverse removal procedure and lubricate spindle bushings (13), pivot bolt (17) and front wheels with SAE 30 oil. Adjust the tie rods as required for ⅛-inch toe-in.

To remove steering gear, remove steering wheel cap, retaining nut and washer, then lift off steering wheel (1) and spacer tube (2). Drive out roll pin and remove shaft extension (3) and washer (6). Raise front and block securely. Disconnect tie rods from ends of rack. Unbolt steering housing (11) from support (10), then withdraw steering unit from bottom. Refer to Fig. MT8 and drive out roll pin securing pinion (3) to steering shaft (1). Withdraw shaft and remove pinion. Remove bolt from one end of two-piece rack (4) and slide rack out of housing (2). Clean and in-

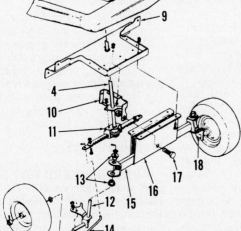

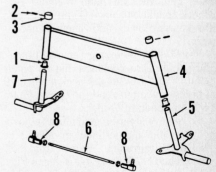

Fig. MT7 — Partially exploded view of front axle and steering system used on Models 440 and 445.

1. Steering wheel
2. Spacer tube
3. Shaft extension
4. Steering shaft
5. Cover
6. Washer
7. Flange bushing
8. Console
9. Frame
10. Support
11. Rack & pinion assy.
12. Spindle R.H.
13. Spindle bushings
14. Tie rod (2 used)
15. Axle main member
16. Axle support
17. Pivot bolt
18. Spindle L.H.

Fig. MT8 — Steering rack and pinion assembly removed from Model 440 or 445.

1. Steering shaft
2. Housing
3. Pinion
4. Rack (2 piece)

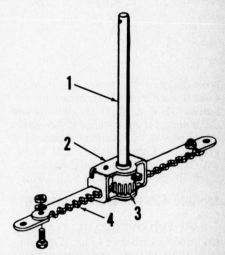

Fig. MT9 — Exploded view of front axle assembly used on Models 460, 465, 470, 475, 480 and 485.

1. Spindle bushings (4 used)
2. Set screw
3. Collar
4. Axle main member
5. Spindle L.H.
6. Tie rod
7. Spindle R.H.
8. Tie rod ends

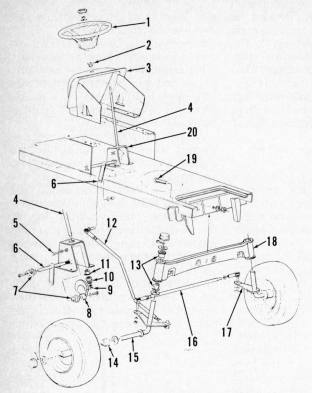

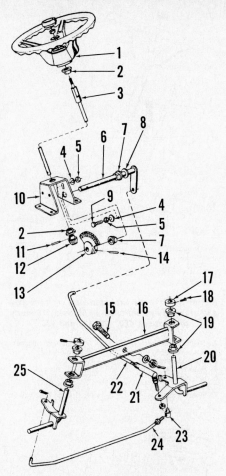

Fig. MT10 — Exploded view of front axle and steering system used on Models 495, 497, 498, 760, 780, 784, 786 and 820.

1. Steering wheel
2. Bushing
3. Dash
4. Steering shaft
5. Adjusting screw
6. Steering arm
7. Bushings
8. Sector gear
9. Retaining nut
10. Pinion gear
11. Bushing
12. Drag link
13. Spindle bushings
14. Bushing
15. Spindle R.H.
16. Tie rod
17. Spindle L.H.
18. Axle
19. Pivot bolt
20. Steering gear housing

Fig. MT11 — Exploded view of front axle and steering gear assembly used on Models 520 and 525.

1. Steering wheel		14. Roll pin
2. Flange bushing		15. Pivot bolt
3. Steering shaft		16. Axle main member
4. Belleville washers		17. Collar
5. Nut		18. Set screw
6. Quadrant shaft		19. Spindle bushings
7. Flange bushings		20. Spindle L.H.
8. Washer		21. Drag link end
9. Cap screw		22. Drag link
10. Support		23. Tie rod end
11. Roll pin		24. Tie rod
12. Pinion gear		25. Spindle R.H.
13. Quadrant gear		

spect all parts and renew any showing excessive wear or other damage. Flange bushing (7 – Fig. MT7) can be removed and a new bushing installed if necessary.

Reassemble by reversing disassembly procedure. Lubricate rack and pinion with multi-purpose lithium grease. Lubricate flange bushing with SAE 30 oil.

Models 460-465-470-475-480-485

R&R AND OVERHAUL. To remove axle main member (4 – Fig. MT9), support front of unit and remove front wheel assemblies. Disconnect drag link from arm on spindle (5) and remove tie rod assembly. Loosen set screws (2), remove collars (3) and lower spindles (5 and 7) from axle. Remove pivot bolt, then remove axle main member (4). Check spindle bushings (1) and spindles for excessive wear and renew as necessary.

Reinstall by reversing removal procedure. Lubricate bushings and all pivot points with SAE 30 oil. Check toe-in and if necessary, adjust ends on tie rod (6) to obtain a toe-in of ⅛-inch.

To remove steering gear, first remove mower unit as outlined in MOWER DECK paragraphs. Raise front of chassis and block securely. Refer to Fig. MT4, unbolt bracket (13) and disconnect drag link (14) from steering rack (12). Drive roll pin (9) from pinion (8). Withdraw steering shaft (7) with steering wheel (4) from top and remove steering rack (12), housing (11) and pinion (8) from bottom. Remove pinion (8) and

two-piece rack (12) from housing. Remove cap (1), nut (2), washer (3) and steering wheel (4) from shaft (7).

Inspect all parts and renew as necessary. When reassembling, oil steering shaft bushings and lubricate pinion (8) and rack (12) with multi-purpose lithium base grease.

Models 495-497-498-760-780-784-786-820

R&R AND OVERHAUL. To remove axle main member (18 – Fig. MT10), raise and support front of unit and remove front wheels. Remove tie rod (16) and disconnect drag link (12) from spindles (15 and 17). Remove retainer from top of spindles, then withdraw spindles and bushings (13) from axle ends. Remove pivot bolt (19) and lower axle from frame.

Clean and inspect parts for excessive wear and renew as needed. Reassemble by reversing removal procedure. Lubricate bushings with SAE 30 oil. Adjust tie rod as required for ⅛-inch toe-in.

To remove steering gear, first raise hood and disconnect battery cables. Remove steering wheel cap, retaining nut and washer, then lift off steering wheel (1 – Fig. MT10) and steering tube (if so equipped). Remove retaining nut (9), washer, pinion gear (10) and flange bushing (11), then withdraw steering shaft (4). Remove retaining bolts from both ends of steering arm shaft (6) and remove sector gear (8) and flange bushings (7). Disconnect drag link (12) from

steering arm (6) and remove steering arm shaft.

Clean and inspect all parts for excessive wear and renew as needed. Reassemble by reversing removal procedure. Lubricate steering gears with light coat of multi-purpose lithium base grease and flange bushings with SAE 30 oil. Adjust backlash between steering gears by turning adjusting bolt (5) in or out, then tighten jam nut.

Models 520-525

R&R AND OVERHAUL. To remove axle main member (16 – Fig. MT11), raise front of unit and block securely. Remove tie rod assembly (23 and 24) and disconnect drag link end (21) from arm on spindle (20). Remove front wheel as-

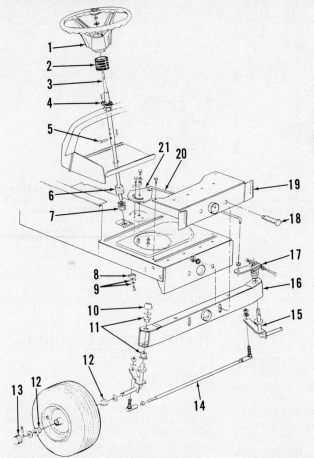

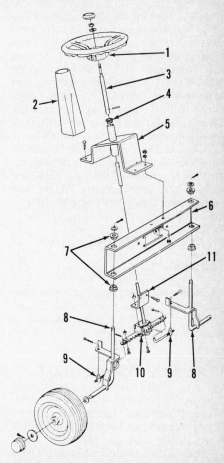

Fig. MT12—Exploded view of front axle and steering system used on Models 630, 632, 638 and 698.

1. Steering wheel
2. Bellow
3. Steering shaft extension
4. Bushing
5. Clevis pin
6. Steering shaft & gear
7. Bushing
8. Spacer
9. Washers
10. Push cap
11. Spindle bushings
12. Wheel bushings
13. Hub cap
14. Tie rod
15. Spindle
16. Axle
17. Steering arm
18. Shoulder bolt
19. Pivot bracket
20. Drag link
21. Gear segment

Fig. MT13—Exploded view of front axle and steering system used on Models 796 and 797.

1. Steering wheel
2. Cover
3. Steering shaft extension
4. Bushing
5. Support bracket
6. Axle
7. Spindle bushings
8. Spindles
9. Tie rod
10. Steering rack assy.
11. Mounting bracket

semblies. Loosen set screws (18), remove collars (17) and lower spindles (20 and 25) from axle. Remove pivot bolt (15), then remove axle main member (16). Inspect spindle bushings (19) and spindles (20 and 25) for excessive wear and renew as necessary.

Reinstall by reversing removal procedure. Lubricate spindle bushings with SAE 30 oil. Check front wheel toe-in and adjust if necessary, to obtain a toe-in of ⅛-inch.

To remove steering gear, tilt hood forward and on Model 525, remove battery cover and disconnect battery cables. Refer to Fig. MT11 and drive out roll pin (11). Withdraw steering wheel (1) and steering shaft (3) from above and remove pinion gear (12) and lower bushing (2) from support (10). Drive out roll pin (14) and remove quadrant gear (13) and inner bushing (7). If necessary to remove quadrant shaft (6) and outer bushing (7), block under front frame securely. Unbolt seat bracket from rear frame. Remove bolts from the front of each fender and unbolt rear frame from front frame. Raise front of rear frame about 4 inches, then slide quadrant shaft (6) out left side.

Clean and inspect all parts and renew any showing excessive wear or other damage. Reassemble by reversing disassembly procedure. Lubricate bushings (2

and 7) with SAE 30 oil and apply multipurpose lithium base grease to teeth of pinion gear (12) and quadrant gear (13). Adjust nuts (5) on cap screw (9) to remove excessive backlash between steering gears.

Models 630-632-638-698

R&R AND OVERHAUL. To remove axle main member (16 – Fig. MT12), raise and support front of unit and remove front wheels. Disconnect tie rod (14) from spindles (15). Unbolt steering arm (17) from left spindle and remove push cap (10) from right spindle, then withdraw spindles and bushings (11) from axle ends. Unbolt and separate axle from frame.

Reassemble by reversing removal procedure. Lubricate bushings and front wheels with SAE 30 oil. Adjust tie rod ends as required for ⅛-inch toe-in.

To remove steering gears, remove steering wheel cap, retaining nut and washer, then lift off steering wheel (1 – Fig. MT12) and bellow (2). Disconnect battery cables. Remove clevis pin (5), then unbolt bushing (4) and withdraw shaft extension (3) and bushing. Disconnect drag link (20) from steering arm (17), then unbolt and remove gear segment (21). Remove retaining bolt from bottom of steering shaft (6) and with-

draw washers (9), spacer (8), flange bushing (7) and steering shaft (6).

Inspect all parts for excessive wear or other damage and renew as needed. Reassemble by reversing disassembly procedure. Lubricate steering gears with multi-purpose lithium base grease and bushings with SAE 30 oil.

Models 796-797

R&R AND OVERHAUL. To remove axle main member (6 – Fig. MT13), support front of unit and remove front wheels. Disconnect tie rods (9) from spindles (8). Remove cotter pins and washers, then withdraw spindles and bushings (7) from axle ends. Unbolt and remove axle from frame.

Inspect for excessive wear and renew as needed. Reassemble by reversing removal procedure and lubricate spindle bushings and front wheels with SAE 30 oil. Adjust tie rods as required for ⅛-inch toe-in.

To remove steering gear, remove steering wheel cap, retaining nut and washer, then lift off steering wheel (1–Fig. MT13) and cover (2). Drive out roll pin and remove shaft extension (3) and flange bushing (4). Disconnect tie rods (9) from rack (10). Unbolt mounting bracket (11) and withdraw steering unit from bottom. Drive out roll pin and withdraw shaft (1–Fig. MT8) and pinion (3). Remove bolt from end of two-piece rack and slide rack (4) out of housing (2).

Clean and inspect parts and renew as needed. Reassemble by reversing removal procedure. Lubricate rack and pinion with multi-purpose lithium base grease and flange bushing with SAE 30 oil.

ENGINE

All Models

For overhaul and repair procedures on engines listed in Specification Table, refer to Small Air Cooled Engines or Large Air Cooled Engines Service Manuals.

REMOVE AND REINSTALL. To remove engine assembly, first disconnect spark plug wire and proceed as follows: On electric start models, disconnect and remove battery, then disconnect starter cable and alternator wires. On recoil start models, pull starter rope out a few inches and tie a knot in rope between starter and starter safety switch. Remove rope handle and feed

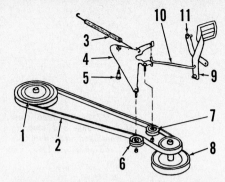

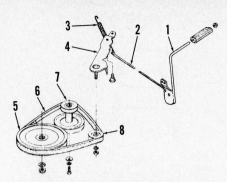

Fig. MT15 — Traction drive clutch used on Models 380, 385, 390, 395, 470, 475, 480 and 485. Clutch used on Models 495 (late production), 497, 498, 520 and 525 is similar.

1. Transmission input pulley	6. Flat idler
2. Traction drive belt	7. Vee idler
3. Clutch spring	8. Engine pulley
4. Idler bracket	9. Clutch pedal
5. Pivot bolt	10. Clutch rod
	11. Clutch lock

Fig. MT16 — Traction drive clutch used on Model 400. Clutch used on Model 410 (early production), 796 and 797 is similar.

1. Clutch pedal	5. Transmission pulley
2. Clutch rod	6. Traction drive belt
3. Clutch spring	7. Engine pulley
4. Idler bracket	8. Idler pulley

rope back through safety switch. Knot will prevent rope from coiling inside starter housing. On all models, disconnect ignition wire from magneto and disconnect throttle cable at engine. On Models 520 and 525, shut off fuel valve and disconnect fuel line. On all models, remove mower as outlined in MOWER DECK paragraphs. Remove drive belt as outlined in TRACTION DRIVE CLUTCH AND DRIVE BELTS paragraphs. Unbolt and remove engine pulley. Remove engine mounting bolts and lift engine from frame.

Reinstall by reversing removal procedure.

TRACTION DRIVE CLUTCH AND DRIVE BELTS

All Models

The traction drive clutch on all models is belt idler type. On models equipped with variable speed drive, variable speed pulley is also clutch idler. Clutch idlers are spring tensioned and no adjustment is required. If drive belt slips during normal operation due to excessive belt wear or stretching, new belts should be installed.

Models 360-362

REMOVE AND REINSTALL. To remove drive belt, first remove mower as outlined in MOWER DECK paragraphs. Depress clutch pedal and engage pedal lock. See Fig. MT14. Spring belt guide away from engine pulley and slip belt off pulley. Remove belt from idler pulley. Unbolt transmission pulley, then slide pulley from shaft and remove belt.

Install new belt by reversing removal procedure.

Models 380-385-390-395-470-475-480-485-495 (Late Production)-497-498-520-525

REMOVE AND REINSTALL. To remove traction drive belt (2–Fig. MT15), first remove mower and mower drive belt as outlined in MOWER DECK paragraphs. Depress clutch pedal and engage pedal lock. Loosen or remove belt guards as needed to remove belt from pulleys. On models equipped with transaxle, remove clutch idler pulley and slip belt out of pulleys. On models equipped with transmission, remove retaining

Fig. MT14 — Bottom view of Model 360 or 362 with blade and mower housing removed, showing transmission clutch idler and main drive belt.

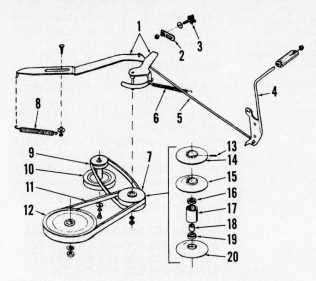

Fig. MT17—Exploded view of clutch and variable speed assembly used on Models 402, 405, 406, 407, 410 (late production) and 412. Clutch and variable drive used on Models 420, 425, 430, 435, 460 and 465 is similar.

1. Clutch and variable speed bracket
2. Speed control stop
3. Knob
4. Clutch pedal
5. Clutch rod
6. Clutch spring
7. Variable speed pulley
8. Variable speed spring
9. Primary drive belt
10. Engine pulley
11. Secondary drive belt
12. Transmission input pulley
13. Roll pin (2 used)
14. Sheave half (top)
15. Movable sheave half
16. Bearing
17. Sleeve
18. Spacer
19. Bearing
20. Sheave half (bottom)

nut on transmission pulley and idler pulley, then slide pulleys off their shafts and remove belt. On some early models, drag link must be disconnected from steering rack to remove belt from chassis.

Install new belt in reverse order of removal. Be sure belt guards and keepers are properly installed.

Models 400-410 (Early Production)-796-797

REMOVE AND REINSTALL. To remove traction drive belt (6–Fig. MT16), first remove mower and mower drive belt as outlined in MOWER DECK paragraphs. Depress clutch pedal and engage pedal lock. Remove engine pulley belt guard. Remove retaining nut from idler pulley and transmission pulley. Slide pulleys off shafts and remove drive belt.

Renew belt by reversing removal procedure.

Models 402-405-406-407-410 (Late Production)-412-420-425-430-435-460-465

REMOVE AND REINSTALL. To remove traction drive belts (9 and 11–Fig. MT17), first remove mower and mower drive belt as outlined in MOWER DECK paragraphs. Depress clutch pedal and engage pedal lock. On Models 460 and 465 disconnect drag link from steering rack. On all models, remove engine pulley belt guard and unhook clutch tension springs. Remove retaining nuts from variable speed pulley and transmission pulley, then slide pulleys off shafts and remove drive belts.

Install new belts by reversing removal procedure. When reassembling transmission pulley, install hub side up.

Models 440-445

REMOVE AND REINSTALL. To remove traction drive belt (5–Fig.

MT18), first remove mower as outlined in MOWER DECK paragraphs. Unhook idler arm tension springs. Remove belt guards as needed to remove belt from pulleys, then slip belt off transmission pulley and engine pulley.

Install new belt by reversing removal procedure.

Model 495 (Early Production)

REMOVE AND REINSTALL. To remove drive belts (2 and 10–Fig. MT19), remove mower as outlined in MOWER DECK paragraphs. Depress clutch pedal and engage pedal lock. Unhook clutch tension spring and remove engine pulley belt guard. Disconnect drag link from steering rack. Remove retaining nut from lower jackshaft pulley. Slide pulley from jackshaft and remove primary drive belt from pulleys and idlers. Unbolt and remove upper frame cover. Loosen transaxle mounting bolts and adjusting nuts at rear of frame. Move transaxle forward and remove secondary drive belt from pulleys.

Install new belts by reversing removal procedure. Adjust secondary drive belt by tightening adjusting nuts evenly until belt deflects about ½-inch when 10 pounds pressure is applied between transaxle pulley and jackshaft pulley. Tighten transaxle mounting bolts.

Models 630-632-638-698

REMOVE AND REPLACE. To remove drive belt, first remove mower as outlined in MOWER DECK paragraphs. Remove engine pulley belt guard. Unbolt and remove idler pulley. Unbolt transaxle pulley belt guard. Remove transaxle shift lever. Remove belt from engine pulley, then lift belt up and over transaxle pulley.

Install new belt by reversing removal procedure.

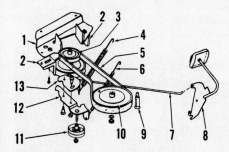

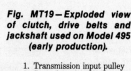

Fig. MT18 — Traction drive clutch assembly used on Models 440 and 445.

1. Engine pulley belt guard	8. Clutch pedal
2. Belt keepers	9. Shoulder bolt
3. Engine pulley	10. Transmission input pulley
4. Spring (heavy)	
5. Traction drive belt	11. Clutch
6. Spring (light)	12. Clutch idler arm
7. Clutch rod	13. Idler pivot bracket

Fig. MT19—Exploded view of clutch, drive belts and jackshaft used on Model 495 (early production).

1. Transmission input pulley
2. Secondary drive belt
3. Jackshaft pulley (upper)
4. Bearing retainer
5. Ball bearing
6. Jackshaft
7. Plate
8. Spacer
9. Jackshaft pulley (lower)
10. Primary drive belt
11. Flat idler
12. Engine pulley
13. Vee idler
14. Pivot bolt
15. Clutch spring
16. Clutch bracket
17. Clutch rod
18. Clutch lock
19. Clutch pedal

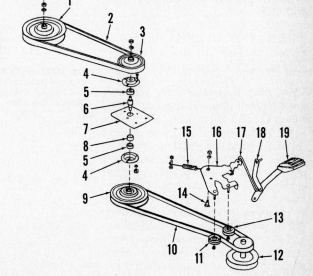

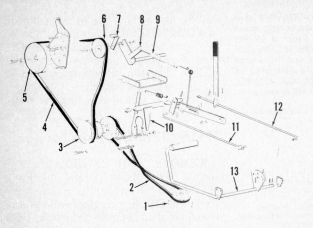

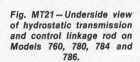

Fig. MT20 — Traction drive clutch assembly, drive belts and jackshaft used on Models 760, 780, 784, 786 and 820.

1. Engine pulley
2. Primary drive belt
3. Jackshaft
4. Secondary belt
5. Transaxle pulley
6. Clutch idler pulley
7. Belt guard
8. Clutch arm
9. Idler spring
10. Jackshaft bracket
11. Brake cam rod
12. Clutch-brake rod
13. Clutch-brake pedal assy.

2337 transaxle. All other models use single speed, forward and reverse Foote transmissions.

Models 760-780-784-786

ADJUSTMENT OF HYDROSTATIC CONTROL. To adjust control linkage, raise both rear wheels off ground and block securely. Loosen nuts on both ends of connecting rod (Fig. MT21). Place hydrostatic control lever in neutral and start engine. With clutch-brake pedal released, adjust connecting rod until rear wheels do not rotate. Shut off engine and tighten nuts.

Models 360-362-380-385-390-395-400-402-405-406-407-410-412-420-425-430-435-440-445-460-465-470-475-480-485-520-525-796-797

REMOVE AND REINSTALL. To remove transmission, first remove mower as outlined in MOWER DECK paragraphs and remove traction drive belt as outlined in previous paragraphs. Unbolt and remove transmission pulley if not removed during belt removal. Disconnect transmission shift linkage, and unbolt and remove engine cover. Disconnect drive chain at master link. Unbolt transmission from frame and remove transmission.

Reinstall by reversing removal procedure.

Models 495-497-498-630-632-638-698-760-780-784-786-820

REMOVE AND REINSTALL. To remove transaxle, remove mower as

Models 760-780-784-786-820

REMOVE AND REINSTALL. To remove drive belts (2 and 4 – Fig. MT20), remove mower as outlined in MOWER DECK paragraphs. Depress clutch pedal and engage pedal lock. Loosen or remove belt guides as needed to remove belts from pulleys. Loosen stop bolt behind jackshaft pulley and pry pulley forward while slipping front belt off jackshaft pulley and engine pulley.

NOTE: Observe twist in front belt. If new belt is installed backwards, the tractor will run backwards.

Note position of clutch idler belt guard, then unbolt and remove idler. On Model 820, remove retaining bolt and remove transaxle pulley and rear belt. On all other models, loosen bolt holding rear axle bracket to frame, then pry frame over about ¼-inch and remove belt.

Install new belts by reversing removal procedure.

TRANSMISSION

All Models

For transmission overhaul procedures, refer to Foote, Peerless or Eaton paragraphs in TRANSMISSION REPAIR section of this manual. Models 390 and 395 are equipped with Peerless Model 501 transmissions. Models 440 and 445 use Peerless Model 515 transmissions. Peerless Model 701 transmissions are used on Models 470, 475, 480, 485, 796 and 797. Models 495 and 497 use Peerless 600 series transaxles. Models 520 and 525 are equipped with Peerless Model 714 transmissions. Peerless Model 813 transaxles are used on Models 498, 630, 632, 638 and 698. Models 760, 780, 784 and 786 are equipped with Eaton hydrostatic transmissions and Peerless Model 1322 transaxles. Model 820 uses Peerless Model

VARIABLE SPEED PULLEY

Models 402-405-406-407-410 (Late Production)-412-420-425-430-435-460-465

R&R AND OVERHAUL. On all models, to remove variable speed pulley (7 – Fig. MT17), remove traction drive belts as outlined in previous paragraphs. Then, with pulley assembly removed, drive out roll pins and separate sheave halves (14 and 20) and movable sheave half (15) from sleeve (17). Press bearings (16 and 19) and spacer (18) from sleeve.

Clean and inspect all parts and renew any showing excessive wear or other damage. Movable sheave half (15) must slide freely on sleeve (17). Reassemble by reversing the disassembly procedure. Apply a dry lubricant to movable sheave half and sleeve. Bearings (16 and 19) are sealed and require no additional lubrication.

Fig. MT21 — Underside view of hydrostatic transmission and control linkage rod on Models 760, 780, 784 and 786.

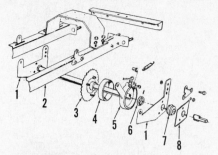

Fig. MT22—Rear axle and drive sprocket assembly used on Models 360 and 362. Axle mounting brackets (1) are also the mower rear height adjusters.

1. Axle mounting brackets
2. Frame
3. Axle & sprocket assy.
4. Brake drum
5. Brake band
6. Collar
7. Axle bearing
8. Bearing plate

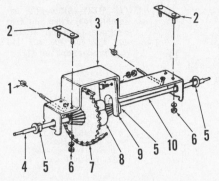

Fig. MT23—Differential assembly used on Models 380, 385, 390 and 395.

1. Adjusting nuts
2. Mounting plates
3. Axle mounting bracket
4. Axle shaft R.H.
5. Axle bearings (3)
6. Locknuts
7. Drive chain
8. Differential assy.
9. Bearing bracket
10. Axle shaft L.H.

outlined in MOWER DECK paragraphs. Disconnect and remove shift linkage. Disconnect brake linkage. Support rear of unit and remove mounting bolts from support braces and frame. Move transaxle forward, then remove belt guides as needed and remove belt from input pulley. Raise rear of unit and roll transaxle rearward.

Reinstall in reverse of removal procedure.

DRIVE CHAIN

Models 360-362

There is no chain adjustment on these models. If chain slack is excessive, check chain and sprockets for wear and renew as necessary.

Models 380-385-390-395-460-465-470-475-480-485-520-525

ADJUSTMENT. Drive chain tension should be checked periodically. Chain should deflect approximately ½-inch when depressed with thumb halfway between sprockets. To adjust chain tension, loosen locknuts (2 each side) which secure axle mounting bracket to main frame. Turn adjusting nuts located at rear of main frame, clockwise to tighten or counter-clockwise to loosen drive chain. Adjust each side equally so rear axle is perpendicular to center line of chassis. When chain tension is correct, tighten mounting locknuts.

Models 400-402-405-406-407-410-412-420-425-430-435

ADJUSTMENT. Drive chain tension should be checked periodically. Chain should deflect about ½-inch when depressed with thumb midway between sprockets. To adjust drive chain, loosen transmission mounting nuts slightly. Turn draw bolt, located under right side

of main frame, clockwise to tighten or counter-clockwise to loosen drive chain. When chain tension is correct, retighten transmission mounting nuts.

Models 440-445-796-797

ADJUSTMENT. The drive chain should deflect approximately ½-inch when depressed with thumb midway between sprockets. To adjust drive chain, loosen chain idler mountings nuts and slide chain idler rearward to tighten or forward to loosen drive chain. When chain tension is correct, retighten idler mounting nuts.

All Models So Equipped

REMOVE AND REINSTALL. To remove drive chain, rotate driven sprocket to locate master link. Disconnect master link and remove drive chain. Clean and inspect chain and sprockets and renew if excessively worn.

Install drive chain and adjust chain tension as outlined in previous paragraphs. Lubricate chain with a light coat of SAE 30 oil.

DIFFERENTIAL, DRIVE SPROCKET AND REAR AXLE

All Models

For differential overhaul procedures, refer to MTD Products paragraphs in DIFFERENTIAL REPAIR section of this manual. On models equipped with transaxles, refer to TRANSMISSION section.

Models 360-362

REMOVE AND REINSTALL. These models are not equipped with a differential. To remove drive sprocket and rear axle assembly (3–Fig. MT22), first

rotate rear axle to locate master link in drive chain, then disconnect drive chain. Disconnect brake rod and spring at brake band and unbolt brake band anchor arm from frame. Support rear of chassis with a hoist, unbolt rear axle mounting brackets (1) from frame, raise rear of frame and roll out wheel and axle assembly. Unbolt and remove wheel assemblies and axle bearing and bracket assemblies from axle. Remove brake band, then unbolt brake drum (4) from sprocket.

Clean and inspect all parts and renew any showing excessive wear or other damage. Drive sprocket and axle is serviced only as an assembly. Reassemble by reversing disassembly procedure. Lubricate drive chain and axle bearings with SAE 30 oil. Adjust brake as necessary.

Models 380-385-390-395

REMOVE AND REINSTALL. To remove rear axle and differential assembly, locate master link in drive chain, then disconnect drive chain (7–Fig. MT23). Remove adjusting nuts (1) from draw bolts and support rear of chassis with a hoist. Remove locknuts (6) and washers, then move axle mounting bracket (3) forward until draw bolts are free from frame. Raise rear of chassis and roll rear axle and differential assembly rearward.

Unbolt and remove rear wheel assemblies and remove outer axle bearings (5). Unbolt bearing bracket (9) and rotate bracket around axle until it is clear of axle mounting bracket (3). Slide axle and differential assembly toward left side until right axle shaft (4) is free of bearing hole in mounting bracket. Then, slide mounting bracket from left axle (10).

Clean and inspect axle bearings (5) and renew as necessary. When reinstalling, flanges on outer axle bearings must face outward (against wheel hubs). Flange on inner axle bearing must be toward differential housing. Balance of reassembly is reverse of disassembly procedure. Lubricate axle bearings and drive chain with SAE 30 oil. Refer to appropriate DRIVE CHAIN paragraph and adjust chain tension.

Models 400-402-405-406-407-410-412 (Late Production)-412

REMOVE AND REINSTALL. To remove differential and rear axle assembly, raise rear of unit and support securely. Locate master link in drive chain and disconnect chain. Remove rear wheel assemblies. Refer to Fig. MT24 and unbolt axle bearing brackets (4, 8 and 9) from main frame. Remove differential and axle assembly, being

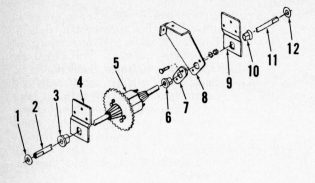

Fig. MT24 — Differential assembly, rear axles and axle bearings used on Models 400, 402, 405, 406, 407, 410 (late production) and 412.

1. Washer
2. Axle shaft R.H.
3. Axle bearing (outer)
4. Bearing bracket
5. Differential assy.
6. Axle bearing (center)
7. Bearing plate
8. Bearing bracket
9. Bearing bracket
10. Axle bearing (outer)
11. Axle shaft L.H.
12. Washer

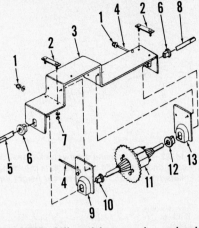

Fig. MT27 — Differential, rear axles and axle bearings used on Models 460, 465, 470, 475, 480 and 485.

1. Adjusting nuts
2. Mounting plates
3. Axle mounting bracket
4. Draw bolts
5. Axle shaft R.H.
6. Axle bearings (outer)
7. Locknut (4 used)
8. Axle shaft L.H.
9. Bearing bracket
10. Axle bearing
11. Differential assy.
12. Axle bearing
13. Bearing bracket

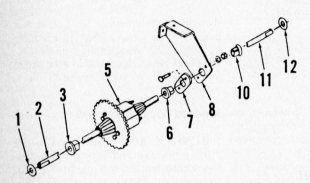

Fig. MT25 — Differential assembly used on Models 420, 425, 430 and 435.

1. Washer
2. Axle shaft R.H.
3. Axle bearing (outer)
5. Differential assy.
6. Axle bearing (center)
7. Bearing plate
8. Bearing bracket
10. Axle bearing (outer)
11. Axle shaft L.H.
12. Washer

careful not to damage brake caliper assembly. Clean all paint, rust or burrs from axle shafts and remove axle bearings and brackets.

Clean and inspect axle bearings and renew as necessary. When reinstalling, flanges on outer axle bearings (3 and 10) must face outward (against wheel hubs). Flange on inner axle bearing (6) must be toward differential housing. Lubricate axle bearings with SAE 30 oil. If drive chain requires adjustment, refer to DRIVE CHAIN paragraphs.

Model 410 (Early Production)

REMOVE AND REINSTALL. To remove rear axle and drive sprocket assembly, support rear of unit and remove rear wheel assemblies. Locate master link in drive chain and disconnect chain. Disconnect brake rod, then unbolt and remove brake caliper assembly. Withdraw axle bearings from each side of frame. Remove axle and sprocket assembly by sliding unit first to the left and then to the right until free of frame.

Clean and inspect all parts and renew any showing excessive wear or other damage. Drive sprocket and axle is serviced only as an assembly. Reassemble by reversing disassembly procedure. Lubricate drive chain and axle bearings with SAE 30 oil. If drive chain requires adjustment, refer to DRIVE CHAIN paragraphs.

Models 420-425-430-435

REMOVE AND REINSTALL. To remove differential and rear axle assembly, raise rear of unit and support securely. Unbolt and remove rear wheel assemblies. Locate master link in drive chain and disconnect chain. Disconnect brake rod, then unbolt and remove brake caliper assembly. Remove washers (1 and 12 – Fig. MT25) and outer axle bearings (3 and 10) from axles and frame. Unbolt bearing bracket (8) from frame, then remove differential and rear axle assembly by sliding unit first to the left and then to the right out of frame. Clean all paint, rust or burrs from left axle shaft (11) and slide center bearing (6) and bearing bracket from axle.

Clean and inspect axle bearings and renew as necessary. When reinstalling, flanges on outer axle bearings (3 and 10) must face outward (against wheel hubs). Flange on center axle bearing (6) must be toward differential housing. Lubricate axle bearings and drive chain with SAE 30 oil. If drive chain requires adjustment, refer to DRIVE CHAIN paragraphs.

Models 440-445

REMOVE AND REINSTALL. To remove differential and rear axle assembly, place mower lift and blade clutch lever in disengaged position. Remove belt keepers at engine pulley and slip mower drive belt from engine pulley. Raise rear of unit and support securely. Locate master link in drive chain and disconnect chain. Remove rear wheel assemblies and disconnect brake rod. Unbolt and remove left axle bearing and bracket (8 and 10 – Fig. MT26). Unbolt right and center axle bearing brackets (4 and 8) from main frame. Move differential assembly forward to clear brake caliper unit, then remove differential and rear axle assembly out toward right side. Clean all rust, paint or burrs from axle shafts and remove axle bearings and brackets.

Clean and inspect axle bearings and renew as necessary. When reinstalling, flanges on outer axle bearings (3 and 10) must face outward (against wheel hubs).

Fig. MT26 — Differential and rear axle assembly used on Models 440 and 445.

1. Washer
2. Axle shaft R.H.
3. Axle bearing (outer)
4. Bearing bracket
5. Differential assy.
6. Axle bearing (center)
8. Bearing bracket
9. Bearing bracket
10. Axle bearing (outer)
11. Axle shaft L.H.
12. Washer
13. Brake disc

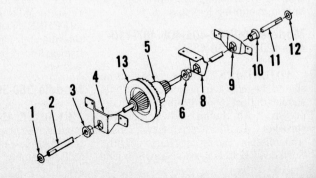

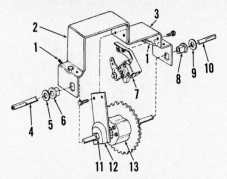

Fig. MT28 — Differential, rear axles and axle bearings used on Models 520 and 525.

1. Draw bolts
2. Axle mounting bracket (right half)
3. Axle mounting bracket (left half)
4. Axle shaft R.H.
5. Washer
6. Axle bearing (outer)
7. Brake caliper assy.
8. Axle bearing (outer)
9. Washer
10. Axle shaft L.H.
11. Axle bearing (center)
12. Bearing bracket
13. Differential assy.

Fig. MT29 — Exploded view of differential, rear axles and bearings used on Models 796 and 797.

1. Spacer
2. Mounting bracket
3. Axle bearing (outer)
4. Washer
5. Differential assy.
6. Brake assy.
7. Axle bearing (inner)
8. Support bracket
9. Brake rod
10. Drive chain
11. Chain idler
12. Idler arm

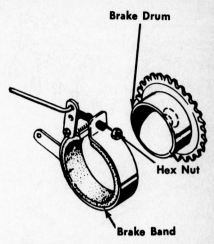

Fig. MT30 — Band type brake used on Models 360 and 362. Brake drum is bolted to drive chain sprocket.

Flange on center axle bearing (6) must be toward differential housing. Balance of reassembly is reverse of disassembly procedure. Lubricate drive chain and axle bearings with SAE 30 oil. If drive chain requires adjustment, refer to DRIVE CHAIN paragraphs.

Models 460-465-470-475-480-485

REMOVE AND REINSTALL. To remove differential and rear axle assembly, raise rear of unit and block securely. Locate master link in drive chain and disconnect chain. Remove locknuts (7 — Fig. MT27), remove adjusting nuts (1) and washers, then move axle mounting bracket (3) forward until draw bolts (4) clear main frame. Disconnect brake return spring. Raise rear of unit and roll differential and rear axle assembly with mounting bracket (3) and bearing brackets (9 and 13) rearward. Unbolt and remove rear wheel assemblies and remove outer axle bearings (6). Unbolt bearing brackets (9 and 13) and rotate bracket (13) around axle until it is clear of bracket (3). Slide axle and differential assembly toward left side until right axle shaft (5) is free of bearing hole in mounting bracket. Then, slide mounting bracket from left axle shaft (8). Clean all paint, rust or burrs from axle shafts and slide axle bearings (10 and 12) with brackets (9 and 13) from axle shafts.

Clean and inspect axle bearings and renew as necessary. When reinstalling, flanges on outer axle bearings must face outward (against wheel hubs). Flanges on inner axle bearings (10 and 12) must be toward differential housing. Balance of reassembly is the reverse of disassembly procedure. Lubricate drive chain and axle bearings with SAE 30 oil. Refer to appropriate DRIVE CHAIN paragraph and adjust chain tension.

Models 495-497-498-630-632-638-698-760-780-784-786-820

Transmission gears, shafts, differential and axle shafts are contained in one case and removed as an assembly. To remove transaxle assembly, refer to TRANSMISSION paragraphs.

Models 520-525

REMOVE AND REINSTALL. To remove differential and rear axle assembly, remove grass catcher, then raise rear of unit and block securely. Locate master link in drive chain and disconnect chain. Remove rear wheel assemblies, washers (5 and 9 — Fig. MT28) and outer axle bearings (6 and 8). Disconnect brake springs, then unbolt and remove brake caliper assembly (7). Remove adjusting nuts at rear of frame and locknuts from mounting plates at each side of frame. Unbolt bearing bracket (12) from axle mounting brackets, separate bracket halves (2 and 3) and withdraw differential and axle assembly out from left side. Remove paint, rust or burrs from axle shaft (4) and slide center bearing (11) with bracket (12) from axle.

Clean and inspect axle bearings and renew as necessary. When reinstalling, flanges on outer axle bearings (6 and 8) must face outward (against wheel hubs). Flange on center bearing (11) must be toward differential housing. Balance of reassembly is reverse of disassembly procedure. Lubricate drive chain and axle bearings with SAE 30 oil. Refer to appropriate DRIVE CHAIN paragraph and adjust chain tension.

Models 796-797

REMOVE AND REINSTALL. To remove differential and rear axle assem-

bly (5 — Fig. MT29), raise rear of unit and block securely. Disconnect drive chain at master link. Unbolt axle support bracket (8). Unbolt axle bearing brackets (2) and lower and remove differential assembly from frame. Remove rear wheels and spacers (1). Remove paint, rust or burrs from axle and slide bearings off axle.

Reinstall in reverse order of removal.

GROUND DRIVE BRAKE

All Models

The brake used on all models except 360 and 362 is caliper disc type. Band type brake is used on Models 360 and 362. Brake on all models is operated by pedal on right side of riding mower.

Models 360-362

ADJUSTMENT. To adjust brake, turn adjusting hex nut (Fig. MT30) clockwise ½-turn. Recheck brake operation and repeat adjustment if necessary.

R&R AND OVERHAUL. Normal overhaul consists of renewing brake band and lining assembly. To remove brake band, refer to Fig. MT30 and remove adjusting hex nut. Disconnect brake rod and unbolt band anchor arm. Slide band off brake drum. Renew brake band by reversing removal procedure and adjust brake as outlined in previous paragraph.

Models 380-385-390-395-400-402-405-406-407-410-412-420-425-430-435-440-445-460-465-470-475-480-485-520-525-796-797

ADJUSTMENT. To adjust brake, disengage traction drive clutch and depress brake pedal, then attempt to push unit.

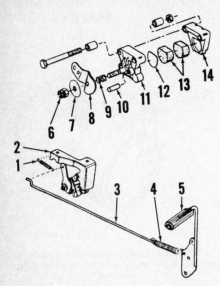

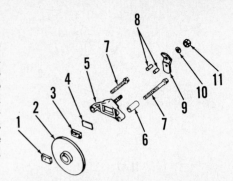

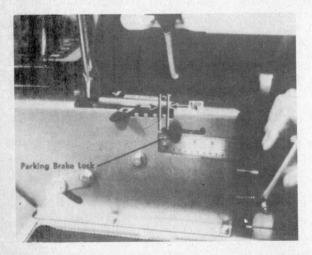

Models 495-497-498

ADJUSTMENT. To check brake adjustment, depress brake pedal by hand until slight resistance is felt. At this point, spring in brake linkage begins to stretch and parking brake lock should have moved approximately ¼-inch forward. See Fig. MT32. If adjustment is incorrect, tighten or loosen brake adjusting nut on actuating lever until correct dimension is obtained.

NOTE: DO NOT OVERTIGHTEN. Overtightening will reduce effective braking action.

R&R AND OVERHAUL. Disconnect brake return spring and brake rod from cam actuating lever (9–Fig. MT33). Unscrew cap screws (7), then remove brake parts (3 through 11). Remove adjusting nut (11) and separate cam lever (9), actuating pins (8), back-up plate (4) and outer brake pad (3) from carrier (5). Slide brake disc (2) from shaft and remove inner brake pad (1) from holder slot in transmission case.

Clean and inspect all parts for excessive wear or other damage. Renew parts as required and reassemble by reversing removal procedure.

Models 630-632-638-698

ADJUSTMENT. To adjust brake, loosen outside jam nut on disc brake actuating lever and tighten inside adjusting nut ½-turn. Check brake operation and repeat adjustment as needed. Retighten jam nut while holding adjusting nut.

R&R AND OVERHAUL. Disconnect brake spring and brake rod from actuating lever. Unbolt and remove brake pad holder. Slide brake disc off transaxle shaft and remove inner brake pad from holder slot in transaxle housing.

Clean and inspect parts and renew any showing excessive wear or other dam-

Fig. MT33—Exploded view of disc brake assembly used on Models 495, 497 and 498. Other models with separate brake disc (2) are similar.

1. Brake pad (inner)	
2. Brake disc	7. Cap screws
3. Brake pad (outer)	8. Adjusting pins
4. Back-up plate	9. Cam lever
5. Carrier	10. Washer
6. Spacer	11. Adjusting nut

age. Reinstall by reversing removal procedure.

Models 760-780-784-786-820

ADJUSTMENT. To adjust brake, loosen jam nut (Fig. MT34) and turn adjusting bolt all the way in, then unscrew bolt one turn. Tighten jam nut.

R&R AND OVERHAUL. Disconnect brake spring and brake rod from actuating lever. Unbolt and remove disc brake assembly from transaxle. Separate housings and remove brake pads.

Clean and inspect parts for excessive wear or damage and renew as necessary. Reinstall by reversing removal procedure.

MOWER DRIVE BELTS

Models 360-362

REMOVE AND REINSTALL. To remove mower drive belt, raise front of

Fig. MT31—Exploded view of typical caliper brake. Drive chain sprocket is also brake disc.

1. Return spring	8. Cam lever
2. Caliper assy.	9. Spring
3. Brake rod	10. Actuating pin (2)
4. Brake tension spring	11. Carrier (outer)
5. Brake pedal	12. Back-up plate
6. Adjusting nut	13. Brake pads
7. Washer	14. Carrier (inner)

If rear wheels rotate, release brake and tighten adjusting nut on cam lever ½-turn. Recheck brake and repeat adjustment as required.

R&R AND OVERHAUL. To remove brake caliper assembly, disconnect return spring (1–Fig. MT31), and brake rod (3). Unbolt mounting bracket from frame and remove caliper assembly. Remove adjusting nut (6), washer (7), cam lever (8), spring (9) and actuating pins (10). Remove through-bolts and separate brake pads (13) and back-up washer (12) from carriers (11 and 14).

Clean and inspect all parts for wear or other damage and renew as necessary. Reassemble by reversing removal procedure. Be sure actuator pins (10) are installed with rounded end toward cam lever (8).

Fig. MT32—On Models 495, 497 and 498, move brake pedal forward by hand until slight resistance is noted. At this time, parking brake lock should have moved about ¼-inch forward.

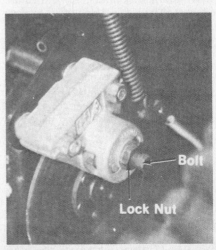

Fig. MT34—View of caliper brake and adjusting bolt used on Models 760, 780, 784, 786 and 820.

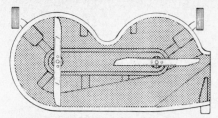

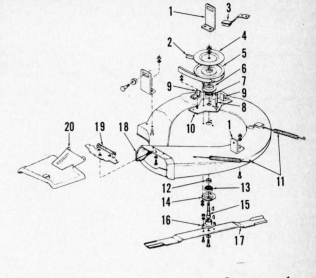

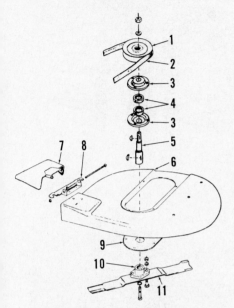

Fig. MT35 — Underside view of twin blade mower unit showing correct position of blades when properly timed.

Fig. MT37 — Exploded view of typical mower unit used on Models 380, 385, 400, 402, 405, 406, 407, 410 (early production), 420 and 425. On front engine models, tension springs (11) are connected to rear of mower deck.

1. Lift brackets
2. Mower drive belt
3. Blade brake pad
4. Blade brake disc
5. Mower pulley
6. Bearing housing
7. Ball bearing
8. Shoulder bolt
9. Belt keepers
10. Plate
11. Mower tension springs
12. Spacer
13. Ball bearing
14. Bearing housing
15. Blade spindle
16. Adapter
17. Blade
18. Mower housing
19. Deflector bracket
20. Deflector

Fig. MT36 — Exploded view of mower unit used on Models 360 and 362.

1. Mower pulley	7. Deflector
2. Mower drive belt	8. Deflector bracket
3. Bearing housings	9. Inspection plate
4. Ball bearings	10. Adapter
5. Blade spindle	11. Blade
6. Mower housing	

Fig. MT38 — Exploded view of typical twin blade mower used on Models 390, 395, 430, 435, 460, 465, 470, 480, 485, 495, 497, 498, 630, 632, 638 and 698.

1. Blade brake pad
2. Blade brake disc
3. Mower pulley
4. Mower drive belt
5. Bearing housings
6. Ball bearings
7. Blade spindle
8. Mower tension springs
9. Mower housing
10. Belt guards
11. Plate
12. Adapter
13. Blade
14. Deflector bracket
15. Deflector
16. Gage wheel
17. Pivot bar
18. Adjusting lever
19. Wheel bracket

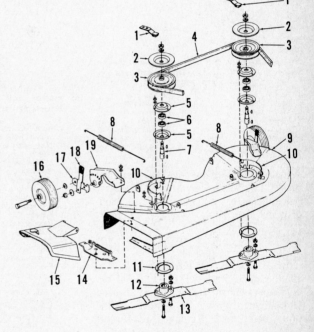

unit and block securely. Disconnect spark plug wire. Remove center bolt from mower blade and remove blade and adapter. Remove six hex nuts and lock-washers from bottom of mower housing and remove housing. Fully depress clutch pedal and engage lock. Spring belt guide away from engine pulley, then remove main drive belt from pulley. Remove belt guard from mower pulley. With blade clutch lever in disengaged position remove mower drive belt.

Install new belt by reversing removal procedure. Make certain belt guard and belt keepers are properly installed and belt alignment is correct.

Models 380-385-390-395-420-425-460-465-470-475-480-485-495-497-498-630-632-638-698-760-784-786

REMOVE AND REINSTALL. To remove mower drive belt, disconnect spark plug wire and proceed as follows: Place lift and blade clutch lever in dis-

engaged position. Remove belt keeper and shoulder bolt at engine pulley, then slip mower drive belt from engine pulley. Move lift and blade clutch lever fully forward in engaged position and unhook both tension springs from mower housing. Unpin lift arms from mower housing and remove mower unit from under right side. Unbolt and remove belt keepers from mower housing and remove mower drive belt.

Install new belt by reversing removal procedure. Make certain belt keepers and shoulder bolts are properly installed.

Models 400-402-405-406-407-410 (Early Production)

REMOVE AND REINSTALL. To remove mower drive belt, disconnect spark plug wire and proceed as follows: Place lift and blade clutch lever in en-

gaged position. Unbolt and remove belt keepers at engine pulley. Unbolt and remove belt guards, belt keepers or shoulder bolts at mower pulleys. Move lift and blade clutch lever to disengaged position and remove belt from pulleys.

Install new belt by reversing removal procedure. Make certain belt keepers, belt guards and shoulder bolts are properly installed.

Models 410 (Late Production)-412-520-525

REMOVE AND REINSTALL. To remove mower drive belt, disconnect spark plug wire and proceed as follows: Place blade clutch lever in disengaged position. Unbolt and remove belt guards or shoulder bolts at engine pulley. Remove blade pulley belt keeper. Unbolt and remove idler pulley, then remove belt from mower.

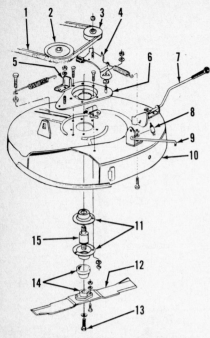

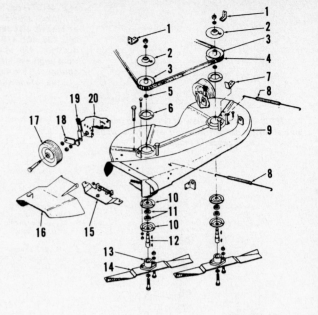

Fig. MT40—Exploded view of twin blade mower unit used on Models 440 and 445.

1. Blade brake pad
2. Blade brake disc
3. Mower pulley
4. Mower belt
5. Spacer
6. Plate
7. Belt keeper
8. Mower tension springs
9. Mower housing
10. Bearing housings
11. Ball bearings
12. Blade spindle
13. Adapter
14. Blade
15. Deflector bracket
16. Deflector
17. Gage wheel
18. Pivot bar
19. Adjusting lever
20. Wheel bracket

Fig. MT39—Exploded view of typical single blade mower unit used on Models 410 (late production), 412, 796 and 797.

1. Blade belt
2. Blade pulley
3. Idler pulley
4. Idler & brake arm
5. Belt keeper
6. Spindle mounting plate
7. Clutch lever
8. Safety switch
9. Stabilizer rod
10. Mower deck
11. Bearing housings
12. Blade
13. Center bolt
14. Blade adapter
15. Spindle & bearing assy.

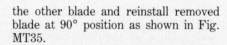

Fig. MT41—Exploded view of mower unit used on Models 520 and 525.

1. Rear discharge grass chute
2. Clutch arm & blade brake
3. Blade brake spring
4. Clutch idler pulley
5. Mower pulley
6. Belt keeper
7. Blade clutch control lever
8. Mower drive belt
9. Clutch tension spring
10. Control cable bracket
11. Spacer
12. Bearing housing
13. Ball bearing
14. Plate
15. Lift bracket
16. Mower housing
17. Spacer
18. Ball bearing
19. Bearing housing
20. Blade spindle
21. Adapter
22. Blade

Install new belt by reversing removal procedure. Make certain belt keepers, belt guards or shoulder bolts are properly installed.

Model 760

REMOVE AND REINSTALL. To remove deck drive belt (6 – Fig. MT42), disconnect spark plug wire and proceed as follows: Loosen or remove engine pulley belt guards. Remove three screws holding mower pulley upper belt guard (1A) to lower belt guard (1), then slip belt out of pulleys. To remove blade belt (3), remove belt guards from mower pulleys. Push idler pulley (5) towards center of deck and slip belt off idler pulley. Remove rear mounting bolt from right hand hanger bracket. Remove belt from pulleys and slip underneath hanger brackets.

Install new belts in reverse order of removal procedure.

Models 780-784-820

ADJUSTMENT. To adjust blade belt tension, loosen nuts on spindle plates (11 – Fig. MT43) on bottom of mower deck. Tighten adjusting nut (8) until spacer (9) on adjustment screw (10) is tight. Tighten spindle plate nuts.

TIMING BLADES. Remove blade retaining bolt from either blade. Rotate the other blade and reinstall removed blade at 90° position as shown in Fig. MT35.

REMOVE AND REINSTALL. Disconnect spark plug wire. Loosen or remove engine pulley belt guides. Remove blade drive pulley belt guard (5 – Fig. MT43) and slip deck drive belt (4) off pulleys. To remove blade drive belt (7), unbolt and remove deck drive pulley (6) and belt cover (3). Loosen spindle plate (11) nuts on bottom of mower deck. Loosen tension adjusting nut (8), then slip belt off pulleys.

Install new belts in reverse order of removal and adjust belt tension.

Models 796-797

REMOVE AND REINSTALL. To remove mower drive belt, remove spark plug wire and proceed as follows: Place blade clutch lever in engaged position. Loosen mower deck front mounting bolts and remove rear mounting bolts. Unbolt rear axle inner bearing support bracket. Loosen engine mounting bolts, then raise engine for clearance and slide support bracket and bearing away from engine pulley. Remove belt keeper and idler pulley from mower deck, then slip belt off pulleys.

Install new belt in reverse order of removal.

MOWER SPINDLE

All Models

R&R AND OVERHAUL. Spindle overhaul procedure is similar on all models. Refer to Figs. MT36 through MT43 for parts identification and ar-

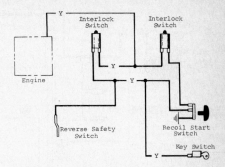

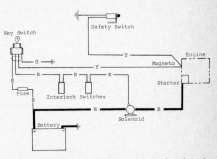

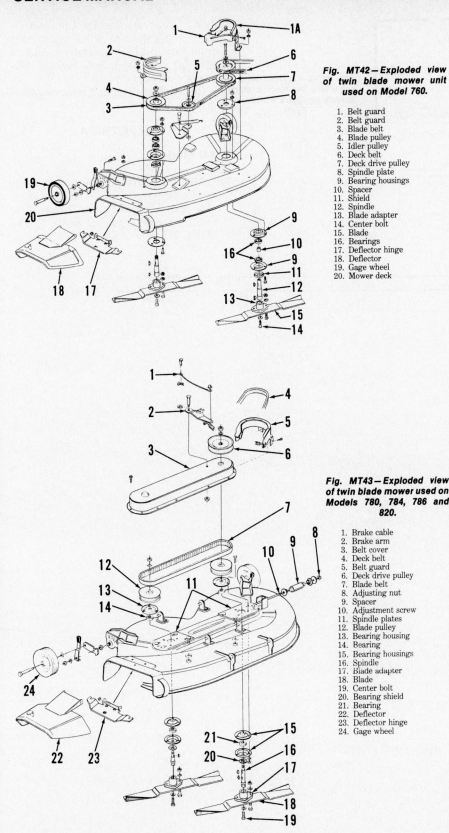

Fig. MT42 — Exploded view of twin blade mower unit used on Model 760.

1. Belt guard
2. Belt guard
3. Blade belt
4. Blade pulley
5. Idler pulley
6. Deck belt
7. Deck drive pulley
8. Spindle plate
9. Bearing housings
10. Spacer
11. Shield
12. Spindle
13. Blade adapter
14. Center bolt
15. Blade
16. Bearings
17. Deflector hinge
18. Deflector
19. Gage wheel
20. Mower deck

Fig. MT44 — Typical wiring schematic for recoil start models. "Y" wires are yellow.

Fig. MT45 — Typical wiring schematic for electric start models without headlights.

1. R-Red
2. G-Green
3. Y-Yellow

Fig. MT43 — Exploded view of twin blade mower used on Models 780, 784, 786 and 820.

1. Brake cable
2. Brake arm
3. Belt cover
4. Deck belt
5. Belt guard
6. Deck drive pulley
7. Blade belt
8. Adjusting nut
9. Spacer
10. Adjustment screw
11. Spindle plates
12. Blade pulley
13. Bearing housing
14. Bearing
15. Bearing housings
16. Spindle
17. Blade adapter
18. Blade
19. Center bolt
20. Bearing shield
21. Bearing
22. Deflector
23. Deflector hinge
24. Gage wheel

move and separate spindle, bearing housings and bearings.

Clean and inspect all parts and renew any showing excessive wear or other damage. Reassemble by reversing removal procedure. Torque blade center bolt to 35 ft.-lbs. and blade adapter bolts to 20 ft.-lbs. Spindle bearings are sealed and require no additional lubrication.

MOWER DECK

Models 360-362

REMOVE AND REINSTALL. Disconnect spark plug wire and raise front of unit and block securely. Remove blade center bolt and remove blade and adapter. Remove six hex nuts from bottom of mower housing and remove mower deck. Spring belt guide away from engine pulley and slip drive belt from pulley, then slide mower out right side.

Reinstall by reversing removal procedure.

All Other Models

REMOVE AND REINSTALL. Disconnect spark plug wire and raise front of unit and block securely. Unbolt and remove belt guides at engine pulley. Place blade clutch lever in disengaged position, then slip mower belt off engine pulley. Unhook mower deck tension springs and disconnect safety starting

rangement. To remove spindle, disconnect spark plug wire and remove mower as outlined in MOWER DECK paragraphs. If equipped with mower belt idler pulley, unbolt and remove pulley. On all models, unbolt and remove mower pulley belt guards, belt keepers and

shoulder bolts, then remove mower belt. Remove retaining nut or bolt and lift off mower pulley and blade brake disc (if equipped). Remove blade center bolt and remove blade and adapter from bottom of spindle. Remove bolts securing spindle bearing housings to mower deck. Re-

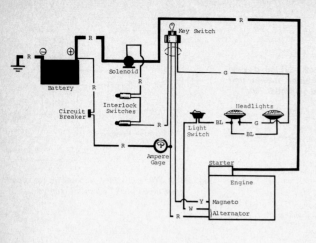

Fig. MT46 — Typical wiring schematic for electric start models equipped with headlights.

1. R-Red
2. G-Green
3. W-White
4. Y-Yellow
5. BL-Blue

switch wires on models so equipped. On Models 520 and 525 unbolt grass chute from mower deck. On all models, disconnect lift arms from deck brackets and remove mower.

Reinstall by reversing removal procedure.

ELECTRICAL

All Models

All models are equipped with safety starting interlock system to prevent engine from starting with traction or mower drives engaged. Refer to Fig. MT44, MT45 or MT46.

MURRAY

MURRAY OHIO MANUFACTURING CO.
P.O. Box 268
Brentwood, TN 37027

Model	Make	Engine Model	Horsepower	Cutting Width, In.
2503	B&S	130000	5	25
2513	B&S	170000	7	25
25501	B&S	130000	5	25
25502	B&S	170000	7	25
3013	B&S	170000	7	30
3033	B&S	170000	7	30
3043	B&S	190000	8	30
3063	B&S	190000	8	30
3233	B&S	190000	8	32
3235	B&S	190000	8	32
3633	B&S	190000	8	36
30501	B&S	170000	7	30
30502	B&S	190000	8	30
31501	B&S	190000	8	31
36503	B&S	250000	11	36
39001	B&S	250000	11	39

FRONT AXLE

Models 2503-2513-3013-3033-3043-3063

The axle member (18–Fig. M1) is of the non-pivoting type and is also the front frame member. To remove the axle member, raise front of unit and block securely. Remove front wheel assemblies and unbolt and remove tie rod (21). Remove cotter pins and washers from top of spindles (20 and 22), then remove the spindles. On electric start models, disconnect battery cables. On all models, loosen locknut and unscrew clutch and brake pedal pad(s). Unbolt and raise console, then unbolt and remove front apron on Models 2503, 2513, 3013 and 3033. On all models, disconnect clutch and brake rod(s), then unbolt and remove axle member.

Clean and inspect all parts and renew as necessary. Reassemble by reversing the disassembly procedure.

To remove the steering shafts, disconnect battery cables on electric start models. On Models 3043 and 3063, remove the lower bolt through sleeve (4–Fig. M1) and withdraw steering wheel, upper shaft, sleeve and coupling (1 thru 5) as an assembly. Remove second bolt through sleeve and separate sleeve and coupling from upper shaft. Drive out roll pin (2) and remove steering wheel from shaft. On Models 2503, 2513, 3013 and 3033, remove the screw from collar (8). Slide collar downward on sleeve (9) and remove the bolt through upper end of sleeve. Remove steering wheel (6) and sleeve (9) with collar (8) from upper shaft (7). Drive out the two roll pins and remove the upper shaft. On all models,

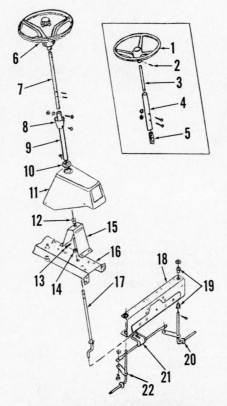

Fig. M1 – Exploded view of front axle and steering system used on Models 2503, 2513, 3013, 3033, 3043 and 3063. Items (1 through 5) in inset are used on Models 3043 and 3063.

1. Steering wheel
2. Roll pin
3. Steering shaft (upper)
4. Sleeve
5. Coupling
6. Steering wheel
7. Steering shaft (upper)
8. Collar
9. Sleeve
10. Nylon bearing
11. Console
12. Nylon bearing
13. Cotter pin
14. Washer
15. Steering support
16. Frame
17. Steering shaft (lower)
18. Axle (front frame) member
19. Nylon bearings
20. Spindle L.H.
21. Tie rod
22. Spindle R.H.

unbolt and remove tie rod (21). Unbolt and raise console (11), remove cotter pin (13), raise front of unit and withdraw lower steering shaft (17).

Clean and inspect all parts for excessive wear or other damage. Remove nylon bearings (10 and 12) and renew if necessary. Lubricate bearings with SAE 30 oil. Reassemble by reversing the disassembly procedure.

Models 3233-3235-3633

To remove the axle main member (24–Fig. M2), support front of unit and remove front wheel assemblies. Disconnect tie rod ends (19) from arms on spindles (22 and 25). Remove cotter pins and washers from top of spindles, then remove the spindles. Remove axle pivot bolt (23) and slide axle main member out from under side.

Clean and inspect all parts and renew any showing excessive wear or other damage. Reinstall by reversing the removal procedure. Lubricate pivot bolt and nylon spindle bearings with SAE 30 oil.

To remove the steering shafts and gears, first disconnect battery cables. Remove the lower bolt through sleeve (5) and withdraw steering wheel (1), collar (3), upper shaft (4), sleeve (5) and coupling (6) as an assembly. Remove screw from collar, slide collar downward on sleeve, drive out roll pin (2) and remove steering wheel. Remove the remaining bolt through sleeve and separate sleeve, shaft and coupling. Disconnect tie rods from lower steering shaft (18). Unbolt console (8) and lay console to the side, taking care not to damage electrical wiring. On Model 3235, remove cotter pin from lower

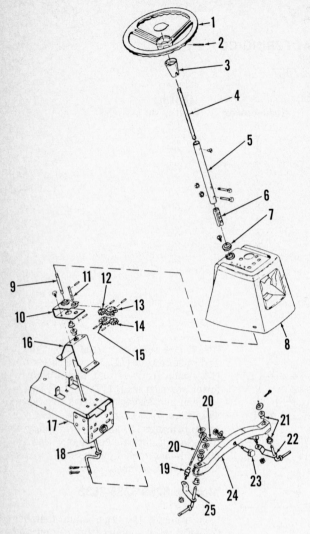

move front wheel assemblies (19). Disconnect tie rod ends (16) and linkage rod (7) from arms on spindles (15 and 17). Remove cotter pins from top of spindles, then remove spindles. Remove axle pivot bolt (20) and slide axle main member out from under side.

Clean and inspect all parts for excessive wear or any other damage and renew as needed. Reinstall by reversing the removal procedure. Lubricate pivot bolt and nylon spindle bearings with SAE 30 oil.

To remove the steering shafts and sector assembly, first disconnect battery cables on Model 30502. Remove lower bolt through steering shaft (2) and pinion shaft (4) and withdraw steering wheel (1) and steering shaft (2) as an assembly. Unbolt sector assembly (6) from axle mounting hanger (9) and axle support assembly (11), then withdraw unit.

Clean and inspect all parts for excessive wear or any other damage and renew as needed. Reassemble by reversing the disassembly procedure. Lubricate nylon bearings and lower steering shaft pivot points with SAE 30 oil. Apply a light coat of lithium grease to pinion gear (4) and sector assembly gear (6).

steering shaft above frame. Raise front of unit and withdraw steering shaft (18) from bottom of frame (17). On Models 3233 and 3633, drive roll pin from pinion gear (12) and remove extension shaft (9) and gear (12). Drive roll pin from sector gear (15), raise front of unit and withdraw steering shaft (18) from bottom of frame (17). Remove sector gear (15). Sector gear (13), pinion gear (14) and idler shaft (11) can be removed after driving out the remaining roll pins. Unbolt gear support (10) and remove nylon bearings from gear support and steering support (16).

Clean and inspect all parts and renew any showing excessive wear or other damage. Reassemble by reversing the disassembly procedure. Lubricate nylon bearings and lower steering shaft pivot points with SAE 30 oil. Apply a light coat of lithium grease to pinion gears (12 and 14) and sector gears (13 and 15).

Models 25501, 25502, 30501 and 30502

To remove axle main member (10 – Fig. M2A), support front of unit and re-

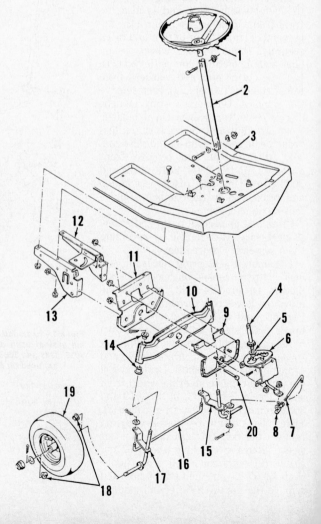

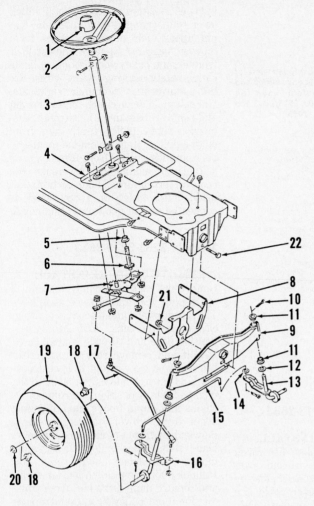

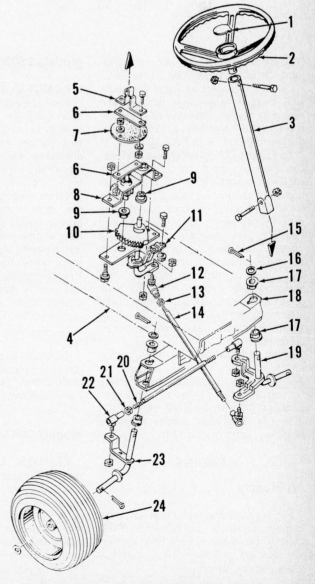

steering gear assembly (7) to frame assembly (4), then withdraw assembly.

Clean and inspect all parts for excessive wear or any other damage and renew as needed. Reassemble by reversing the disassembly procedure. Lubricate bearing (5) with SAE 30 oil. Apply a light coat of lithium grease to pinion gear (6) and steering gear (7).

Model 39001

Front axle frame (18–Fig. M2C) can be removed as a complete unit with removal of front axle hanger. Steering components may be removed individually. To remove steering components, raise front of unit and securely block. Remove front wheel assemblies (24). Remove nuts securing steering linkage rod ball joints (12 and 22) to steering gear arm (10) and spindle arms (19 and 23), then withdraw linkage rods. Remove cotter keys (15), then slide steering spindles (19 and 23) from axle frame

Fig. M2B—Exploded view showing front axle and steering components used on Models 31501 and 36503.

1. Cover
2. Steering wheel
3. Steering post
4. Frame
5. Bearing
6. Pinion gear & shaft
7. Steering gear assy.
8. Axle support
9. Axle main member
10. Cotter key
11. Bearing
12. Washer
13. Spindle assy., L.H.
14. Washer
15. Tie rod
16. Spindle assy., R.H.
17. Drag link assy.
18. Bearing
19. Wheel assy.
20. Washer
21. Nut
22. Bolt

Models 31501-36503

To remove axle main member (9–Fig. M2B), support front of unit and remove front wheel assemblies (19). Disconnect drag link (17) from steering arm on steering gear assembly (7). Remove mounting bolts securing axle support (8). Remove nut (21) from mounting bolt (22). Remove axle support (8), then while supporting axle member (9) withdraw mounting bolt (22). Lower axle assembly clear of tractor frame.

Complete disassembly of unit with reference to Fig. M2B. Clean and inspect all parts for excessive wear or any other damage and renew as needed. Reinstall by reversing the removal procedure. Lubricate spindle bearings (11) with SAE 30 oil.

To remove the steering post and sector assembly, first disconnect battery cables from battery post. Remove lower bolt through steering post (3) and pinion shaft (6), then remove any other components that will obstruct steering post removal. Lift steering wheel (2) and steering post (3) out as an assembly. Disconnect drag link (17) from steering arm on steering gear assembly (7). Remove three mounting bolts securing

Fig. M2C—Exploded view showing front axle and steering components used on Model 39001.

1. Cover
2. Steering wheel
3. Steering post
4. Frame
5. Upper steering coupling
6. Coupling plate
7. Steering disc
8. Steering gear bracket assy.
9. Bearing
10. Steering gear assy.
11. Steering bracket
12. Ball joint
13. Locknut
14. Drag link sleeve
15. Cotter key
16. Washer
17. Bearing
18. Axle frame
19. Spindle assy., L.H.
20. Tie rod sleeve
21. Locknut
22. Ball joint
23. Spindle assy., R.H.
24. Wheel assy.

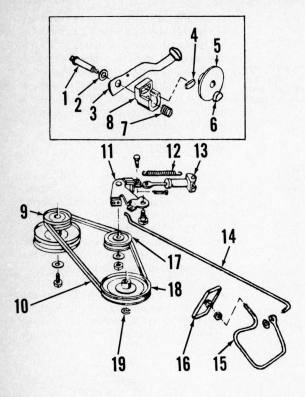

Fig. M3—Exploded view of typical main drive clutch and brake system used on Models 2503, 2513, 3013 and 3033.

1. Shoulder bolt
2. Washer
3. Parking brake cam lever
4. Brake pad (outer)
5. Brake disc
6. Brake pad (inner)
7. Spring
8. Carrier
9. Engine pulley
10. Traction drive belt
11. Clutch idler arm
12. Clutch tension spring
13. Air cylinder
14. Clutch rod
15. Pedal
16. Pedal pad
17. Clutch idler pulley
18. Transmission input pulley
19. Retaining ring

(18). Remove front axle hanger mounting bolts to withdraw axle frame (18).

Clean and inspect all parts for excessive wear or any other damage and renew as needed. Reinstall by reversing the removal procedure. Lubricate spindle bearings (17) with SAE 30 oil.

To remove the steering post and sector assembly, first disconnect battery cables from battery post. Remove lower bolt through steering post (3) and upper steering coupling (5), then remove any other components that will obstruct steering post removal. Lift steering wheel (2) and steering post (3) out as an assembly. Disconnect drag link arm (14) from steering arm on steering gear assembly (10). Remove mounting bolts securing steering gear assembly (10) and steering gear bracket assembly (8) to frame assembly (4), then withdraw assembly.

Clean and inspect all parts for excessive wear or any other damage and renew as needed. Reassemble by reversing the disassembly procedure. Lubricate bearings (9) with SAE 30 oil. Apply a light coat of lithium grease to pinion gear and steering gear (10).

ENGINE

All Models

For overhaul and repair procedures on engines listed in Specification Table refer to Small Air Cooled Engines Service Manual or Large Air Cooled Engines Service Manual.

Models 2503-2513-3013-3033

REMOVE AND REINSTALL. To remove the engine assembly, disconnect spark plug wire, ignition wire and throttle control cable. On Model 3033, disconnect battery cables, starter cable and alternator wires. On all models, place mower unit in lowest position and move blade clutch lever to disengaged (stop) position. Disconnect lift chains and unpin upper ends of scissor arms from mower hangers. Remove mower belt from engine pulley and disconnect blade clutch safety starting switch wires. Remove mower unit from under right side. Unhook main drive clutch tension spring and remove main drive belt from transmission input pulley first, then from engine pulley. Remove engine mounting bolts and lift engine assembly from frame.

Reinstall engine by reversing the removal procedure. Make certain that blade clutch safety starting switch is connected and in good operating condition.

Models 3043-3063

REMOVE AND REINSTALL. To remove the engine assembly, disconnect spark plug wire and proceed as follows: On Model 3043, unbolt and remove the seat. On Model 3063, tilt seat forward and disconnect battery cables, starter cable and alternator wires. On both models, disconnect ignition wire and throttle control cable. Place mower unit in lowest position and move blade clutch lever to disengaged (stop) position. Unpin upper ends of scissor arms from mower hangers. Remove mower belt from engine pulley and disconnect blade clutch safety starting switch wires. Remove mower unit from under right side. Unhook main drive clutch tension spring and remove main drive belt from transmission input pulley first, then from engine pulley. Remove engine mounting bolts and lift engine from frame.

Reinstall engine by reversing the removal procedure. Make certain that blade clutch safety starting switch is connected and in good operating condition.

Models 3233-3235-3633

REMOVE AND REINSTALL. To remove the engine assembly, disconnect spark plug wire and proceed as follows: Tilt seat forward and disconnect battery cables, starter cable, alternator wires, ignition wire and throttle control cable. Unbolt and remove rear shroud crossbrace. Place mower unit in lowest position and move blade clutch lever to disengaged (stop) position. Disconnect lift chains and unpin upper ends of scissor arms from mower hangers. Remove mower belt from engine pulley, remove blade clutch idler pulley, then remove mower unit from under right side. Unhook main drive clutch tension spring and remove main drive belt from transmission input pulley first, then from engine pulley. Unbolt and remove engine assembly from frame.

Reinstall engine by reversing the removal procedure.

Models 25501, 25502, 30501 and 30502

REMOVE AND REINSTALL. To remove engine assembly, first remove all protective shields and obstructing parts as needed. Disconnect spark plug wire and throttle cable. On Model 30502, disconnect battery cables and on all models disconnect engine wiring. Mark engine wiring as needed for reassembly. Lower mower deck as needed to allow access to drive belts. Remove traction drive belt and blade drive belt. Unbolt engine from frame mounting assembly, then withdraw engine.

Reinstall engine by reversing the removal procedure. Reattach linkage and cables and check for correct operation. Complete reassembly in reverse order of disassembly.

Models 31501-36503-39001

REMOVE AND REINSTALL. To remove engine assembly, disconnect spark plug wire and proceed as follows:

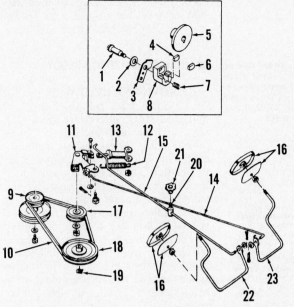

Fig. M4 — Exploded view of typical main drive clutch and brake system used on Models 3043, 3063, 3233, 3235 and 3633.

Fig. M4 — Exploded view of typical main drive clutch and brake system used on Models 3043, 3063, 3233, 3235 and 3633.

1. Shoulder bolt
2. Washer
3. Cam lever
4. Brake pad (outer)
5. Brake disc
6. Brake pad (inner)
7. Spring
8. Carrier
9. Engine pulley
10. Traction drive belt
11. Clutch idler arm
12. Clutch tension spring
13. Air cylinder
14. Clutch rod
15. Brake rod
16. Pedal pad
17. Clutch idler pulley
18. Transmission input pulley
19. Retaining ring
20. Park brake lock
21. Knob
22. Brake pedal
23. Clutch pedal

Remove engine cover assembly. Disconnect battery cables, starter cable, alternator wires, ignition wire and throttle control cable. Mark engine wiring as needed for reassembly. Lower mower deck as needed to allow access to drive belts. Remove traction drive belt and blade drive belt. Unbolt engine from frame mounting assembly, then withdraw engine.

Reinstall engine by reversing the removal procedure. Reconnect electrical wires and throttle cable, then check for correct operation. Complete reassembly in reverse order of disassembly.

TRACTION DRIVE CLUTCH AND DRIVE BELT

Rear Engine Models

The traction drive clutch used on all models is of the belt idler type. On Models 2503, 2513, 3013 and 3033 clutch is operated by a pedal on right side. On all other models, the clutch pedal is on the left side. On all models when clutch pedal is fully depressed, all tension is removed from drive belt (10 – Fig. M3, M4 or M4A) and engine pulley (9) is allowed to rotate freely within the belt. At this time, on models so equipped a brake pad on clutch idler arm contacts the transmission input pulley. This transmission braking action will assist in stopping.

CAUTION: This brake is ineffective when transmission is in neutral position.

On Models 25501, 25502, 30501 and 30502 adjust drive clutch by turning adjustment nut (11 – Fig. M4A). If complete adjustment is used and belt slippage still occurs, then belt must be renewed. On all other models there is no adjustment on the traction drive clutch.

If the drive belt slips during normal operation due to excessive belt wear or stretching, renew the belt as outlined under the following REMOVE AND RENEW paragraphs.

Models 2503-2513-3013-3033-3043-3063

REMOVE AND RENEW. To remove the traction drive belt (10 – Fig. M3 or M4), disconnect spark plug wire and proceed as follows: Place mower unit in lowest position and move blade clutch lever to disengaged (stop) position. On Models 2503, 2513, 3013 and 3033, disconnect the lift chains. On all models, unpin upper ends of scissor arms from mower hangers. Remove mower belt from engine pulley and disconnect blade clutch safety starting switch wires. Remove mower unit from under right side. Unhook main drive clutch tension spring (12) and remove clutch idler pulley (17). Remove traction drive belt from transmission input pulley (18) first, then from engine pulley (9).

Install new belt by reversing the removal procedure. Make certain that blade clutch safety starting switch is connected and in good operating condition.

Models 3233-3235-3633

REMOVE AND RENEW. To remove the traction drive belt (10 – Fig. M4), disconnect spark plug wire and proceed as follows: Place mower unit in lowest position and move blade clutch lever to disengaged (stop) position. Disconnect lift chains and unpin upper ends of scissor arms from mower hangers. Remove the blade clutch idler pulley, slip mower belt from engine pulley and remove mower unit from under right side. Unhook main drive clutch tension spring (12) and remove clutch idler pulley (17). Remove traction drive belt from transmission pulley (18) first, then from engine pulley (9).

Install new belt and reinstall mower unit by reversing the removal procedure.

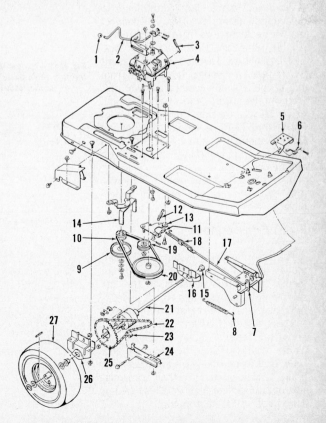

Fig. M4A — Exploded view of traction drive parts used on Models 25501, 25502, 30501 and 30502.

1. Knob
2. Shift lever
3. Spring
4. Transmission assy.
5. Pedal
6. Spring
7. Lever assy.
8. Spring
9. Engine pulley assy.
10. Drive belt
11. Adjustment nut
12. Spring
13. Idler arm assy.
14. Belt guide assy.
15. Axle bearing
16. Axle mounting bracket
17. Linkage rod
18. Spring
19. Idler pulley
20. Transmission drive pulley
21. Differential
22. Chain
23. Roller
24. Chain adjustment bracket
25. Sprocket
26. Axle mounting bracket
27. Wheel assy.

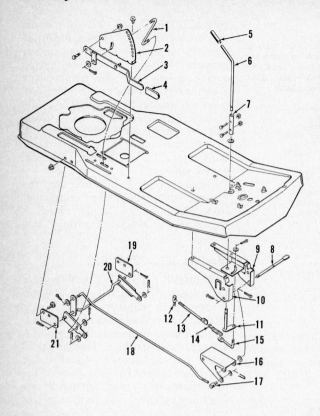

occurs, then belt must be renewed. Renew drive belt as outlined in the following REMOVE AND RENEW paragraphs.

Models 31501-36503-39001

REMOVE AND RENEW. To remove the traction drive belt (6–Fig. M4C, M4D and M4E), disconnect spark plug wire and proceed as follows: Place mower unit in lowest position and move blade clutch lever to disengaged (stop) position. Disconnect mower deck hangers from deck lifter assembly and front hanger bracket. Disconnect mower clutch idler engagement rod, then remove mower belt from engine pulley and remove mower unit. Loosen tension on traction drive belt idler pulley and remove all belt guides and protective shields as needed to allow access to drive belt. Remove drive belt from drive pulleys.

Inspect all drive pulleys for excessive wear or any other damage and renew as needed. Install new drive belt and reinstall mower unit by reversing the removal procedure.

MOWER CLUTCH AND DRIVE BELT

Rear Engine Models

Models 2503, 2513, 25501, 25502, 30501 and 30502 are equipped with

Fig. M4B—Exploded view of mower deck control components and blade engagement components for Models 25501, 25502, 30501 and 30502.

1. Lift rod
2. Index plate
3. Lift lever
4. Grip
5. Grip
6. Blade lever
7. Tube
8. Pin
9. Left suspension plate
10. Right suspension plate
11. Rod and arm assy.
12. Adjustment nut
13. Linkage rod
14. Spring
15. Front link
16. Front suspension bracket
17. Adjustment nut
18. Linkage rod
19. Left mounting plate
20. Deck lifter assy.
21. Right mounting plate

Models 25501, 25502, 30501 and 30502

REMOVE AND RENEW. To remove the traction drive belt (10–Fig. M4A), disconnect spark plug wire and proceed as follows: Place mower unit in lowest position and move blade clutch lever to disengaged (stop) position. Disconnect mower deck hangers from deck lifter assembly (20–Fig. M4B) and front suspension plates. Disconnect adjustment nut (12), then remove mower belt from engine pulley and remove mower unit from under right side. Unhook main drive clutch tension spring (12–Fig. M4A) and remove clutch idler pulley (19). Remove traction drive belt from transmission pulley (20) first, then from engine pulley (9).

Install new belt and reinstall mower unit by reversing the removal procedure.

Front Engine Models

The traction drive clutch used on all models is of the belt idler type. On all models the clutch pedal is on the left side. On all models when clutch pedal is fully depressed, all tension is removed from drive belt (6–Fig. M4C, M4D and M4E) and engine pulley is allowed to rotate freely within the belt. At this time, brake pads come into contact with brake disc located on either transmission or transaxle assembly. This braking action will assist in stopping.

CAUTION: This brake is ineffective when transmission is in neutral position.

Adjustment of drive clutch is done by turning adjustment nut (27). If complete adjustment is used and belt slippage still

Fig. M4C—Exploded view of traction drive parts used on Model 31501.

1. Shift yoke
2. Shift lever
3. Knob
4. Transmission assy.
5. Spring
6. Drive belt
7. Frame
8. Idler bracket assy.
9. Hairpin clip
10. Spring
11. Transmission input pulley
12. Idler pulley
13. Chain idler bracket
14. Axle bearing
15. Bearing plate
16. Chain
17. Roller
18. Master chain link
19. Sprocket
20. Differential assy.
21. Wheel assy.
22. Brake link assy.
23. Brake rod
24. Clutch rod
25. Clutch & brake pedal
26. Lever assy.
27. Adjusting nut
28. Spring

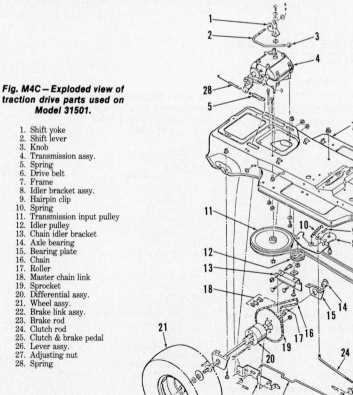

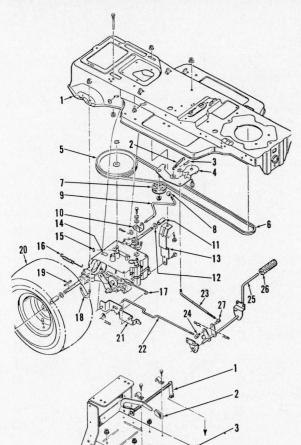

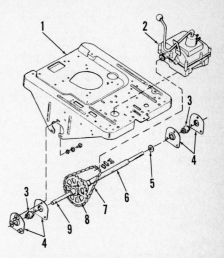

Fig. M4D — Exploded view of traction drive parts used on Model 36503.

1. Frame
2. Spring
3. Hairpin clip
4. Idler bracket assy.
5. Transaxle input pulley
6. Drive belt
7. Idler pulley
8. Knob
9. Shift lever
10. Shift yoke
11. Axle spacer
12. Washer
13. Transaxle support
14. Transaxle assy.
15. Woodruff key
16. Spring
17. Spring
18. "U" bolt
19. Square key
20. Wheel assy.
21. Brake link assy.
22. Brake rod
23. Clutch rod
24. Adjusting nut
25. Lever assy.
26. Clutch & brake pedal
27. Adjusting nut

Fig. M5 — View of typical transmission, drive chain, differential, rear axles and axle bearings used on all models.

1. Rear frame section
2. Transmission
3. Axle bearings
4. Bearing retainers
5. Washer
6. Sleeve
7. Drive chain
8. Sprocket
9. Rear axle & differential assy.

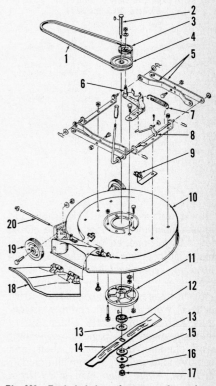

Fig. M4E — Exploded view of traction drive parts used on Model 39001.

1. Handle assy.
2. Grip
3. Frame
4. Idler bracket assy.
5. Transaxle bracket
6. Drive belt
7. Clutch rod
8. Parking brake arm
9. Belt guide
10. Engine pulley
11. Idler pulley
12. Clutch idler pulley
13. Idler bracket
14. Spring
15. Backside idler pulley
16. Pulley
17. Transaxle bracket
18. Belt guide
19. Adjusting nut
20. Adjusting rod
21. Spring
22. Link plate
23. Adjusting nut
24. Brake rod
25. Lever assy.
26. Clutch & brake pedal
27. Adjusting nut
28. Grip
29. Disc brake assy.
30. Brake disc
31. Knob
32. Transaxle assy.
33. Extruded washer
34. Square key
35. Wheel hub
36. "E" clip
37. Wheel assy.

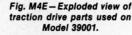

Fig. M6 — Exploded view of mower unit used on Models 2503 and 2513.

1. Mower drive belt
2. Blade shaft
3. Clutch idler pulley
4. Mower pulley
5. Scissor arms
6. Blade clutch idler arm
7. Clutch tension spring
8. Clutch lever & linkage
9. Safety starting switch
10. Mower housing
11. Bearing & housing assy.
12. Blade adapter
13. Cushion washers
14. Blade
15. Flat washer
16. Shake proof washer
17. Nut
18. Deflector
19. Gage wheel
20. Pin

single blade rotary mowers. All other models are equipped with twin blades.

CAUTION: Always disconnect spark plug wire before performing any inspection, adjustment or other service on the mower.

Make certain that safety starting switches are connected and in good operating condition before returning mower to service.

On all models a belt idler type blade clutch is used. On Models 25501, 25502, 30501 and 30502 adjust drive clutch by turning adjustment nut (12 – Fig. M4B).

If complete adjustment is used and belt slippage still occurs, then belt must be renewed. On all other models the clutch idler is spring loaded and requires no adjustment. If mower drive belt slips during normal operation, due to excessive belt wear or stretching, renew belt as outlined in the following paragraphs.

Models 2503-2513

REMOVE AND RENEW. To remove the mower drive belt (1 – Fig. M6), disconnect spark plug wire and remove

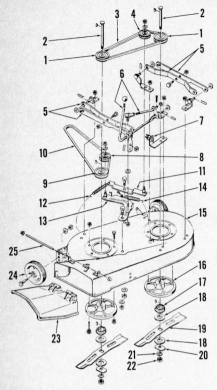

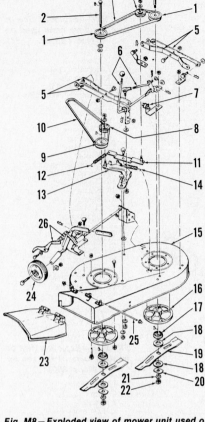

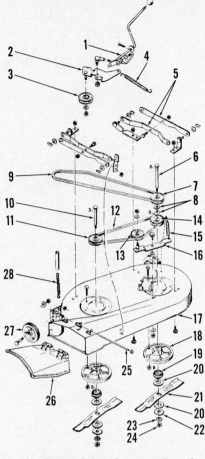

Fig. M7 — Exploded view of mower unit used on Models 3013 and 3033.

1. Mower pulleys	13. Blade clutch idler arm
2. Blade shafts	14. Clutch tension spring
3. Blade cross belt	15. Mower housing
4. Idler pulley	16. Bearing & housing assy.
5. Scissor arms	17. Blade adapter
6. Clutch lever & linkage	18. Cushion washers
7. Safety starting switch	19. Blade
8. Clutch idler pulley	20. Flat washer
9. Mower drive pulley	21. Lockwasher
10. Mower drive belt	22. Nut
11. Idler arm	23. Deflector
12. Idler spring	24. Gage wheel
	25. Pin

Fig. M8 — Exploded view of mower unit used on Models 3043 and 3063.

1. Mower pulleys	14. Clutch tension spring
2. Blade shafts	15. Mower housing
3. Blade cross belt	16. Bearing & housing assy.
4. Idler pulley	17. Blade adapter
5. Scissor arms	18. Cushion washers
6. Clutch lever & linkage	19. Blade
7. Safety starting switch	20. Flat washer
8. Clutch idler pulley	21. Lockwasher
9. Mower drive pulley	22. Nut
10. Mower drive belt	23. Deflector
11. Idler arm	24. Gage wheel
12. Idler spring	25. Pin
13. Blade clutch idler arm	26. Mower height adjuster

Fig. M9 — Exploded view of mower unit used on Models 3233, 3235 and 3633.

1. Clutch lever & linkage	14. Mower pulley
2. Blade clutch idler arm	15. Idler spring
3. Clutch idler pulley	16. Idler arm
4. Clutch tension spring	17. Mower housing
5. Scissor arms	18. Bearing & housing assy.
6. Blade clutch idler arm	19. Blade adapter
7. Mower drive pulley	20. Cushion washers
8. Spacer washers	21. Blade
9. Mower drive belt	22. Flat washer
10. Blade shaft	23. Lockwasher
11. Mower pulley	24. Nut
12. Blade cross belt	25. Pin
13. Idler pulley	26. Deflector
	27. Gage wheel
	28. Lift chain

mower unit as follows: Place mower in lowest position and move blade clutch lever to disengaged position and move blade clutch lever to disengaged (stop) position. Disconnect lift chains and unpin upper ends of scissor arms (5) from mower hangers. Remove mower drive belt (1) from engine pulley and disconnect wires from blade clutch safety starting switch (9). Remove mower unit from under right side. Unbolt and remove idler pulley (3) from clutch idler arm (6). Move idler arm away from mower pulley (4) and remove belt (1).

Install new belt by reversing the removal procedure. Make certain that blade clutch safety starting switch is connected and in good operating condition.

Models 3013-3033-3043-3063

REMOVE AND RENEW. To remove the mower drive belt (10 – Fig. M7 or M8) and blade cross belt (3), disconnect spark plug wire and remove mower unit as follows: Place mower in lowest position and move blade clutch lever to disengaged (stop) position. On Models 3013 and 3033, disconnect lift chains. On all models, unpin scissor arms (5) from mower hangers. Remove mower drive belt (10) from engine pulley and disconnect wires from blade clutch safety starting switch (7). Remove mower unit from under right side. Unhook idler spring (12) and remove blade cross belt (3). Remove nut, washer and idler pulley (8) from clutch idler arm (11). Remove mower drive belt (10).

Install new belts by reversing the removal procedure. Make certain that blade clutch safety starting switch is connected and in good operating condition.

Models 3233-3235-3633

REMOVE AND RENEW. To remove the mower drive belt (9 – Fig. M9) and blade cross belt (12), disconnect spark plug wire and remove mower unit as follows: Place mower unit in lowest position and move blade clutch lever to disengaged (stop) position. Disconnect lift chains (28) and unpin upper ends of scissor arms (5) from mower hangers. Remove nut, washer and idler pulley (3) from clutch idler arm (2). Slip mower drive belt (9) from engine pulley and remove mower unit from under right side. Remove mower drive belt, unhook idler spring (15) and remove blade cross belt (12).

Install new belts by reversing the removal procedure.

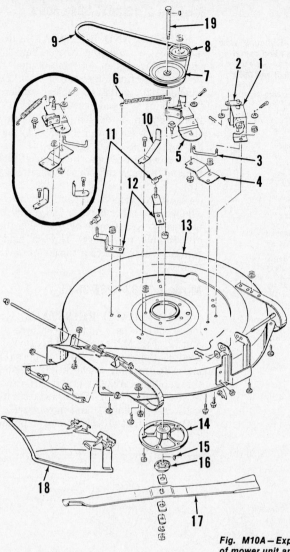

Fig. M10 – Exploded view of mower unit and operating components on Models 25501, 25502, 30501 and 30502. Parts shown in inset apply to Models 25501 and 25502.

1. Bracket assy.
2. Pivot rod
3. Idler link
4. Idler support bracket
5. Idler assy.
6. Spring
7. Pulley
8. Idler pulley
9. Drive belt
10. Belt guide
11. Adjustment nut
12. Rear bracket
13. Deck assy.
14. Bearing & housing assy.
15. Woodruff key
16. Blade adapter
17. Blade
18. Deflector chute
19. Blade shaft

Models 25501, 25502, 30501 and 30502

REMOVE AND RENEW. To remove mower drive belt (9-Fig. M10), disconnect spark plug wire and proceed as follows: Place mower unit in lowest position and move blade clutch lever to disengaged (stop) position. As needed disconnect mower deck hangers from deck lifter assembly (20 – Fig. M4B) and front suspension plates. Disconnect adjustment nut (12), then remove mower belt from engine pulley, idler pulley and mower pulley.

Install new belt and reinstall mower unit by reversing the removal procedure.

Front Engine Models

Model 31501 is equipped with a single blade rotary mower. Models 36503 and 39001 are equipped with twin blade rotary mowers.

CAUTION: Always disconnect spark plug wire before performing any inspec-

Fig. M10A – Exploded view of mower unit and operating components used on Model 31501.

1. Adjusting rod
2. Adjusting nut
3. Lifter bracket
4. Pulley cover
5. Jackshaft
6. Pulley
7. Pulley
8. Idler pulley
9. Drive belt
10. Control rod
11. Adjusting nut
12. Pivot lever
13. Spring
14. Spacer
15. Rear support bracket
16. Adjusting nut
17. Deck link
18. Belt guide
19. Link retainer
20. Brake & idler assy.
21. Spring
22. Belt guide
23. Skid flat stock, R.H.
24. Skid flat stock, L.H.
25. Idler mount assy.
26. Housing assy.
27. Chute deflector
28. Jackshaft housing assy.
29. Woodruff key
30. Blade adapter
31. Washer
32. Blade
33. Washer
34. Blade washer
35. Lockwasher
36. Nut
37. Hairpin clip
38. Hanger rod
39. Hanger

tion, adjustment or other service on the mower.

Make certain that safety starting switches are connected and in good operating condition before returning mower to service.

On all models a belt idler type blade clutch is used. Adjust drive clutch by turning adjustment nut (11 – Fig. M10A and M10B). If adjustment limit is reached and belt slippage still occurs, then belt must be renewed. Renew drive belt as outlined under the following REMOVE AND RENEW paragraphs.

Models 31501-36503-39001

To remove mower drive belt (9 – Fig. M10A and M10B), disconnect spark plug wire and proceed as follows: Place mower unit in lowest position and move blade clutch lever to disengaged (stop) position. Disconnect mower deck hangers from deck lifter assembly and front hanger bracket. Disconnect mower clutch idler engagement rod, then remove mower belt from engine pulley and remove mower unit.

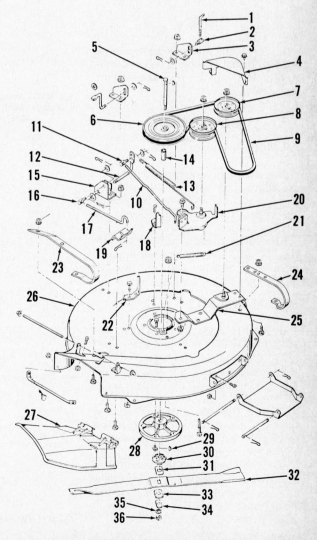

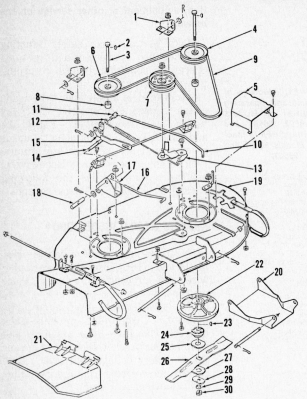

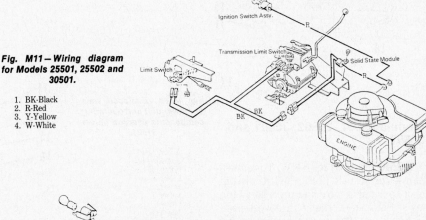

Models 3013-3033-3043-3063

REMOVE AND REINSTALL. Remove mower unit and mower belts (3 and 10 – Fig. M7 or M8) as outlined in previous section. Remove nuts (22), washers (20 and 21), cushion washers (18), blades (19) and blade adapters (17) from bottom of shafts (2). Withdraw shafts (2) with mower pulleys (1) and drive pulley (9) from above. Note location of belt guide bolt, then unbolt and remove bearing and housing assemblies (16).

Clean and inspect all parts for excessive wear or other damage and renew as necessary. Reassemble by reversing the disassembly procedure. Tighten blade retaining nuts (22) to a torque of 15-18 ft.-lbs.

Models 3233-3235-3633

REMOVE AND REINSTALL. Remove mower unit and mower belts (9 and 12 – Fig. M9) as outlined in previous section. Remove nuts (24), washers (22 and 23), cushion washers (20), blades (21) and blade adapters (19) from bottom of shafts (6 and 10). Withdraw shaft (6) with drive pulley (7) and mower pulley (14), then shaft (10) with mower pulley

Fig. M10B – Exploded view of mower unit and operating components used on Models 36503 and 39001.

1. Lifter bracket
2. Woodruff key
3. Jackshaft
4. Pulley
5. Pulley cover
6. Pulley
7. Idler pulley
8. Spacer
9. Drive belt
10. Control rod
11. Adjusting nut
12. Spring
13. Idler bracket assy.
14. Pivot lever
15. Brake bar
16. Link rod
17. Rear hanger bracket
18. Adjusting bar
19. Housing assy.
20. Front hanger
21. Chute deflector
22. Jackshaft housing assy.
23. Woodruff key
24. Blade adapter
25. Blade washer
26. Blade
27. Blade washer
28. Washer
29. Lockwasher
30. Nut

Inspect all drive pulleys for excessive wear or any other damage and renew as needed. Install new drive belt and reinstall mower unit by reversing the removal procedure.

MOWER DECK

All Models

The following paragraphs outline the procedures for removing the blades, blade shafts and bearing and housing assemblies. Bearings and housings are available only as assemblies.

Lubricate all linkage pivot points with SAE 30 oil. All idler pulley bearings and blade shaft bearings are pre-lubricated and require no lubrication.

Models 2503-2513

REMOVE AND REINSTALL. Remove mower unit and mower drive belt (1 – Fig. M6) as outlined in previous section. Remove nut (17), washers (15 and 16) cushion washers (13), blade (14) and adapter (12) from bottom of shaft (2). Withdraw shaft (2) with mower pulley (4) from above. Note location of belt guide bolt, then unbolt and remove bearing and housing assembly (11).

Clean and inspect all parts and renew as necessary. Reassemble by reversing the disassembly procedure. Tighten blade retaining nut (17) to a torque of 15-18 ft.-lbs.

Fig. M11 – Wiring diagram for Models 25501, 25502 and 30501.

1. BK-Black
2. R-Red
3. Y-Yellow
4. W-White

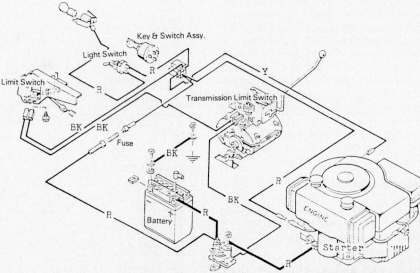

Fig. M12 – Wiring diagram for Model 30502. Refer to legend in Fig. M11 for identification of wires.

(11) from above. Unbolt and remove bearing and housing assemblies (18).

Clean and inspect all parts and renew any showing excessive wear or other damage. Reassemble by reversing the disassembly procedure. Tighten blade retaining nuts (24) to a torque of 15-18 ft.-lbs.

Models 25501, 25502, 30501 and 30502

REMOVE AND REINSTALL. Remove mower unit and mower drive belt (9 – Fig. M10) as outlined in previous section. Remove blade shaft (19) retaining nut, lockwasher, washers, blade (17), and adapter (16) from bottom of deck. Withdraw shaft (19) with mower pulley (7) from above. Be sure not to lose Woodruff keys located in shaft (19). Unbolt and remove bearing and housing assembly (14).

Clean and inspect all parts and renew as needed. Reassemble by reversing the disassembly procedure.

Model 31501

REMOVE AND REINSTALL. Remove mower unit and mower drive belt (9 – Fig. M10A) as outlined in previous section. Remove jackshaft (5) retaining nut (36), lockwasher (35), washers (34 and 33), blade (32), washer (31) and adapter (30) from bottom of deck. Withdraw jackshaft (5) with mower pulley (6) from above. Be sure not to lose Woodruff keys located in shaft. Unbolt and remove bearing and housing assembly (28).

Clean and inspect all parts and renew as needed. Reassembly by reversing the disassembly procedure.

Models 36503-39001

REMOVE AND REINSTALL. Remove mower unit and mower drive belt (9 – Fig. M10B) as outlined in previous section. Remove jackshaft (3) retaining nut (30), lockwasher (29), washers (28 and 27), blade (26), washer (25) and adapter (24) from bottom of deck. Withdraw jackshaft (3) with mower pulley (4 or 6) from above. Be sure not to lose Woodruff keys located in shaft. Unbolt and remove bearing and housing assembly (22).

Clean and inspect all parts and renew as needed. Reassemble by reversing the disassembly procedure.

ELECTRICAL

Models 25501-25502-30501-30502-31501-36503-39001

Shown in electrical diagrams Figs.

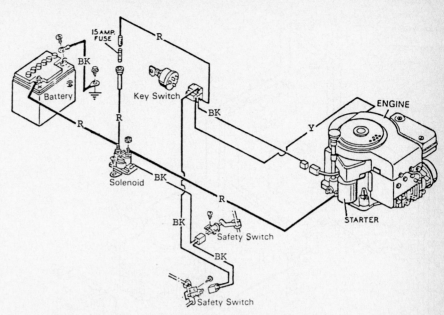

Fig. M13 – Wiring diagram for Model 31501. Refer to legend in Fig. M11 for identification of wires.

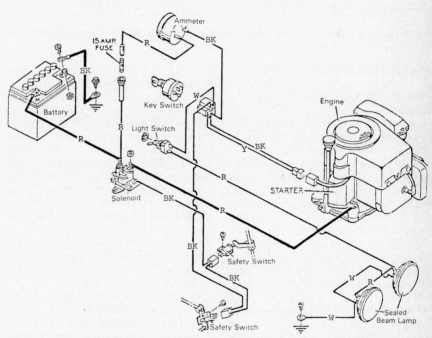

Fig. M14 – Wiring diagram for Model 36503. Refer to legend in Fig. M11 for identification of wires.

M11, M12, M13, M14 and M15 are electrical components and connecting wires. For component and wire identification refer to Fig. M11 for Models 25501, 25502 and 30501, Fig. M12 for Model 30502, Fig. M13 for Model 31501, Fig. M14 for Model 36503 and Fig. M15 for Model 39001.

TRANSMISSION

All Models

The transmission used on all models except 25501, 25502, 30501, 30502 and 31501 is manufactured by the J.B. Foote Foundry Co. A Peerless model transmission is used on Models 25501, 25502, 30501, 30502 and 31501. A Peerless model transaxle is used on Models 36503 and 39001.

For transmission or transaxle overhaul procedures, refer to the Foote or Peerless section in the TRANSMISSION REPAIR service section.

Refer to the following paragraphs for removal and installation of transmission or transaxle.

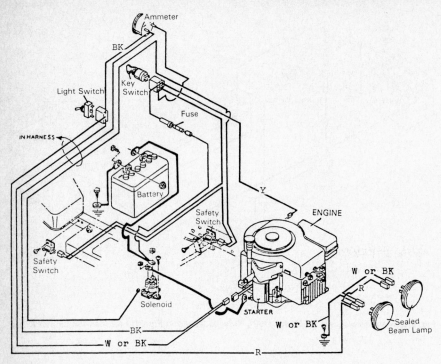

Fig. M15 — Wiring diagram for Model 39001. Refer to legend in Fig. M11 for identification of wires.

Models 2503-2513-3013-3033

REMOVE AND REINSTALL. To remove Foote 2010 transmission, disconnect spark plug wire and remove the mower unit as follows: Place mower unit in lowest position and move blade clutch lever to disengaged (stop) position. Disconnect lift chains and unpin upper ends of scissor arms from mower hangers. Remove mower belt from engine pulley and disconnect blade safety starting switch wires. Remove mower unit from under right side. Unhook main drive clutch tension spring (12 – Fig. M3) and remove main drive belt (10) from transmission input pulley (18). Remove retaining ring (19) and pulley from transmission input shaft. Locate master link in drive chain and disconnect drive chain. Unbolt and remove seat assembly and engine shroud. Disconnect transmission safety starting switch wires and unhook brake lever spring. Remove mounting bolts and lift transmission from frame.

Reinstall transmission by reversing the removal procedure. Make certain that transmission and blade clutch safety starting switches are connected and in good operating condition.

Models 3043-3063

REMOVE AND REINSTALL. To remove Foote 2010 transmission, disconnect spark plug wire and proceed as follows: Place mower unit in lowest position and move blade clutch lever to disengaged (stop) position. Unpin upper

ends of scissor arms from mower hangers. Remove mower belt from engine pulley and disconnect blade clutch safety starting switch wires. Remove mower unit from under right side. Unhook main drive clutch tension spring (12 – Fig. M4) and remove main drive belt (10) from transmission pulley (18). Remove retaining ring (19) and pulley from transmission input shaft. Disconnect brake rod (15) from brake cam lever (3). Locate master link in drive chain and disconnect drive chain. Unbolt and remove seat assembly and engine shroud assembly. Disconnect transmission safety starting switch wires, then unbolt and remove transmission assembly.

Reinstall transmission by reversing the removal procedure. Make certain that transmission and blade clutch safety starting switches are connected and in good operating condition.

Models 3233-3235-3633

REMOVE AND REINSTALL. To remove Foote 2010 (Models 3233 and 3633) or Foote 2600 (Model 3235) transmission, disconnect spark plug wire and proceed as follows: Place mower unit in lowest position and move blade clutch lever to disengaged (stop) position. Disconnect lift chains and unpin upper ends of scissor arms from mower hangers. Remove the blade clutch idler pulley, slip mower belt from engine pulley and remove mower unit from under right side. Unhook main drive clutch tension spring (12 – Fig. M4) and remove main drive belt (10) from transmission pulley (18).

Remove retaining ring (19) and pulley from transmission input shaft. Disconnect brake rod (15) from brake cam lever (3). Locate master link in drive chain and disconnect drive chain. Unbolt and remove seat assembly and engine shroud. Disconnect transmission safety starting switch wires, then unbolt and remove transmission assembly.

Reinstall transmission by reversing the removal procedure. Make certain that transmission safety starting switch is connected and in good operating condition.

Models 25501, 25502, 30501 and 30502

REMOVE AND REINSTALL. To remove Peerless 500 (Model 25501) or Peerless 700 (Models 25502, 30501 and 30502) transmission, disconnect spark plug wire and proceed as follows: Place mower unit in lowest position and move blade clutch lever to disengaged (stop) position. Disconnect mower deck hangers from deck lifter assembly (20 – Fig. M4B) and front suspension plates. Disconnect adjustment nut (12), then remove mower belt from engine pulley and remove mower unit from under right side. Unhook main drive clutch tension spring (12 – Fig. M4A) and remove clutch idler pulley (19). Remove traction drive belt from transmission pulley (20) first, then from engine pulley (9). Remove retaining ring and pulley from transmission input shaft. Locate master link in drive chain and disconnect drive chain. Unbolt and remove seat assembly and engine shroud. Disconnect transmission safety starting switch wires and unhook brake lever spring. Remove mounting bolts and lift transmission from frame.

Reinstall transmission by reversing the removal procedure. Make certain that transmission and blade clutch safety starting switches are connected and in good operating condition.

Model 31501

REMOVE AND REINSTALL. To remove the transmission, disconnect spark plug wire and proceed as follows: Place mower unit in lowest position and move blade clutch lever to disengaged (stop) position. Disconnect mower deck hangers from deck lifter assembly and front hanger bracket. Disconnect mower clutch idler engagement rod, then remove mower belt from engine pulley and remove mower unit. Loosen tension on traction drive belt idler pulley and remove all belt guides and protective shields as needed to allow access to drive belt. Remove drive belt from drive pulleys. Remove retaining ring and pulley from transmission input shaft. Locate master link in drive chain and disconnect drive chain. Remove seat deck and seat assembly. Disconnect transmission safety starting switch wires and unhook brake lever spring. Remove mounting bolts and lift transmission assembly from frame.

Reinstall transmission by reversing the removal procedure. Make certain that transmission and blade clutch safety starting switches are connected and in good operating condition.

OVERHAUL. For overhaul procedures, refer to Peerless transmission Series 700 in the TRANSMISSION REPAIR section of this manual.

Models 36503-39001

REMOVE AND REINSTALL. To remove the transaxle assembly, disconnect spark plug wire and battery ground cable from battery terminal. Remove mower deck and drive belt as outlined in MOWER DECK section. Remove traction drive belt from transaxle input pulley. Disconnect brake linkage. Remove cap screws retaining transaxle to frame, then on Model 36503 remove "U" bolts securing axle housings to frame and on Model 39001 remove cap screws securing frame brackets to axle housings. Raise rear of tractor and remove shift lever as needed, then remove transaxle assembly from tractor.

Reinstall by reversing the removal procedure. Adjust clutch and brake linkage as required.

OVERHAUL. For overhaul procedures, refer to Peerless transaxle Series 800 on Model 36503 and Series 2300 on Model 39001.

DRIVE CHAIN

Rear Engine Models

ADJUSTMENT. On all models except 25501, 25502, 30501 and 30502 adjust drive chain (7 – Fig. M5) as follows: Loosen the three cap screws on each side of frame securing bearing retainers (4) to frame (1). Move rear axle assembly rearward (equal distance on both sides) to tighten drive chain. Drive chain should deflect approximately ½-inch when about five pounds pressure is applied on chain. When chain tension is correct, tighten bearing retainer mounting cap screws. On Models 25501, 25502, 30501 and 30502 drive chain is adjusted by sliding adjustment bracket (24 – Fig. M4A) in slots on frame assembly. To adjust, loosen securing nut and slide bracket until approximately ½-inch deflection is attained at center of drive chain when about five pounds pressure

is applied. When chain tension is correct, retighten securing nut.

REMOVE AND REINSTALL. To remove drive chain, rotate differential sprocket to locate master link. Disconnect master link and remove chain. Clean and inspect drive chain and renew if excessively worn.

Install drive chain and adjust chain tension as outlined in the preceding paragraph. Lubricate chain with a light coat of SAE 30 oil.

Model 31501

ADJUSTMENT. Drive chain is adjusted by sliding adjustment bracket (13 – Fig. M4C) in slots on frame assembly. To adjust, loosen securing nuts and slide bracket until approximately ½-inch deflection is attained at center of drive chain when about five pounds pressure is applied. When chain tension is correct, retighten securing nuts.

REMOVE AND REINSTALL. To remove drive chain, rotate differential sprocket to locate master link. Disconnect master link and remove chain. Clean and inspect drive chain and renew if excessively worn.

Install drive chain and adjust chain tension as outlined in the preceding paragraph. Lubricate chain with a light coat of SAE 30 oil.

DIFFERENTIAL

Rear Engine Models

REMOVE AND REINSTALL. For assistance in removal of differential and rear axle assembly refer to Fig. M4A for Models 25501, 25502, 30501 and 30502 and Fig. M5 for all other models. To remove, raise rear of unit and securely support. Remove rear wheel assemblies. Rotate differential sprocket to locate the master link in drive chain, then disconnect the chain. Unbolt axle mounting brackets and remove differential assembly.

Reinstall differential and rear axle assembly by reversing the removal procedure. Adjust drive chain tension as outlined in previous section. Lubricate axle bearings with SAE 30 oil.

OVERHAUL. The differential used on all models except 25501, 25502, 30501 and 30502 is manufactured by Indus Wheel Company, Division of Carlisle Corp. A Peerless model differential is used on Models 25501, 25502, 30501 and 30502.

For differential overhaul procedures, refer to the Indus or Peerless section in the DIFFERENTIAL REPAIR service section.

Model 31501

REMOVE AND REINSTALL. For assistance in removal of differential and rear axle assembly refer to Fig. M4C. To remove, raise and securely support rear of unit. Remove rear wheel assemblies. Rotate differential sprocket to locate the master link in drive chain, then disconnect the chain. Unbolt axle mounting brackets and remove differential assembly.

Reinstall differential and rear axle assembly by reversing the removal procedure. Adjust drive chain tension as outlined in previous section. Lubricate axle bearings with SAE 30 oil.

OVERHAUL. A Peerless model differential is used. For overhaul procedures, refer to the Peerless section in the DIFFERENTIAL REPAIR service section.

GROUND DRIVE BRAKE

All Models

Traction drive brake is applied by depressing clutch/brake pedal on single pedal models or by depressing brake pedal on dual pedal models. All models except Models 25501, 25502, 30501, 30502, 31501, 36503 and 39001 are equipped with a brake pad on clutch idler arm (11 – Fig. M3 or M4) which contacts the transmission input pulley (18) and assists in stopping.

CAUTION: This brake is ineffective when transmission is in neutral position.

Models 2503, 2513, 3013 and 3033 are also equipped with a hand operated disc brake located on the transmission output shaft. See items (1 thru 8 – Fig. M3). This brake can be used as an emergency brake, parking brake or in conjunction with the brake on clutch idler.

Models 3043, 3063, 3233, 3235 and 3633 are equipped with a foot operated disc brake. Brake disc (5 – Fig. M4) is located on the transmission output shaft. Use this brake in conjunction with the brake on clutch idler, or as an emergency brake or parking brake.

Models 25501, 25502, 30501, 30502, 31501, 36503 and 39001 are equipped with a foot operated disc brake. Brake disc is located on the transmission output shaft on Models 25501, 25502, 30501, 30502 and 31501 and transaxle brake shaft on Models 36503 and 39001.

On models equipped with clutch idler brake pad, depress clutch pedal before fully applying the disc brake. There is no adjustment on the disc brakes. When applying, the disc brake and brake lever travels the full length of its slot, brake

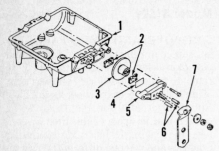

Fig. M16—Exploded view of traction brake assembly used on Models 25501, 25502, 30501 and 30502.

1. Lower case housing
2. Pads
3. Disc
4. Pad plate

5. Pad holder
6. Dowel pin
7. Lever

pads must be renewed. Refer to the following paragraphs for procedures.

Models 2503-2513-3013-3033

REMOVE AND RENEW. To renew the brake pads (4 and 6–Fig. M3), disconnect the brake spring and remove knob from brake lever (3). Remove shoulder bolt (1) with washer (2), cam lever (3), carrier (8), outer pad (4) and spring (7). Slide brake disc (5) outward on transmission shaft and remove inner brake pad (6) from holding slot in transmission housing. Install new brake pads and reassemble by reversing the disassembly procedure.

Models 3043-3063-3233-3235-3633

REMOVE AND RENEW. To renew the brake pads (4 and 6–Fig. M4), dis-

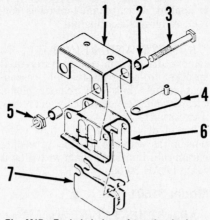

Fig. M17—Exploded view of traction brake assembly used on Model 39001.

1. Bracket assy.
2. Spacer
3. Bolt
4. Lever

5. Locknut
6. Bracket
7. Pads

connect the brake rod (15) from cam lever (3). Tilt seat forward and unscrew shoulder bolt (1). Remove shoulder bolt with washer (2), cam lever (3), carrier (8), outer brake pad (4) and spring (7). Slide brake disc (5) outward on transmission shaft and remove inner brake pad (6) from holding slot in transmission housing. Install new brake pads and reassemble by reversing the disassembly procedure.

Models 25501-25502-30501-30502-31501

REMOVE AND RENEW. To renew brake pads (2–Fig. M16), remove all parts as needed to attain access to disc

brake components. Remove jam nut, nut and washer from pad holder stud (5), then withdraw lever (7). Remove pad holder mounting bolts, then withdraw pad holder, pads and brake disc.

Install new brake pads and reassemble by reversing the disasembly procedure.

Model 36503

REMOVE AND RENEW. Disc brake assembly on Model 36503 is similar to type shown in Fig. M16, except unit is mounted on brake shaft and secured to transaxle housing.

To renew brake pads, remove all parts as needed to attain access to disc brake components. With reference to components shown in Fig. M16, remove jam nut, nut and washer from pad holder stud (5), then withdraw lever (7). Remove pad holder mounting bolts, then withdraw pad holder, pads and brake disc.

Install new pads and reassemble by reversing the disassembly procedure.

Model 39001

REMOVE AND RENEW. To renew brake pads (7–Fig. M17), remove all parts as needed to attain access to disc brake components. Unhook actuating spring from lever (4), then remove bracket assembly (1) mounting cap screws. Remove locknuts (5), then withdraw bolts (3) and slide pads (7) out of bracket (6).

Install new brake pads and reassemble by reversing the disassembly procedure.

MUSTANG

MOWETT SALES CO., INC.
110 W. Mason
P.O. Box 218
Odessa, MO 64076

| | | Engine | | Cutting |
Model	Make	Model	Horsepower	Width, In.
245	B&S	130000	5	24
248	B&S	190000	8	24

FRONT AXLE

All Models

R&R AND OVERHAUL. Front end assembly (11–Fig. MU1) is secured to frame rails (7 and 8). To remove, raise front of unit until front wheel assemblies (10) are clear of ground. Loosen four bolts retaining mounting plate over steering shaft. Remove two bolts securing steering wheel shaft (2) to steering post, then lift steering wheel and shaft assembly from unit. Unbolt front end assembly from frame rails, then withdraw assembly from mower unit.

Complete disassembly and inspect all components for excessive wear or any other damage. Renew all parts as needed, then reassemble in reverse order of disassembly. After installation if steering is too tight, loosen four bolts holding metal plate over steering post. If steering is too loose, tighten bolts as needed. Nylon bearings in front spindles and oil impregnated bearings in front wheel assemblies do not require lubrication.

ENGINE

All Models

For overhaul and repair procedures on engines listed in Specification Table refer to Small Air Cooled Engines Service Manual.

Refer to the following paragraphs for removal and installation procedures.

REMOVE AND REINSTALL. To remove engine, first raise unit to allow access to underneath side. Unbolt blade retaining bolt and withdraw washers, spacers (26–Fig. MU1) and blade (30). Remove four bolts securing mower deck (22) to frame rails (7 and 8), then withdraw deck clear of unit. Remove drive belt (23) from pulleys, then withdraw clutch (25) from engine crankshaft. Be sure not to lose square key used between crankshaft keyway and clutch keyway.

Loosen four bolts retaining mounting plate over steering shaft. Remove two bolts securing steering wheel shaft (2) to steering post, then lift steering wheel and shaft assembly from unit.

Remove all bolts as needed to withdraw hood (3) from frame rails. Remove wiring, cables and any obstructing components to allow removal of engine assembly. Remove engine mounting bolts, then lift engine assembly from mower unit.

After repair of engine reverse removal procedure to reinstall engine.

TRACTION DRIVE CLUTCH

All Models

Figure MU2 shows a view of traction drive clutch components. Actuating foot pedal will tighten idler pressure on drive belt which will engage traction power train. Excessive pressure on foot pedal will cause drive belt to over-stretch and slip on pulleys. If slippage occurs during engagement, then belt must be renewed. Unit speed is regulated by throttle control.

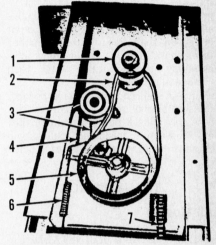

Fig. MU2 – View of traction drive assembly.

1. Clutch assy.	
2. Belt	5. Transmission pulley
3. Idler assy.	6. Spring
4. Idler rod	7. Chain

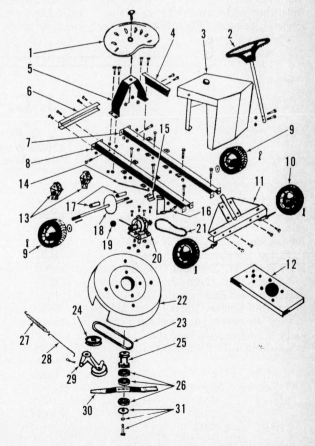

Fig. MU1 – Exploded view of complete mower unit.

1. Seat
2. Steering wheel & shaft
3. Hood
4. Chain guard
5. Seat support
6. Rear channel
7. Left frame rail
8. Right frame rail
9. Rear wheel assy.
10. Front wheel assy.
11. Front end assy.
12. Mounting plate
13. Rear axle bearing assy.
14. Idler rod
15. Foot pedal bearing
16. Foot pedal
17. Axle spacer
18. Rear axle assy.
19. Knob
20. Transmission assy.
21. Chain
22. Mower deck
23. Belt
24. Pulley
25. Clutch
26. Spacers
27. Spring
28. Idler rod
29. Idler assy.
30. Blade
31. Bolt and washers

To renew clutch idler or drive belt refer to the following paragraphs.

REMOVE AND RENEW. Unhook ignition wire from spark plug, then invert mower assembly so it is resting on the rear wheels and seat. Unbolt blade retaining bolt and withdraw washers, spacers and blade. Remove four bolts securing mower deck to frame rails, then withdraw deck clear of unit. Unhook spring (6 – Fig. MU2) and idler rod (4) from idler assembly (3), then remove drive belt from pulleys. Remove snap ring retaining transmission input pulley on shaft, then withdraw pulley from shaft. Remove bolt holding idler assembly to mounting plate, then remove idler assembly.

Inspect and renew all parts as needed. Reassemble in reverse order of disassembly. Check pulley alignment during reassembly, misaligned pulleys will cause premature damage to drive belt.

TRANSMISSION

All Models

REMOVE AND REINSTALL. Unhook ignition wire from spark plug, then invert mower assembly so it is resting on the rear wheels and seat. Unbolt blade retaining bolt and withdraw washers, spacers and blade. Remove four bolts securing mower deck to frame rails, then withdraw deck clear of unit. Unhook spring (6 – Fig. MU2) and idler rod (4) from idler assembly (3), then remove drive belt from pulleys. Remove snap ring retaining transmission input pulley on shaft, then withdraw pulley from shaft.

Loosen all bolts that extend through left and right frame rails. Loosen bolt that extends through steering post support bracket and engine mounting plate. Push mounting plate assembly downward to release tension on chain, then remove chain from transmission sprocket. Remove transmission mounting bolts and lift transmission assembly from mounting plate.

After repair, reassemble in reverse order of disassembly. Pry upward on engine mounting plate to tighten chain tension. Check chain alignment after installation. If chain tries to jump off rear axle sprocket, then axle position must be changed to correct alignment. Move washers on outside of rear wheels from side to side to change axle position.

OVERHAUL. After transmission is removed, refer to FOOTE TRANSMISSION REPAIR section of this manual for detailed service procedures.

CHAIN

All Models

ADJUSTMENT. Loosen all bolts that extend through left and right frame rails. Loosen bolt that extends through steering post support bracket and

engine mounting plate. Pry upward on mounting plate until correct chain tension is reached, then retighten all bolts. Chain is equipped with one half link that may be removed to provide additional chain adjustment.

REMOVE AND RENEW. Loosen all bolts that extend through left and right frame rails. Loosen bolt that extends through steering post support bracket and engine mounting plate. Push mounting plate assembly downward to release tension on chain, then remove chain from transmission sprocket.

Lower mower unit, then raise rear wheel assembly clear of ground. Unbolt rear axle bearing assembly (13 – Fig. MU1) from frame rails, then withdraw rear axle assembly from mower unit.

Inspect and renew all components as needed. Reassemble in reverse order of disassembly. Adjust chain tension as outlined in ADJUSTMENT section. Lubricate chain with a good grade of oil after installation.

MOWER CUTTING HEIGHT

All Models

Mower blade cutting height may vary from 1½-3½ inches off the ground.

ADJUSTMENT. Unbolt blade retaining bolt and reposition spacers (26 – Fig. MU1) to raise or lower blade height. Reinstall blade retaining bolt and washer after adjustment.

J. C. PENNEY

J. C. PENNEY CO., INC.
11800 West Burleigh Street
Milwaukee, Wisconsin 53201

The following J. C. Penney riding mowers were manufactured for J. C. Penney Co., Inc., by Murray Ohio Manufacturing, Brentwood, Tennessee and by MTD Products, Cleveland, Ohio. Service procedures for these J. C. Penney models will not differ greatly from those given for similar Murray and MTD models. However, parts are not necessarily interchangeable and should be obtained from J. C. Penney Co., Inc.

J.C. Penney Model	Murray Model	MTD Model
1905	2503	....
1907	2503	....

J.C. Penney Model	Murray Model	MTD Model
1907A	2503	....
1908	2513	....
1909	2513	....
1910	2513	....
1824	39001	....
1834*	36503	....
1841**	31501	....
1842**	31501	....
1820	....	525
1831	....	520
1832	....	362
1835***	....	498
1839†	....	495
1840††	....	495
1844	....	820

J.C. Penney Model	Murray Model	MTD Model
1845	....	497
1846	....	497
1847	....	498
1848	....	497

*Model 1834 uses a four spindle mower.
**Models 1841 and 1842 use a two spindle mower.
***Model 1835 uses a three spindle mower and a Foote Series 4000 transaxle.
†Model 1839 uses a Foote Series 2010 transmission.
††Model 1840 uses a Peerless Series 700 transmission.

RIDE KING

SWISHER MOWER & MACHINE CO.
P.O. Box 67
333 East Gay St.
Warrensburg, MO 64093

Model	Make	Engine Model	Horsepower	Cutting Width, In.
A-32	Tecumseh	V60	6	32
R-32	Tecumseh	V60	6	32

STEERING SYSTEM

Model A-32

Steering is controlled by turning steering wheel (1–Fig. RK1) which rotates steering shaft and sprocket (20). Steering shaft sprocket meshes with steering gear (21) located on gearbox. Steering gear (21) will allow front wheel to pivot 360 degrees. For reference in disassembly and repair refer to Fig. RK1.

Model R-32

Steering is controlled by turning steering wheel (1–Fig. RK2) which rotates steering shaft and sprocket (20). Steering shaft sprocket rotates steering gear (21) by use of chain (37). Front wheel assembly has a 360 degree turning radius. For reference in disassembly and repair refer to Fig. RK2.

ENGINE

All Models

For overhaul and repair procedures on engines listed in Specification Table refer to Small Air Cooled Engines Service Manual.

Refer to the following paragraphs for removal and installation procedures.

REMOVE AND REINSTALL. Disconnect spark plug lead from spark plug. Remove front hood assembly. On Model A-32 use caution as fuel tank is removed with hood assembly. Remove traction drive belt cover, then withdraw drive belt(s) from pulleys. Remove mower drive belt inspection covers, then loosen tension on belt and slip drive belt off pulleys. Remove steering wheel (1–Fig. RK1 and RK2). Inspect and remove as needed any part or parts that will obstruct removal of engine assembly. Loosen and remove engine securing bolts and nuts, then lift engine assembly clear of mounting plate and mower unit. Place engine assembly to

the side for inspection and repair.

After repair, reinstall in reverse order of disassembly.

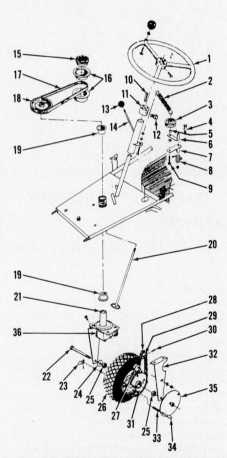

Fig. RK1—Exploded view showing steering and traction drive components for Model A-32.

1. Steering wheel	20. Steering shaft & sprocket
2. Idler spring	21. Steering gear
3. Pulley	22. Axle bolt
4. Shoulder bolt	23. Lockpin lever
5. Spacer	24. Spacer washer
6. Belt release finger	25. Bearing
7. Idler arm	26. Wheel assy.
8. Spring	27. Chain idler
9. Bolt	28. Chain
10. Key	29. Sprocket
11. Hub	30. Pin
12. Set screw	31. Drive sprocket & hub assy.
13. Knob	32. Bracket
14. Idler control assy.	33. Spring
15. Starter cup	34. Lockpin
16. Pulley	35. Chain guard
17. Drive belt	36. Gearbox assy.
18. Pulley	
19. Bearing	

Fig. RK2—Exploded view showing steering and traction components for Model R-32.

1. Steering wheel	21. Steering gear
2. Control rod	22. Axle bolt
3. Pulley	23. Lockpin lever
4. Shoulder bolt	24. Spacer washer
5. Spacer	25. Bearing
6. Belt release finger	26. Wheel assy.
7. Idler arm	27. Chain idler
8. Spring	28. Chain
9. Bolt	29. Sprocket
13. Knob	30. Pin
14. Idler control assy.	31. Drive sprocket & hub assy.
15. Starter cup	32. Bracket
16. Pulley	33. Spring
17. Drive belt	34. Lockpin
18. Pulley	35. Chain guard
19. Bearing	36. Gearbox assy.
20. Steering shaft & sprocket	37. Chain

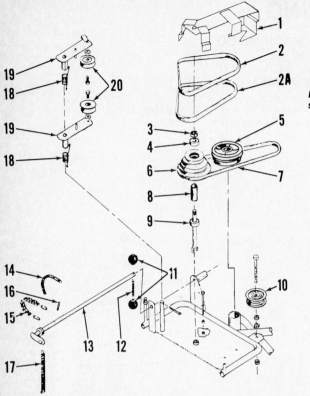

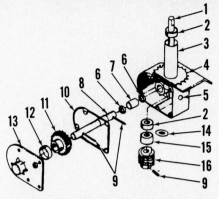

Fig. RK3 – Exploded view showing dual speed drive unit.

1. Cover
2. Low speed belt
2A. High speed belt
3. Nut
4. Bearing
5. Pulley
6. Pulley
7. Drive belt
8. Spacer
9. Cap screw
10. Pulley
11. Knob
12. Stud bolt
13. Shift lever
14. Spring
15. Chain
16. Pin
17. Spring
18. Spring
19. Idler arm
20. Pulley

Fig. RK4 – Exploded view showing traction drive gearbox assembly.

1. Worm shaft
2. Ball bearing
3. Gear case
4. Drive gear
5. Oil plug
6. Seal
7. Bearing
8. Gear shaft & spacer
9. Roll pin
10. Gasket
11. Gear
12. Retainer ring
13. Gear case cover
14. Spacer washer
15. Thrust bearing
16. Worm gear

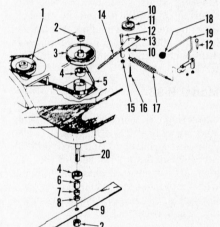

Fig. RK5 – Exploded view showing mower blade drive unit and drive belt engagement system.

1. Engine pulley
2. Locknut
3. Blade pulley
4. Bearing
5. Drive belt
6. Spacer (1-inch)
7. Spacer (½-inch)
8. Spacer (¼-inch)
9. Blade
10. Nut
11. Idler pulley
12. Shoulder bolt
13. Control arm
14. Spring
15. Nut
16. Bolt
17. Idler control rod & spring
18. Knob
19. Idler control assy.
20. Shaft

TRACTION DRIVE BELT

Single Speed

Single traction drive belt (17 – Fig. RK1 and RK2) is engaged and disengaged by control lever (14) which controls the operation of idler pulley (3). Moving control lever (14) forward will engage drive belt and pulling lever rearward will disengage drive belt. If drive belt slips when engaged on Model A-32 inspect condition of idler spring (2 – Fig. RK1) and drive belt (17). Renew idler spring if excessively stretched or damaged in any way. Renew drive belt if frayed, cracked, excessively stretched or damaged in any way. Check for correct operation after repair and reassembly.

Dual Speed

Figure RK3 shows an exploded view of a dual speed drive unit. Turning idler control lever (13 – Fig. RK3) forward will engage low speed belt (2) and turning lever rearward will engage high speed belt (2A). Center position is disengaged or neutral position. If drive belt slips when engaged check for correct operation of idler system. Idler pulley should fully engage drive belt when control lever is turned to engaged position. Renew drive belts if frayed, cracked, excessively stretched or damaged in any way. Inspect all pulleys, springs and levers for excessive wear or any other damage and renew as needed. After

repair and reassembly check for correct operation.

TRACTION DRIVE GEARBOX

All Models

Shown in Fig. RK4 is an exploded view of a traction drive gearbox assembly. To remove gearbox for disassembly proceed as follows: Raise and support front end of mower unit. Loosen chain tensioner (27 – Fig. RK1 and RK2). Remove locknut from axle bolt (22), then withdraw chain guard (35). Unbolt and remove left hand support bracket (32). Remove lockpins (34), then lift drive sprocket and hub assembly (31) from wheel assembly (26) along with drive chain (28). Remove axle bolt (22) and wheel assembly (26). Remove front hood assembly. On Model A-32 use caution as fuel tank is removed with hood assembly. Slip traction drive belt (17) off pulley (18), then withdraw pulley (18) off gearbox worm shaft (1 – Fig. RK4). Slide gearbox assembly out of frame housing, then place gearbox assembly to the side for disassembly and repair.

Inspect bearings 19 and 25 for roughness, binding or any other damage. Inspect all other parts for excessive wear or any other damage and renew parts as needed.

Disassemble gearbox assembly with reference to Fig. RK4. Inspect all parts for excessive wear, binding or any other damage. Inspect drive gears for chipped

or missing teeth and renew all parts as needed. Reassemble in reverse order of disassembly. Manufacturer recommends using a high pressure gear oil with a high lead content in gearbox. Add gearbox oil until level is even with bottom of plug hole (5).

Reinstall gearbox assembly in mower unit and reassemble in reverse order of disassembly.

MOWER DRIVE BELT

All Models

REMOVE AND RENEW. Remove left and right mower deck inspection plate securing nuts. Lift inspection plates off mower deck and place to the side. Release tension on idler control arm as needed, then slip drive belt off engine and blade drive pulleys. Install new drive belt, then adjust idler control rod (17 – Fig. RK5) as needed. Drive belt should not slip on pulleys when idler control rod is engaged and mower blades should not turn when control rod is in

disengaged position. Reinstall inspection plates after completing adjustment.

MOWER BLADE DRIVE

All Models

Mower blade is belt driven, for renewing or adjusting tension on drive belt refer to previous section. Figure RK5 shows an exploded view of mower blade drive unit and drive belt engagement system. Periodically inspect all parts for excessive wear or any other damage and renew all parts as needed. To adjust mower blade cutting height, remove blade locknut (2) and blade (9). Reposition spacers (6, 7 and 8) above or below mower blade until desired cutting height is attained.

ROPER

ROPER SALES CORPORATION
1905 West Court Street
Kankakee, IL 60901

Model	Make	Engine Model	Horsepower	Cutting Width, In.
K511	B&S	130902	5	26
K521	B&S	130902	5	26
K522	B&S	130905E	5	26
K831	B&S	190702	8	32
K832	B&S	190705E	8	32
K852	B&S	190705E	8	32
L711	Tecumseh	V70	7	26
L721	Tecumseh	V70E	7	26
L722	Tecumseh	V70E	7	26
L821	B&S	191707E	8	36
L861	Tecumseh	VM80	8	36
L863	B&S	191707E	8	36

K – prefix models: Rear engine, "Sprint" riders
L – prefix models: Tractor-style, "Mini-Brute", "Rally" riders

STEERING SYSTEM

Models K-511, K-521, K-522, K-831, K-832 and K-852

REMOVE AND REINSTALL. These models are fitted with a non-adjustable steering system as shown in Fig. R1. Tube portion of steering column may be shifted to a high or low position for comfort and convenience of operator.

If front system shows evidence of wear, hard turning or stiff operation, raise and block up under frame just forward of cutting deck so that wheels are clear of surface. Unbolt and remove axle cross member (16 – Fig. R1) from frame without separate removal of king pins (13), wheels or tie bar (14). Slot in steering plate (15) will disengage from arm of steering shaft (6) when assembly is lowered. Thoroughly examine all parts and renew any which are damaged or worn.

If condition of steering column is questionable, unbolt steering shaft tube (3) with steering wheel and remove upward from steering shaft (6), then lower steering shaft out of support bracket and evaluate grommets which support shaft at wear points. Renew as needed.

If equipped with headlight and electric start, it is advisable to remove control console cover (four self-tapping metal screws) for access to steering shaft after steering column tube has been removed.

When reassembly is complete, lubricate steering linkage, shaft grommets, bushings and wheel bearings with SAE 30 engine oil. Wipe off excess to prevent dirt accumulation.

Models L-711, L-721, L-722 L-821, L-861 and L-863

REMOVE AND REINSTALL. All front engine riders are equipped with automotive type, gear sector and pinion steering as shown in Figs. R2 and R3.

Raise and block unit under frame between front edge of cutting deck and steering tie rod so that wheels are clear of work surface. On Models L-711, L-721 and L-722, remove cotter pins from ends of steering link (13 – Fig. R2) then unbolt front member from chassis

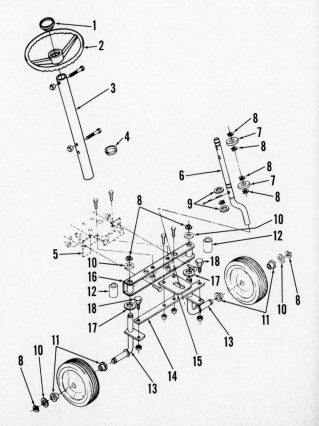

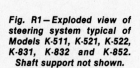

Fig. R1 – Exploded view of steering system typical of Models K-511, K-521, K-522, K-831, K-832 and K-852. Shaft support not shown.

1. Logo cap
2. Steering wheel
3. Shaft tube & bolts
4. Grommet
5. Frame channel
6. Steering shaft
7. Shaft ring (2)
8. "E" ring, ⅝-in. (9)
9. Flat washer, 0.64 (2)
10. Flat washer, 0.63 (4)
11. Wheel bearings
12. Spindle bearings
13. Kingpin/spindle, RH, LH
14. Tie bar
15. Steering plate
16. Cross member
17. Special washer (2)
18. Shoulder bolt, 5/16-24 (2)

platform and roll steering assembly forward from under unit for inspection and further disassembly as needed. On Models L-821, L-861 and L-863, remove nuts from ball joint studs of drag link (17—Fig. R3) and remove drag link, then remove front pivot bolt (24) and withdraw axle. Parts of steering system should now be carefully inspected and renewed if damaged or worn.

Before proceeding with disassembly of steering shaft, pinion or sector, front axle should be reinstalled and unit lowered to rest on all wheels. Drag link is left disconnected.

On Models L-711, L-721 and L-722, remove hood, then unbolt and remove dash panel from chassis cover. Unbolt steering shaft mount (11—Fig. R2) from chassis, withdraw the unit and disassemble as needed. Thoroughly clean all parts and renew any which are damaged or worn. Pay special attention to condition of shoulder bolt (4) and key (7). Renew steering shaft mount (11) if it shows any signs of wear. Lubricate all parts during reassembly.

On Models L-821, L-861 and L-863, swing hood forward to uncover engine compartment and unbolt steering bracket (9—Fig. R3) from frame. Clean, inspect and determine need for further disassembly and renewal of parts.

NOTE: If it is necessary to remove pinion gear (5) from steering shaft (4) or gear sector (11) from steering rod (10), use caution and buck shafts when driving out pins.

Pay particular attention to bushings (3, 8 and 16).

When damaged parts are renewed, lubricate with chassis grease and reassemble. Use SAE 30 engine oil at all wear points in steering linkage before returning unit to service.

ENGINE

All Models

For overhaul and repair procedures on engines listed in Specification Table refer to Small Air Cooled Engines Service Manual.

Refer to the following paragraphs for removal and installation procedures.

Models K-511, K-521, K-522, K-831, K-832 and K-852

REMOVE AND REINSTALL. To remove engine, disconnect spark plug lead and on electric start models, disconnect battery cables from starter motor and battery.

Remove operator's seat and disconnect bowden cable (throttle control wire) at carburetor end.

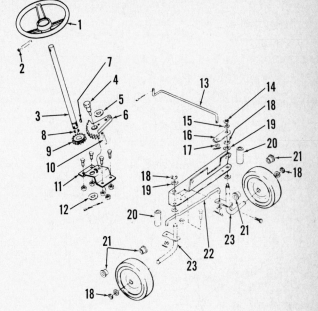

Fig. R2—View of steering system parts as used on Models L-711, L-721 and L-722.

1. Steering wheel
2. Lock pin ¼X1½
3. Steering shaft
4. Shoulder bolt (spec.)
5. Flat washer, 0.515 bore
6. Gear sector
7. Woodruff key, #61
8. "E" ring, ¾-inch
9. Pinion gear
10. Roll pin, 3/16X1
11. Steering shaft mount
12. Flat washer, 0.765
13. Steering link
14. Nylok nut, ⅜-24
15. Flat washer, 0.40 bore
16. Steering arm
17. Flat washer, 0.515 bore
18. "E" ring, ⅝-inch (2)
19. Flat washer, 0.640 bore (4)
20. Spindle sleeve
21. Wheel bearings (4)
22. Tie rod
23. Kingpin/spindle RH, LH

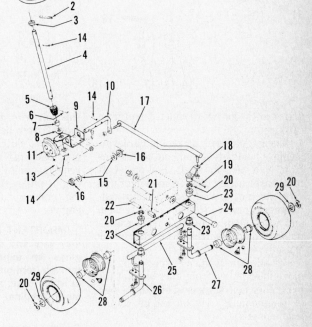

Fig. R3—Exploded view of steering system typical of that used on Models L-821, L-861 and L-863. Upper end of steering shaft is supported by console.

1. Steering wheel
2. Lock pin ¼X1½
3. Bushing
4. Steering shaft
5. Pinion gear
6. Spring pin, ¼X1½
7. Shaft spacer
8. Bushing
9. Steering bracket
10. Steering rod
11. Gear sector
13. Lock pin, ¼X1½
14. Spring pin, 3/16X1 (2)
15. Flat washer, 0.656 (2)
16. Bushing
17. Drag link
18. Steering arm
19. Spring pin, ¼X1¼
20. "E" ring, ¾-inch (4)
21. Rocker pivot
22. Pivot spacer
23. Kingpin bearing (4)
24. Bolt, ½-13X3
25. Tie rod
26. Kingpin/spindle, RH
27. Kingpin/spindle, LH
28. Wheel bearings (4)
29. Flat washer, 0.765 bore (2)

Disengage blade clutch. At front of cutting deck, remove cotter pins from suspension rod, drive out rod at one end and remove spacers. Back off nut, washer and spring at rear of blade brake rod, then remove rod from cutting deck. Disconnect sway chains at rear of cutting deck on Models K-511, K-521 and K-522 or unbolt housing retainer on Models K-831, K-832 and K-852, then slip deck rearward to relieve pressure on drive belt and work belt from lower sheave of engine pulley. Pull cutting deck from under chassis.

Unhook clutch idler spring (16—Fig. R4) at end which is anchored to frame.

Remove bolt (18) and idler pulley (19) from idler arm (17). Remove traction drive belt from sheaves of engine and transmission pulleys, then unbolt engine from rear pan (1) and lift out of chassis. Reverse this sequence to reinstall.

Models L-711, L-721 and L-722

REMOVE AND REINSTALL. To remove engine, unbolt and remove hood. Remove spark plug lead, then disconnect battery cables (L-721 and L-722) and remove cable from starter motor. Disconnect leads from ignition switch

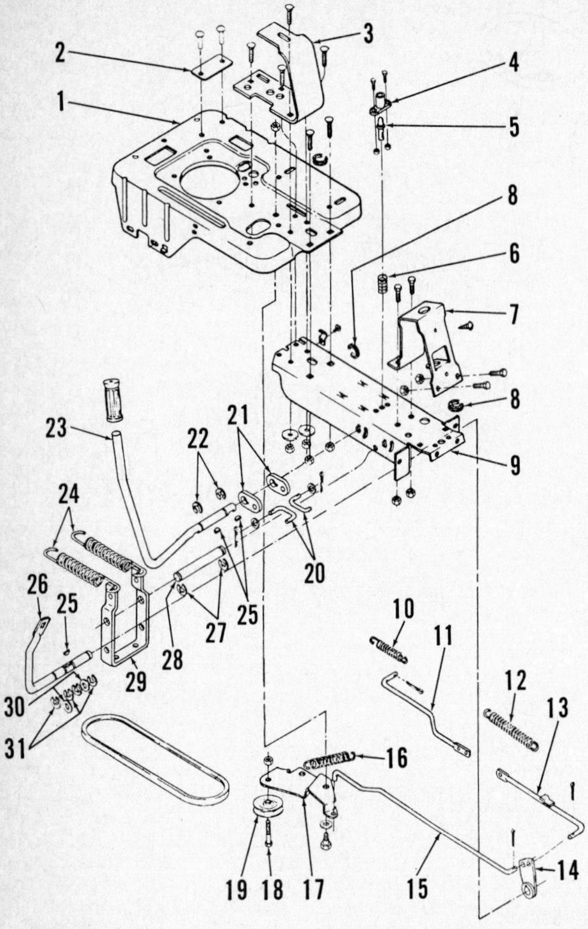

TRACTION DRIVE CLUTCH

All Models

All models use a conventional, pedal-operated, spring-loaded belt idler to apply or relieve belt tension between output pulley of engine and input pulley of transmission or transaxle. No adjustment is provided. If belt becomes worn, stretched or glazed to the point that operation is unsatisfactory, a new belt must be installed.

To renew drive belt or clutch components refer to the following paragraphs for procedures.

Models K-511, K-521, K-522, K-831, K-832 and K-852

REMOVE AND REINSTALL. To remove drive belt, remove mower deck from under chassis, as previously outlined. Disconnect spark plug lead as a safety precaution. Unhook idler spring (16 – Fig. R4) from chassis, then remove nut from idler pulley shaft (18) and remove pulley (19) and drive belt. Pulley and its bearing should be very carefully checked for condition such as worn or cracked sheaves and rough or noisy bearing. Pulley and bearing are serviced as an assembly.

Reverse procedure to reinstall belt and idler. Be sure that long side of idler pulley bearing is upward.

Models L-711, L-721, L-722, L-821, L-861 and L-863

REMOVE AND REINSTALL. Remove mower deck from under unit as previously outlined, and disconnect spark plug lead. Disconnect idler spring (1 – Fig. R5 or 16 – Fig. R6) from frame. Unbolt and remove idler pulley for thorough examination and carefully check condition of idler bracket and all linkage. Note that idler pulley and bearing are serviced as an assembly. When removing belt from transaxle equipped models

Fig. R4 – Exploded view of rear engine chassis to show arrangement of clutch and brake controls typical of all model numbers having a K prefix. Model K-832 is shown.

1. Rear platform
2. Identification plate
3. Seat mount spring
4. Parking brake post
5. Parking brake plunger
6. Parking brake spring
7. Steering shaft support
8. Steering shaft grommet
9. Frame channel
10. Spring (light duty)
11. Rear brake rod
12. Brake spring (main)
13. Front brake rod
14. Clutch arm
15. Clutch rod
16. Clutch idler spring
17. Idler bracket
18. Pulley shaft 3/8-16X1½
19. Idler pulley
20. Control links
21. Clutch pivot (2)
22. "E" rings 5/8-inch
23. Mower clutch lever
24. Rturn spring (2)
25. Woodruff key
26. Pedal shaft
27. "E" rings, 5/8-inch
28. Pivot shaft
29. Hanger bracket
30. Flat washer, 0.640 inch
31. "E" ring, 5/8-inch

and throttle control (bowden cable) from carburetor.

Disengage blade drive clutch lever. Remove blade brake rod, and cotter pins from ends of pivot shaft. Pull out shaft to release cutting deck from hanger and take care not to lose spacers. Shift cutting deck forward to relieve belt tension and work blade drive belt from engine pulley. Remove cutting deck from under chassis.

Unhook clutch idler spring, then remove idler pulley from idler bracket for maximum belt slack and slip drive belt off engine pulley. Unbolt and lift engine from chassis.

Models L-821, L-861 and L-863

REMOVE AND REINSTALL. To remove engine, unlatch and tilt hood forward. Disconnect spark plug cable and battery cables, and on Models L-821 and L-863, remove cable from starter motor. Disconnect ignition primary wires from engine, then unclamp and separate fuel line at filter. Plug or cap open line. Disconnect throttle cable.

Disengage blade drive clutch lever and remove blade brake rods from brackets on chassis. Now, remove cotter pins which secure parallel links to clutch arm assembly at each side, shift mower deck forward to ease tension on drive belt,

slip belt from engine pulley, then slide mower from beneath unit.

Unhook clutch idler spring at chassis end, back off nut on pulley shaft and remove pulley. Work drive belt off engine pulley. Recheck for any interfering connections, then unbolt and remove engine.

IMPORTANT NOTE: Design of muffler may vary and on some TECUMSEH engines an exhaust tube may extend through front grille just below hood hinge; in which case, remove muffler before removing engine.

Fig. R5 – Exploded view of drive clutch and mower clutch controls typical of Models L-711, L-721 and L-722.

1. Idler spring
2. Idler arm
3. Clutch rod
4. Idler pulley
5. Pivot bolt, 5/16-24
6. Pedal sleeve
7. Pedal/shaft
8. Woodruff keys
9. Springs
10. Clutch-brake levers
11. "E" ring, 5/8-inch
12. Corner gussets
13. Hanger
14. Shoulder bolt, 5/16-18
15. Link bar
16. Pivot bolt 5/16-18
17. Clutch pivot lever
18. Shoulder bolt 5/16-18
19. Mower control handle
20. Grip

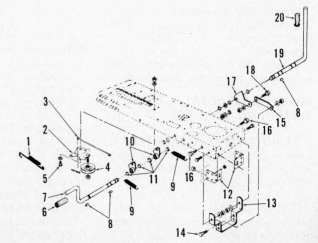

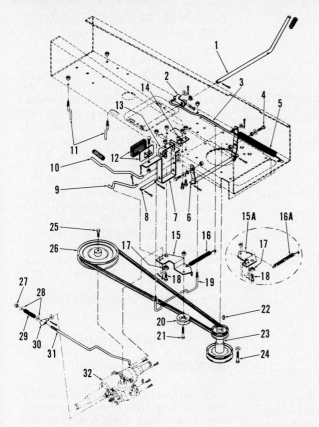

Fig. R6—Exploded view of drive clutch, brake and mower clutch controls typical of Models L-821, L-861 and L-863. Inset view shows different idler bracket for Models L-861 and L-863.

1. Blade clutch control handle
2. Blade clutch arm
3. Blade clutch rod
4. Pivot bolt, 5/16-18X¾
5. Spring
6. Clutch arm assy.
7. Pedal arm
8. Pedal shaft
9. Drive clutch rod
10. Park brake rod
11. Belt guides
12. Pedal & pad
13. Pedal bracket
14. Switch bracket
15/15A. Idler bracket
16/16A. Idler spring
17. Bushing
18. Pivot bolt, 5/16-18X¾
19. Belt guide
20. Idler pulley
21. Shaft bolt (⅜-16X1½)
22. Woodruff key (crankshaft)
23. Engine pulley
24. Cap screw ⅜-24X1⅜ w/Nylon washer
25. Cap screw, ¼-20X¾
26. Transaxle pulley
27. Locknut ⅜-16
28. Flat washer, 0.406 (2)
29. Brake spring
30. Trunnion
31. Brake rod
32. Transaxle assy.

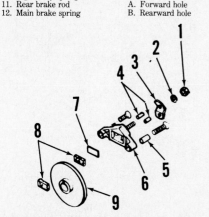

Fig. R7 — When service brake performance is unsatisfactory on rear-engine (K) models, shift spring (12) from hole A to hole B in brake rod (11) to increase tension. Model K-511 linkage is shown. See text.

10. Light duty spring
11. Rear brake rod
12. Main brake spring
13. Forward brake rod
A. Forward hole
B. Rearward hole

Fig. R8—Exploded view of transaxle-mounted disc brake assembly as used on Models L-821, L-861 and L-863.

1. Locknut
2. Flat washer
3. Brake lever
4. Dowel pins
5. Spacer
6. Pad holder
7. Brake pad plate
8. Brake pads
9. Brake disc

(L-821, L-861 and L-863) it will be necessary to remove shift lever plate and to insert a loop of drive belt through shift lever opening. Belt guides (11—Fig. R6) can be loosened so their positions can be shifted to make room for belt to clear sheaves. This also applies to snubbers under chassis.

Reverse procedure to reinstall belt and idler pulley.

TRANSMISSION

All Models

Single-speed transmission for Model K-511 is covered in this section. Refer to TRANSMISSION REPAIR section for service details on other models: FOOTE Model 2010 transmission is used on

Fig. R9—Exploded view of single-speed, reversible transmission used on Model K-511. Brake assembly included.

1. Brake bolt (special)
2. Brake lever
3. Brake jaw
4. Brake pads
5. Spring
6. Front case half
7. Gasket set
8. Input shaft bearing
9. Input shaft & pinion
10. Brake disc
11. "E" ring - ⅝-inch
12. Thrust washer
13. Bearing
14. Bevel gears
15. Clutch
16. Woodruff key
17. Output shaft/sprocket
18. Morton key (clutch)
19. Rear case half
20. Button plug
21. Drive lock pins
22. Clutch yoke
23. Lever extension
24. Grip
25. Lever plate
26. Shift lever

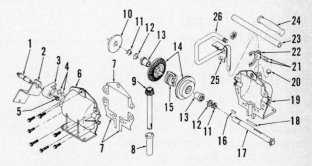

Models K-521, K-522, K-831, K-832, L-711, L-721 and L-722. FOOTE 2600 type is used on Model K-852. Model 649 PEERLESS transaxle (600 series) is used on Models L-821, L-861 and L-863.

Model K-511

R&R AND OVERHAUL. To remove transmission, remove operator's seat and seat mount. Disconnect safety interlock leads at transmission. Disconnect light duty spring and brake rod from brake lever. Raise and block rear of unit so rear wheels can roll, disconnect master link and remove drive chain from transmission output sprocket. Disconnect clutch idler spring from frame, remove idler pulley and pull main drive belt clear of transmission pulley. Remove snap ring at bottom of transmission input shaft and remove pulley. Unbolt transmission from chassis and remove.

Reverse procedure to reinstall. Check adjustment of drive chain before returning unit to service.

To disassemble removed transmission, back out seven cap screws which hold front case half (6 – Fig. R9) to rear case half (19). Brake assembly (parts 1 through 5) will remain with front case half (6) when cases are separated. Shafts, gears and shifter will remain in rear case half (19). Note that there are no spring-loaded parts to pop out of place. Lift output shaft assembly from case half. Use a light-duty puller and remove brake disc (10) from shaft, taking care not to lose key (16). Remove "E" ring (11) at brake disc end of shaft and slip off thrust washer (12), bearing (13) and a bevel gear (14). Clutch (15) is a sliding fit over key (18). When clutch is bumped from shaft and key (18) is removed, balance of parts can be slipped off shaft end. Separate input shaft and pinion from bearing (8). Soak removed parts in solvent for thorough cleaning and easier inspection. Peel old gasket set (7) from case and prepare a new gasket by coating with a non-hardening

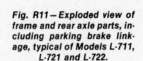

Fig. R10—Exploded view of body sheet metal parts typical of Models L-711, L-721, L-722, L-821, L-861 and L-863. Model L-721 shown.

1. Seat
2. Seat spring
3. Remote throttle control
4. Fender, LH
5. Running board pad (2)
6. Hood
7. Dash panel
8. Spacer (4)
9. Running board, LH
10. Running board, RH
11. Chassis cover
12. Fender, RH
13. Grille support
14. Front grille
15. Number plate
16. Park brake decal

transmission can be removed WITHOUT dismantling chassis cover, as follows:

Remove cutting deck as previously outlined. Disconnect lead from spark plug and, on electric-start models, disconnect battery leads. Unhook anchored end of clutch idler spring (1–Fig. R5) from frame and remove idler pulley (4). Pull drive belt off transmission pulley. Remove snap ring at bottom end of transmission input shaft and remove transmission pulley. Now, unbolt seat spring (2–Fig. R10) from rear of chassis and lift off seat and spring together. Disconnect safety interlock leads and rear brake rod from brake lever on transmission; then raise one rear wheel, disconnect master link and remove drive chain. Unbolt and remove transmission from chassis.

To reinstall, reverse order of removal.

DRIVE CHAIN

All Models So Equipped

To adjust drive chain, raise unit at rear, block up with wheels clear and remove rear wheels. Loosen retainer bolts at each rear axle bearing plate (17–Fig. R11) and shift axle rearward, using chain adjuster (16) on each side to keep axle square with frame. Proper chain tension calls for about 1/4 to 3/8-inch slack between sprockets. Before retightening bolts, be sure that adjusters (16) are engaged in the same corresponding notches.

When adjustment is satisfactory, lubricate drive chain with SAE 30 engine oil. If drive chain fails due to wear, renew chain; do not attempt to repair.

sealer, Hi-Tack or light grade of Permatex. Clean flange surfaces thoroughly.

Check teeth of bevel gears (14) and pinion (9) for undue wear or damage. Inspect shaft (17), input shaft (9) and bearings (8 and 13) for signs of heavy wear, dry galling or severe wear. Renew as needed.

Inspect clutch (15) with care, paying special attention to its keyway and engagement teeth. Key (18) should be renewed.

If clutch yoke (22) is undamaged, there is no need for removal, however, if it is cracked or broken, use a pin punch to tap pins (21) out part way, then grasp pins with small pliers to pull clear.

Reassemble parts in reverse of disassembly order, lubricating all wear surfaces and contact points of shafts, bearings and gear bores and teeth with a stable, medium weight grease. Manufacturer specifies Shell Darina AX.

Torque all assembly cap screws in an even criss-cross pattern to 6-7 ft.-lbs. Reinstall transmission.

Models K-521, K-522, K-831, K-832 and K-852

REMOVE AND REINSTALL. To remove transmission, first remove operator's seat and on electric start models, disconnect and remove battery. Unbolt and remove seat mount spring. Disconnect electrical connections for safety interlock switch at transmission. Disconnect light duty spring and rear brake rod from lever on brake assembly. Release clutch idler spring from frame, then unbolt and remove idler pulley and work main belt off and clear of transmission

pulley. Remove snap ring at low end of transmission input shaft and remove pulley. With transmission in neutral and one wheel raised, disconnect master link and separate chain from output sprocket of transmission. Unbolt and lift out transmission.

Reverse these steps to reinstall. Check drive chain adjustment before operating unit.

Models L-711, L-721 and L-722

REMOVE AND REINSTALL. Removal of transmission is somewhat involved because of its placement beneath operator's seat under sheet metal chassis cover (11–Fig. R10); however,

Fig. R11—Exploded view of frame and rear axle parts, including parking brake linkage, typical of Models L-711, L-721 and L-722.

1. Bearing bracket
2. Masterlink set
3. Chassis
4. Shield
5. Brake spring
6. Park brake post
7. Plunger
8. Brake rod
9. Spring
10. Clutch-brake decal
11. Belt snubber (2)
12. Housing snubber
13. Axle/differential assy.
14. Spacer (2)
15. Flat washer, 0.756 (2)
16. Chain adjuster (2)
17. Bearing plate
18. Axle bearing (2)
19. Spring

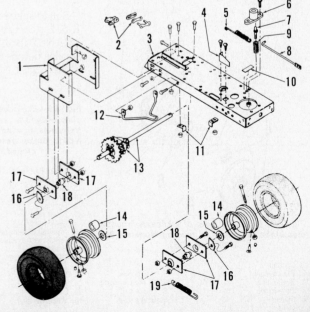

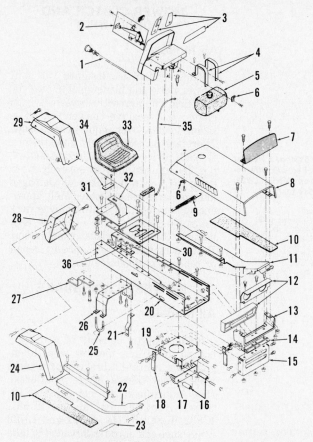

Fig. R12—Exploded view of chassis and body sheet metal parts typical of Models L-821, L-831 and L-863.

1. Choke control
2. Remote throttle control
3. Dash assy.
4. Fuel tank straps
5. Fuel tank
6. Hood cable clevis
7. Hood insert
8. Hood
9. Hood latch spring
10. Running board pad (2)
11. Running board, LH
12. Headlight panel/support
13. Panel frame
14. Hood hinge
15. Lower grille
16. Spacer (2)
17. Axle support
18. Brake rod bracket (2)
19. Engine base
20. Chassis
21. Axle strap
22. Running board, RH
23. Clutch-brake decal
24. Fender, RH
25. Axle clamp
26. Transaxle support
27. Hitch plate
28. Rear cover
29. Fender, LH
30. Chassis cover
31. Cover plate
32. Seat spring
33. Seat
34. Identification plate
35. Hood cable
36. Support

TRANSAXLE

Models L-821, L-831 and L-863

REMOVE AND REINSTALL. To remove transaxle assembly, proceed as follows: Disconnect spark plug lead and remove cutting deck from under unit. Unscrew knob from shift lever, then back out four cap screws and lift off chassis cover plate (30–Fig. R12). Disconnect battery on electric start models, release clutch idler spring (16–Fig. R6) from chassis and remove idler pulley (26). Slip drive belt off transaxle pulley and work forward out of the way. Remove cotter pin from trunnion (30) and release brake rod from brake lever on transaxle.

Raise and block securely under chassis just forward of rear wheels so that wheels are clear then remove nuts from axle clamp (25–Fig. R12) and remove two cap screws which hold transaxle housing in transaxle support (26). Roll transaxle out from under unit on its wheels for further service as needed.

DIFFERENTIAL AND REAR AXLE

All Models

Model K-511 is not equipped with a differential. Final drive sprocket and rear axle are integral.

Models K-521, K-522, K-831, K-832, K-852, L-711 and L-721 are equipped

with a conventional gear-type differential identified as INDUS WHEEL type 73 DP. Refer to DIFFERENTIAL REPAIR section for details.

Limited slip differential, used in Model L-722 only, is covered in this section.

Models L-821, L-831 and L-863 are fitted with PEERLESS transaxle Model 649 which includes differential. Refer to TRANSMISSION REPAIR section.

Models K-511, K-521, K-522, K-831, K-832, K-852, L-711, L-721 and L-722

REMOVE AND REINSTALL. To remove axle and differential, disconnect spark plug lead and remove cutting deck from under unit.

Raise under frame just forward of

rear wheels and block securely with wheels just clear.

With transmission in neutral, disconnect master link and remove drive chain. Removal of rear wheels is optional. Unbolt rear axle bearing plates (17–Fig. R11) from bearing bracket (1) or from frame. Lower axle, wheels and differential out of frame for further disassembly.

Reverse procedure to reinstall.

Model L-722

OVERHAUL. This differential uses friction blocks (pucks) (6–Fig. R14) which are held by spring pressure against a drive disc within housing (4). To disassemble, after wheels and bearing plates have been removed, set axle and sprocket (8) up in a padded vise and carefully unbolt housing (4), releasing spring pressure gradually. When case is opened, inspect all parts for undue wear, paying particulr attention to condition of drive disc (7), friction blocks (6) and pressure springs (5).

Set up slip axle (1) in vise so that springs (5) can easily be inserted into their sockets in housing (4) and reassemble.

IMPORTANT: DO NOT LUBRICATE. Bearing in housing (4) is pre-lubricated, and no other lubricant should be introduced.

Torque case bolts to 20-24 ft.-lbs. in alternating sequence. Reinstall axle bearings, bearing plates and wheels and reassemble. Adjust chain.

GROUND DRIVE BRAKE

Construction details of each disc brake appear in illustrations in applicable TRANSMISSION sections. Parking brakes are essentially a locking device to keep service brakes applied.

Models K-511, K-521, K-522, K-831, K-832 and K-852

ADJUSTMENT. Refer to Fig. R7 and/or to Fig. R4. Unhook light duty spring (10–Fig. R4) at frame, then re-

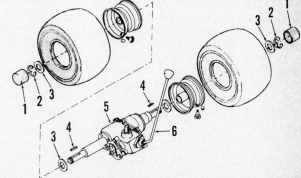

Fig. R13—Transaxle and rear wheels of Models L-821, L-831 and L-863. Transaxle is PEERLESS model 649.

1. Hub cap
2. "E" ring, ¾-inch
3. Washer, 0.765
4. Axle key, 3/16X2
5. Transaxle assy.
6. Shift lever

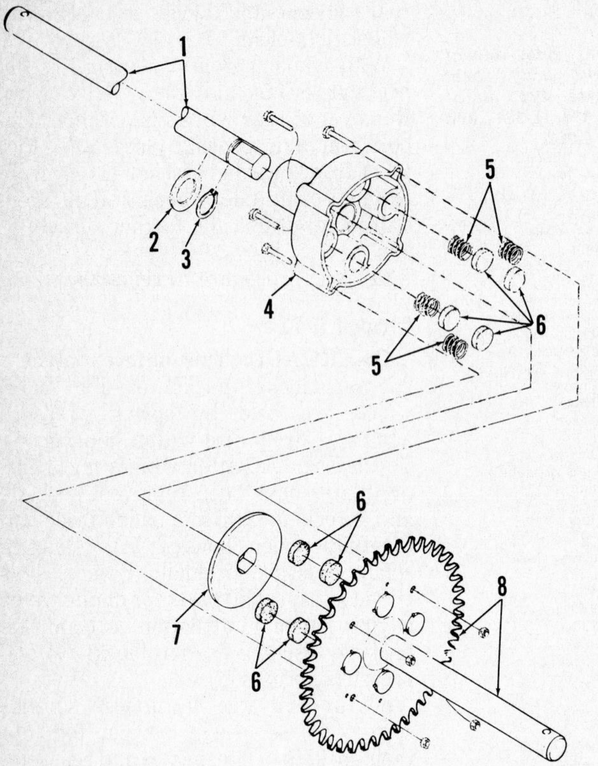

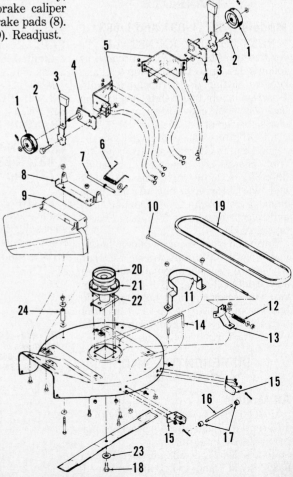

Fig. R14—Exploded view of Stewart limited slip differential used on Model L-722. See text.

1. Slip axle
2. Thrust washer, 0.751
3. Snap ring, 3/4-inch
4. Housing w/bearing
5. Springs (4)
6. Friction blocks (8)
7. Drive disc
8. Axle & sprocket assy.

MOWER CLUTCH AND DRIVE BELT

All Models

Blade clutch is designed to shift mower deck and blade spindle pulley(s) into firm contact with drive belt from engine by positive movement of a spring-loaded, over-center type linkage. There are no engagement adjustments. If blade drive becomes unsatisfactory, possible causes such as severely worn drive belt or worn, bent or damaged linkage parts will be immediately apparent upon inspection. Entire mower deck control assemblies are open for easy manual check and visual examination. Restore mower performance by repair or renewal of defective parts.

MOWER BLADE BRAKE

All Models

Blade brake(s) consists of a rod anchored to a frame member so as to engage and operate a brake shoe against blade pulley sheave. There is no brake shoe adjustment to compensate for wear. However, on Models K-831, K-832, K-852, L-711, L-721 and L-722 brake shoe end of rod is threaded to take

move brake rod (11) from actuating lever on brake. Shift main brake spring (12) to engage rearmost hole in brake rod, then reattach brake rod and light duty spring (10). If this does not satisfactorily restore service braking, check condition of brake disc which may be contaminated with oil or grease, and consider renewal of friction pads, ROPER part numbers 364259 or 370104.

If parking brake does not hold, check condition of plunger (5), spring (6) and parking brake tang on forward brake rod (13). Renew as necessary.

Models L-711, L-721 and L-722

Brake adjustment is similar to that outlined for K models in preceding paragraph.

Models L-821, L-861 and L-863

ADJUSTMENT. Brake linkage is shown in Fig. R6. To determine if service brake needs adjustment, observe if brake holds properly when clutch-brake pedal is fully depressed. If stopping force is ineffective, proceed as follows:

Be sure parking brake rod (10—Fig. R6) is unlocked, then turn locknut (27) on brake rod (31) until about 1/8-inch play can be measured between rearmost flat washer (28) and spring (29). If this adjustment alone does not restore braking, due to wear on brake pads, take up locknut (1—Fig. R8) against brake lever (3) so that brake pads (8) are closer to brake disc (9) but with no drag when brake is

not applied. Recheck brake rod adjustment.

If these procedures do not restore braking, remove brake rod assembly, then unbolt and remove brake caliper assembly (6); install new brake pads (8). Reinstall over brake disc (9). Readjust.

Fig. R15—Exploded view of cutting deck used on Models L-711, L-721 and L-722. NOTE: Items 20, 21 and 22 are furnished only as a complete "jackshaft assembly". Design is typical of that used on rear engine models.

1. Gage wheels, 5 in.
2. Shoulder bolt, 3/8-16
3. Spring lever assy.
4. Wheel bracket & axle
5. Gage wheel bracket
6. Guard spring
7. Hinge pin
8. Hinge bracket
9. Discharge guard
10. Blade brake rod
11. Belt snubber
12. Spring
13. Brake assy.
14. Belt keeper
15. Housing bracket (2)
16. Lower pivot shaft
17. Spacers (2)
18. Lockscrew, 3/8-24X7/8
19. Belt
20. Pulley
21. Jackshaft plate
22. Blade saddle
23. Washer, special
24. Spacer

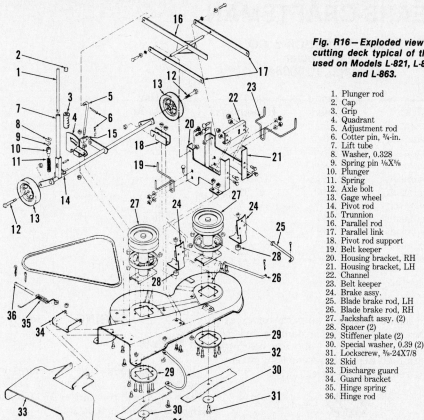

Fig. R16—Exploded view of cutting deck typical of that used on Models L-821, L-861 and L-863.

1. Plunger rod
2. Cap
3. Grip
4. Quadrant
5. Adjustment rod
6. Cotter pin, 3/4-in.
7. Lift tube
8. Washer, 0.328
9. Spring pin 1/8X5/8
10. Plunger
11. Spring
12. Axle bolt
13. Gage wheel
14. Pivot rod
15. Trunnion
16. Parallel rod
17. Parallel link
18. Pivot rod support
19. Belt keeper
20. Housing bracket, RH
21. Housing bracket, LH
22. Channel
23. Belt keeper
24. Brake assy.
25. Blade brake rod, LH
26. Blade brake rod, RH
27. Jackshaft assy. (2)
28. Spacer (2)
29. Stiffener plate (2)
30. Special washer, 0.39 (2)
31. Lockscrew, 3/8-24X7/8
32. Skid
33. Discharge guard
34. Guard bracket
35. Hinge spring
36. Hinge rod

series of holes at front and rear of cutting deck allow for three positions of pivot shaft at front and gage wheels at rear. Cutting height levels are: 1¾, 2⅜ and 3 inches.

Models K-521, K-522, K-831, K-832 and K-852

ADJUSTMENT. Height adjustment is of the quick adjust type. A single lift handle rotates gage wheel axle through five positions for cutting heights which range from 1½ to 3 inches.

Models L-711, L-721 and L-722

ADJUSTMENT. On these models, cutting height is set by manually changing setting of each of the individual selector levers (3 – Fig. R15) to raise or lower gage wheels (1). There is no adjustment for front edge of cutting deck.

Models L-821, L-861 and L-863

ADJUSTMENT. Cutting height is controlled by a single lift handle which raises or lowers both gage wheels simultaneously through a choice of five height positions from 1½ to 3½ inches.

Provision is made for adjusting forward slope of the cutting deck. Measured from a level surface, forward tip of leading blade should be from ⅛ to ¼-inch lower than rearmost tip of rear blade. To adjust, remove cotter pin from stud of trunnion (15 – Fig. R16) and turn trunnion clockwise to lower or counterclockwise to raise rear of deck. Check condition of parallel link (17) and rod assembly (16) and connections at clutch arm. Be sure cutter blades are straight.

up brake spring tension and prevent looseness and rattle in brake rod. This spring (12 – Fig. R15) should be taken up just enough to apply tension to rod (10). Brake assembly (13) must not drag against sheave (20) unless brake is applied.

Brake pad alone is not renewable. If blade brake fails, renew entire assembly.

MOWER CUTTING HEIGHT

Model K-511

ADJUSTMENT. On this model, a

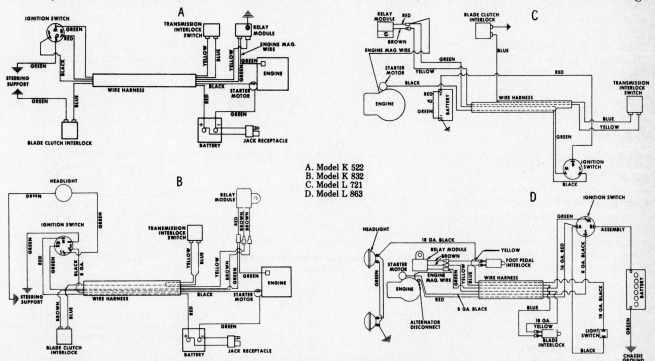

A. Model K 522
B. Model K 832
C. Model L 721
D. Model L 863

Fig. R17 — Convenience views of typical circuit diagrams for electric-start mowers. Units without electric starting and/or lights are basically similar.

SEARS-CRAFTSMAN

SEARS, ROEBUCK & CO.
Sears Tower
Chicago, IL 60684

Model	Make	Engine Model	Horsepower	Cutting Width, In.
502.256011	Tecumseh	V50	5	25
502.256020	Tecumseh	V60	6	25
502.256030	Tecumseh	V70	7	25
502.256040	Tecumseh	V80	8	25
502.256071	Tecumseh	V80	8	30
502.256080	Tecumseh	V100	10	36
502.256091	Tecumseh	V60	6	25
502.256111	Tecumseh	V80	8	30
502.256121	Tecumseh	V80	8	30
502.256130	Tecumseh	V100	10	30
502.256141	B&S	250000	11	30

FRONT AXLE

All Models

REMOVE AND REINSTALL. The axle member (17–Fig. SE1) is non-pivoting type and is also the front frame member. To remove axle member, raise and block front of unit. Remove front wheels, unbolt and remove tie rod and bracket assembly (20) from spindles (19). Remove cotter pins and washers from top of spindles, then remove spindles and bushings (18) from axle ends. Loosen jam nut (8) and unscrew clutch-brake pedal pad (9). Unbolt and raise console (7), then unbolt and remove front apron (22). Disconnect clutch and brake rods (14 and 15), then unbolt and remove axle member.

Clean and inspect all parts and renew as necessary. Reassemble by reversing removal procedure. Lubricate bearings with light coat of SAE 30 oil.

STEERING SHAFT

All Models

REMOVE AND REINSTALL. On Models 502.256011 and 502.256091 unbolt and remove steering handlebar from lower steering shaft (13–Fig. SE1). On all other models, remove screw from collar (2) and slide collar downward on sleeve (3), then remove bolt through upper end of sleeve. Remove steering wheel and sleeve with collar from upper shaft (4), then drive out roll pins (5) and remove upper shaft. On all models, unbolt and remove tie rod and bracket assembly (20) from spindles (19). Unbolt and raise console (7) and remove cotter pin (12) from lower steering shaft (13). Raise front of unit and withdraw steering shaft.

Clean and inspect all parts for excessive wear or any other damage and renew as necessary. Lubricate bearings with SAE 30 oil. Reassemble by reversing removal procedure.

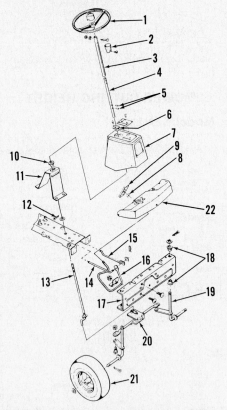

Fig. SE1 – Exploded view of front axle and steering system.

1. Steering wheel	12. Cotter pin
2. Collar	13. Lower steering shaft
3. Sleeve	14. Clutch-brake rod
4. Upper steering shaft	15. Brake rod
5. Roll pins	16. Clutch-brake lever
6. Nylon bearing	17. Axle member
7. Console	18. Spindle bearings
8. Jam nut	19. Spindle
9. Clutch-brake pedal	20. Tie rod & bracket assy.
10. Nylon bearing	21. Wheel
11. Steering support	22. Front apron

ENGINE

All Models

For overhaul and repair procedures on engines listed in Specification Table, refer to Small or Large Air Cooled Engines Service Manuals.

Refer to following paragraphs for removal and installation procedures.

REMOVE AND REINSTALL. Disconnect spark plug wire, ignition wires and throttle control cable. On recoil start models, unbolt and remove seat. On electric start models, tilt seat for-

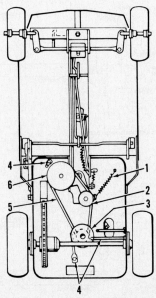

Fig. SE2 – View of typical traction drive clutch and belt. Adjust belt guides to ⅛-inch clearance between guide and belt or pulley.

1. Idler spring	4. Belt guides
2. Idler pulley	5. Drive belt
3. Engine pulley	6. Transmission pulley

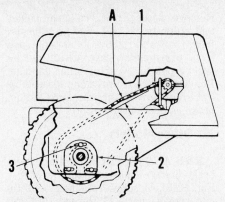

Fig. SE3 — View of drive chain and axle bearing retainer. Chain should deflect approximately ½-inch (A) under light finger pressure.

1. Drive chain
2. Axle bearing retainer
3. Slotted holes

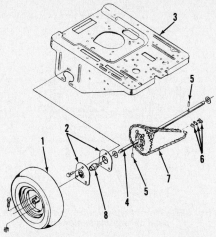

Fig. SE4 — View of typical rear axle assembly used on all models.

1. Rear wheel
2. Bearing retainers
3. Rear frame section
4. Rear axle
5. Roll pins
6. Master link
7. Drive chain
8. Bearing

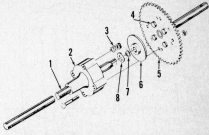

Fig. SE5 — Exploded view of limited slip differential used on Model 502.256011.

1. Slip axle
2. Housing w/bearing
3. Friction block & spring assy.
4. Friction block
5. Axle & sprocket assy.
6. Drive disc
7. Retaining ring
8. Thrust washer

ward and disconnect battery cables and starter cables. On all models, remove mower as outlined in MOWER DECK paragraphs. Unhook traction drive idler spring and slip drive belt off transmission pulley first, then off engine pulley. Remove engine mounting bolts and lift engine from frame.

Reinstall engine by reversing removal procedure.

TRACTION DRIVE CLUTCH AND BELT DRIVE

All Models

All models use a pedal-operated, spring-loaded belt idler to apply or relieve tension on drive belt. No adjustment is provided. If belt slippage occurs during normal operation due to belt wear or stretching, belt must be renewed.

REMOVE AND REINSTALL. Disconnect spark plug wire and remove mower as outlined in MOWER DECK paragraphs. Unhook idler spring (1– Fig. SE2) and unbolt and remove idler pulley (2). Loosen or remove belt guides as needed, and slip drive belt off transmission pulley (6) first, then off engine pulley (3).

Renew belt by reversing removal procedure. Adjust belt guides to ⅛-inch clearance between guide and belt or pulley.

TRANSMISSION

All Models

Models 502.256011 and 502.256091 are equipped with single speed Foote transmissions. All other models are equipped with Peerless model transmissions. For overhaul procedures on all models, refer to TRANSMISSION REPAIR section of this manual.

REMOVE AND REINSTALL. Disconnect spark plug wire and remove mower as outlined in MOWER DECK paragraphs. Unhook traction drive idler spring (1–Fig. SE2). Loosen or remove belt guides as needed, and slip drive belt (5) off transmission pulley (6). Remove retaining ring and pulley from transmission shaft. On Models 502.256011 and 502.256091, remove chain guard and disconnect drive chain. On all other models, remove rear shroud assembly and seat and disconnect drive chain. Disconnect transmission safety switch wires and unhook brake spring and linkage from transmission brake. Remove mounting bolts and lift transmission from frame. Reinstall by reversing removal procedure.

DRIVE CHAIN

All Models

R&R AND ADJUSTMENT. To remove chain (1–Fig. SE3), rotate differential sprocket to locate master link. Disconnect master link and remove chain. Reinstall in reverse order of removal.

To adjust chain tension, loosen bolts securing axle bearing retainer plates (2) and slide axle rearward to tighten chain. Move each side of axle equally to maintain alignment. When tension is correct, chain can be depressed approximately ½-inch with light finger pressure, tighten bearing plate bolts. Lubricate with light coat of SAE 30 oil.

DIFFERENTIAL

All Models

The differential used on all models except Model 502.256011 is manufactured

by Indus Wheel Company. For overhaul procedure, refer to DIFFERENTIAL REPAIR section of this manual. Overhaul procedure for limited slip differential, used in Model 502.256011, is covered in this section.

REMOVE AND REINSTALL. Disconnect spark plug wire and raise and block rear of unit. Remove rear wheels (1–Fig. SE4). Disconnect master link (6) and remove drive chain (7). Unbolt axle bearing plates (2) from rear frame section (3) and lower differential and axle out of frame. Remove bearings (8) from axle ends. Reinstall by reversing removal procedure.

Model 502.256011

OVERHAUL. The differential uses friction blocks (3 and 4–Fig. SE5) which are held by spring pressure against a drive disc (6) within housing (2). To disassemble, carefully unbolt housing from sprocket releasing spring pressure gradually. Separate housing and remove retaining ring (7) and thrust washer (8), then withdraw slip axle (1) from housing.

Inspect all parts for wear or damage and renew as necessary. Reassemble in

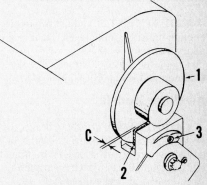

Fig. SE6 — View of typical brake assembly used on Models 502.256011 and 502.256091. Pad to disc clearance (C) should be 0.012-0.020 inch.

1. Brake disc
2. Brake pad
3. Adjustment screw

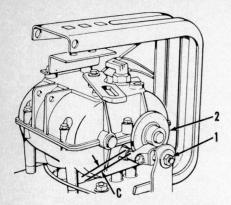

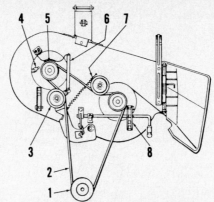

Fig. SE7 — View of typical brake assembly used on all models except 502.256011 and 502.256091. Pad to disc (2) clearance (C) should be 0.010-0.015 inch and is adjusted by turning nut (1).

Fig. SE9 — View of typical drive belt and clutch used on twin blade mowers.

1. Engine pulley
2. Drive belt
3. Idler pulley
4. Belt guide
5. Blade brake
6. Blade brake rod
7. Idler spring
8. Rear hanger bracket

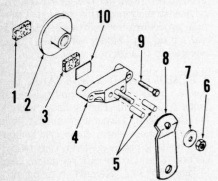

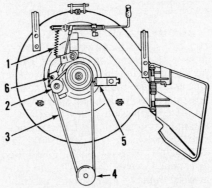

Fig. SE8 — Exploded view of brake assembly used on all models except 502.256011 and 502.256091.

1. Inner brake pad
2. Brake disc
3. Outer brake pad
4. Pad holder
5. Dowel pins
6. Adjusting nut
7. Washer
8. Actuating lever
9. Retaining bolts
10. Back plate

Fig. SE10 — View of typical drive belt and clutch used on single blade mowers.

1. Idler spring
2. Idler pulley
3. Drive belt
4. Engine pulley
5. Belt guide
6. Guide arm

reverse order of disassembly. Torque case bolts to 20-24 ft.-lbs.

NOTE: DO NOT LUBRICATE. Bearing in housing (2) is pre-lubricated and no other lubricant should be used.

GROUND DRIVE BRAKE

Models 502.256011-502.256091

ADJUSTMENT. To adjust brake, disengage parking brake lever. Turn brake adjustment screw (3 – Fig. SE6) to adjust pads for 0.012-0.020 inch clearance between pad (2) and brake disc (1).

All Other Models

ADJUSTMENT. To adjust brake, disengage parking brake lever. Turn brake adjusting nut (1 – Fig. SE7) to adjust pads for 0.010-0.015 inch clearance between pad and brake disc (2).

Models 502.256011-502.256091

R&R AND OVERHAUL. Disconnect brake linkage from actuating lever.

Fig. SE11 — Exploded view of typical single blade mower.

1. Nut
2. Lockwasher
3. Blade washer
4. Washer
5. Blade
6. Washer
7. Blade adapter
8. Mower deck
9. Spindle bearing housing assy.
10. Front slotted bracket
11. Clutch lever & linkage
12. Idler spring
13. Blade clutch idler arm
14. Idler pulley
15. Spindle shaft
16. Drive belt
17. Blade pulley
18. Belt guide
19. Rear slotted bracket

Remove shoulder bolt, brake pad holder and outer brake pad. Slide brake disc from shaft and remove inner brake pad from transmission case. Clean and inspect parts for wear or other damage. Renew parts as required and reassemble by reversing removal procedure. Adjust brake as previously outlined.

All Other Models

R&R AND OVERHAUL. Disconnect brake linkage from actuating lever (8 – Fig. SE8). Unscrew retaining bolts (9) and remove parts (3 through 10). Remove adjusting nut (6) and separate lever (8), actuating pins (5), back-up plate (10) and outer brake pad (3) from carrier (4). Slide brake disc (2) off shaft and remove inner brake pad (1) from transmission case. Clean and inspect parts for excessive wear or any other damage. Renew parts as required and reassemble by reversing removal procedure. Adjust brake as previously outlined.

MOWER DRIVE BELT

Models 502.256071-502.256080

REMOVE AND REINSTALL. Disconnect spark plug wire and remove mower unit as outlined in MOWER DECK paragraphs. On Model 502.256071, remove blade brake rod (6 – Fig. SE9). On Model 502.256080, remove right hanger bracket (8). On both models, unhook idler spring (7) and

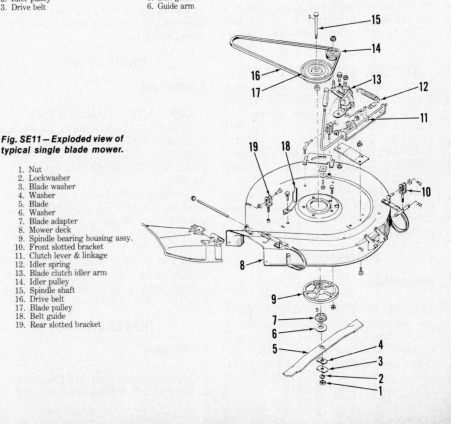

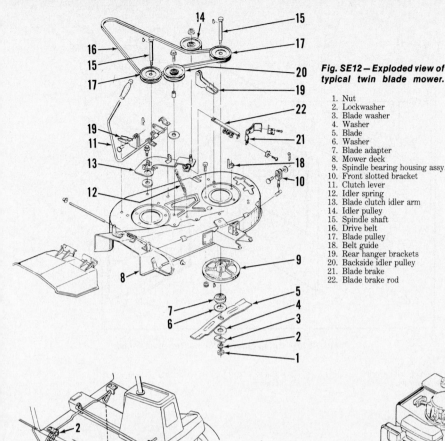

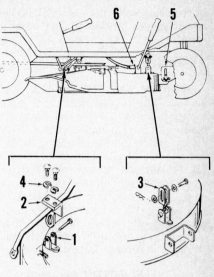

Fig. SE12 — Exploded view of typical twin blade mower.

1. Nut
2. Lockwasher
3. Blade washer
4. Washer
5. Blade
6. Washer
7. Blade adapter
8. Mower deck
9. Spindle bearing housing assy.
10. Front slotted bracket
11. Clutch lever
12. Idler spring
13. Blade clutch idler arm
14. Idler pulley
15. Spindle shaft
16. Drive belt
17. Blade pulley
18. Belt guide
19. Rear hanger brackets
20. Backside idler pulley
21. Blade brake
22. Blade brake rod

Fig. SE14 — View of mower suspension using slotted bracket in front only.

1. Rear mounting bracket
2. Rear hanger bracket
3. Front slotted bracket
4. "U" washers
5. Front suspension bracket
6. Safety switch

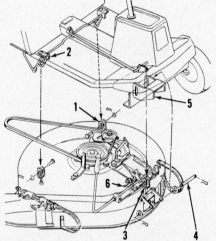

Fig. SE15 — Wiring diagram for recoil start models.

1. Key switch
2. Mower safety switch
3. Transmission safety switch
4. Module
5. R-Red
6. Y-Yellow
7. B-Brown
8. W-White
9. BK-Black

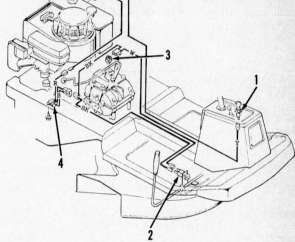

Fig. SE13 — View of mower suspension using front and rear slotted mounting brackets.

1. Rear slotted brackets
2. Rear suspension bracket
3. Front slotted bracket
4. Pin
5. Front suspension bracket
6. Safety switch

moval. Be sure belt is inside belt guide (5) and guide arm (6).

MOWER SPINDLE

All Models

REMOVE AND REINSTALL. Disconnect spark plug wire and remove mower and drive belt as outlined in MOWER DRIVE BELT paragraphs. Remove blade retainer nut (1 – Fig. SE11 or SE12) and remove blade (5), washers, blade holder (7) and Woodruff key from spindle shaft (15). Remove spindle shaft and pulley (17) from spindle bearing housing (9). Unbolt and remove spindle bearing housing from bottom of mower deck.

Clean and inspect parts for excessive wear or any other damage. Renew parts

as required and reinstall in reverse order of removal. Torque blade retainer nut to 30 ft.-lbs.

MOWER DECK

All Models

LEVEL ADJUSTMENT. With unit on level surface, measure distance from blade to ground from side to side and front to rear. Mower should be approximately ⅛-inch lower in front than in back. To adjust front to rear, remove pin from front slotted bracket (3 – Fig. SE13 or SE14) and adjust bracket to raise or lower front of mower. To adjust side to side, some models have adjustable rear slotted brackets (1 – Fig. SE13) to raise or lower sides of mower and other models are adjusted by adding

unbolt and remove idler pulley (3), then remove drive belt (2).

Renew belt in reverse order of removal. Be sure belt is inside belt guides.

All Other Models

REMOVE AND REINSTALL. Disconnect spark plug wire and remove mower unit as outlined in MOWER DECK paragraphs. Unhook idler spring (1 – Fig. SE10) and unbolt and remove idler pulley (2), then remove drive belt (3).

Renew belt in reverse order of re-

or removing washers (4 – Fig. SE14) between hanger bracket (2) and running board.

All Models

REMOVE AND REINSTALL. Disconnect spark plug wire and place mower in lowest position and disengage blade clutch. Disconnect wires from safety switch (6 – Fig. SE13 or SE14). Remove clevis pin from front slotted bracket (3) and remove pin (4) from front suspension bracket (5). Remove clevis pins connecting mower deck to rear suspension brackets (1). Slip blade drive belt off engine pulley and slide mower deck out from under chassis. Reinstall in reverse order of removal.

ELECTRICAL

All Models

All models are equipped with safety interlock switches on transmission shift linkage and blade clutch control linkage. Transmission must be in neutral and blade clutch disengaged before engine will start. Refer to Fig. SE15 or Fig. SE16.

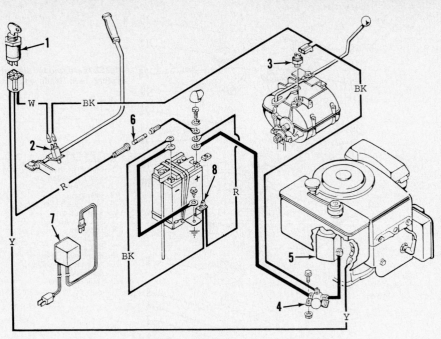

Fig. SE16 — Wiring diagram for electric start models.

1. Key switch
2. Mower safety switch
3. Transmission safety switch
4. Solenoid
5. Starter
6. Fuse
7. Battery charger
8. Charger plug
9. R-Red
10. Y-Yellow
11. W-White
12. BK-Black

SIMPLICITY

SIMPLICITY MFG. CO., INC.
500 North Spring St.
Port Washington, WI 53074

Model	Make	Engine Model	Horsepower	Cutting Width, In.
305	B&S	130202	5	24
315	B&S	130202*	5	26
355	B&S	130202	5	28
808	B&S	190702**	8	30
3005	B&S	130202	5	26
3008-2	B&S	190402	8	30
3008-3	B&S	190707	8	30
3008-FES	B&S	190707	8	36

*130207 on electric start models.
**190707 on electric start models.

FRONT AXLE AND STEERING SYSTEM

Models 305 – 315 – 355 – 3005 – 3008-2

REMOVE AND REINSTALL. The axle main member is also the front frame assembly (10 – Fig. S1). Pivot point is at joint of front and rear frame sections. To disassemble the steering system, support front of unit and remove front wheel assemblies. Unbolt and remove tie rods (14). Remove cotter pins from top of spindles (16 and 19), drive out roll pins (12) and remove both spindles with washers (13) and torsion springs (11). Unbolt and remove steering wheel (8) and cover (7). Remove washer (2), bushing (3), cup (4) and retaining ring (5) from upper end of steering shaft (6). Working through rear opening in front frame, remove cotter pin from steering shaft. Withdraw steering shaft from bottom of frame and remove cup (9), bushing (17) and washer (18).

Clean and inspect all parts and renew any showing excessive wear or other damage. Reassemble by reversing the disassembly procedure. Lubricate spindles and steering shaft bushings with SAE 30 oil.

Models 808 – 3008-3

REMOVE AND REINSTALL. The axle main member is also the front frame assembly (19 – Fig. S2). Pivot point is at joint of front and rear frame sections. To disassemble the steering system, support front of unit and remove front wheel assemblies. Unbolt and remove the tie rod (17) and drag link (11). Remove retaining rings (12) and remove spindles (13 and 18). Remove steering wheel (1), washer (2) and bushing (3). Unbolt and remove steering support and cover (20 and 21). Remove cot-

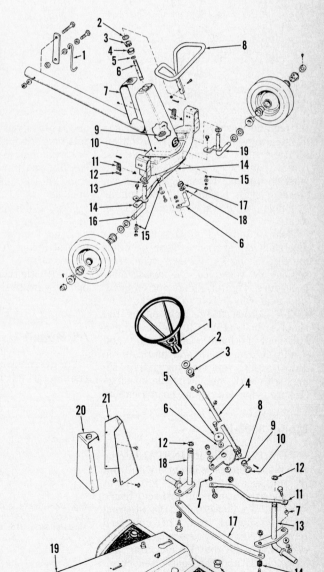

Fig. S1—Exploded view of typical front axle and steering system used on Models 305, 315 and 355. Models 3005 and 3008-2 are similar.

1. Mower hanger
2. Washer
3. Bushing
4. Bushing cup
5. Retaining ring
6. Steering shaft
7. Cover
8. Steering wheel
9. Bushing cup
10. Front frame & axle assy.
11. Torsion spring
12. Roll pin
13. Washer
14. Tie rods
15. Spacers
16. Spindle R.H.
17. Bushing
18. Washer
19. Spindle L.H.

Fig. S2—Exploded view of typical front axle and steering system used on Models 808 and 3008-3.

1. Steering wheel
2. Washer
3. Bushing
4. Steering shaft & pinion
5. Washer
6. Bushing
7. Spacers
8. Quadrant gear
9. Bushing
10. Washer
11. Drag link
12. Retaining ring
13. Spindle R.H.
14. Spring
15. Shoulder bolt
16. Spindle bushings
17. Tie rod
18. Spindle L.H.
19. Front frame & axle assy.
20. Steering support
21. Support cover

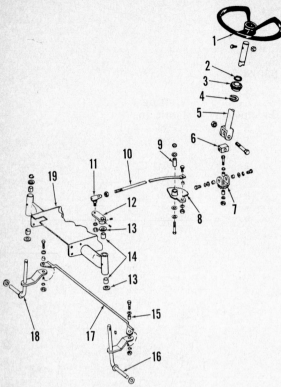

Fig. S3—Exploded view of typical front axle and steering system used on Model 3008-FE3.

1. Steering wheel
2. Washer
3. Bushing
4. Retaining ring
5. Steering shaft
6. "U" joint
7. Gear & yoke assy.
8. Quadrant gear
9. Sleeve
10. Drag link
11. Ball joint end
12. Steering arm
13. Washer
14. Spindle bushings
15. Spacer
16. Spindle L.H.
17. Tie rod
18. Spindle R.H.
19. Front frame & axle assy.

refer to Small Air Cooled Engines Service Manual.

Refer to the following paragraphs for removal and installation procedures.

Models 305 — 315 — 355 — 3005 — 3008-2

REMOVE AND REINSTALL. To remove the engine assembly, unbolt and remove engine hood on all models so equipped. On electric start models, disconnect battery cables and starter wires. On all models, disconnect ignition wire and throttle control cable. Unbolt and remove pulley and belt guard from left side and remove the belt guide. Place the blade clutch lever in disengaged position, depress clutch-brake pedal and remove belts from engine pulley. Unbolt and remove engine assembly. Reinstall engine by reversing the removal procedure.

Models 808 — 3008-3

REMOVE AND REINSTALL. To remove the engine assembly, open the rear frame cover and on electric start models, disconnect battery cables and starter wires. On all models, disconnect throttle cable and ignition wire. On Model 808, push transmission primary drive belt idler forward and remove belt from engine pulley. On Model 3008-3, depress clutch-brake pedal and remove transmission belt from engine pulley. On all models, place blade clutch control lever in disengaged position and remove mower drive belt from engine pulley. Unbolt and remove engine assembly. Reinstall engine by reversing the removal procedure.

Model 3008-FE3

REMOVE AND REINSTALL. To remove the engine assembly, remove the thumb screws at rear of hood and tilt hood forward. Disconnect battery cables, starter wires, ignition wire,

ter pin and washer (10) from lower end of steering shaft (4), then withdraw steering shaft and pinion. Unbolt and remove quadrant gear (8), special washer (5) and bushing (6). Remove steering shaft bushing (9) and spindle bushings (16), if need for renewal is indicated.

Clean and inspect all parts and renew any showing excessive wear or other damage. Using Fig. S2 as a guide, reassemble by reversing the disassembly procedure. Tighten flange nut on quadrant gear center bolt to a torque of 60 ft.-lbs. Torque flange nuts on tie rod shoulder bolts (15) to 30 ft.-lbs. Apply a light coat of lithium grease to pinion and quadrant gear teeth and lubricate tie rod, drag link and steering shaft bushings with SAE 30 oil. Spindle bushings (16) are nylon and require no lubrication.

Model 3008-FE3

REMOVE AND REINSTALL. The axle main member is integral with front frame assembly (19 – Fig. S3). Pivot point is at joint of front and rear frame sections. To disassemble the steering system, support front of unit and remove front wheel assemblies. Unbolt and remove tie rod (17) and drag link assembly (10 and 11). Loosen set screws and remove steering arm (12), washers (13) and left spindle (16). Remove retaining ring, washers and right spindle (18). Unbolt and remove steering wheel (1) and unbolt pinion and yoke (7) from frame. Pull upward on steering shaft

and remove washer (2), bushing (3) and retaining ring (4). Unbolt and remove quadrant gear (8) and sleeve (9). Remove steering shaft and "U" joint assembly (5, 6 and 7). Any further disassembly will be obvious.

Clean and inspect all parts and renew any showing excessive wear or other damage. Lubricate spindle bushings (15) and steering gear teeth with lithium grease. Use SAE 30 oil on bushing (3), sleeve (9), tie rod and drag link ends and on "U" joint bushings.

ENGINE

All Models

For overhaul and repair procedures on engines listed in Specification Table

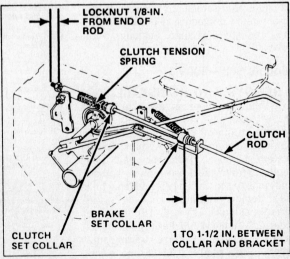

Fig. S4—Brake and clutch linkage adjustment on Models 305, 315 and 355.

LOCKNUT 1/8-IN. FROM END OF ROD

CLUTCH TENSION SPRING

CLUTCH ROD

BRAKE SET COLLAR

CLUTCH SET COLLAR

1 TO 1-1/2 IN. BETWEEN COLLAR AND BRACKET

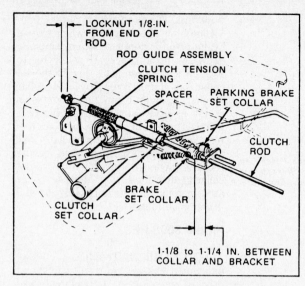

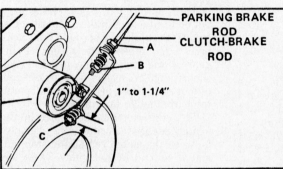

Fig. S5 — Brake and clutch linkage adjustment on Models 3005 and 3008-2.

throttle cable and fuel line. Place blade clutch control lever in disengaged position and remove mower drive belt from engine pulley. Depress clutch-brake pedal and remove main drive belt from engine pulley. Unbolt and lift engine from frame. Reinstall engine by reversing the removal procedure.

MAIN DRIVE CLUTCH AND BRAKE

The main drive clutch on all models is the belt idler type. Band and drum type brake is used on all models.

Models 305 — 315 — 355 — 3005 — 3008-2

ADJUSTMENT. To adjust the clutch and brake linkage, remove rear hitch plate, refer to Figs. S4 and S5 and proceed as follows: With clutch-brake pedal in fully up (clutch engaged) position, adjust the brake set collar to a distance of 1 to 1½ inches (305, 315 and 355) or 1⅛ to 1¼ inches (3005 and 3008-2) from bracket as shown. On all models, adjust locknut to ⅛-inch from end of clutch rod. On Models 305, 315 and 355 with clutch-brake pedal in up position, adjust the clutch set collar so that clutch tension spring is preloaded 1/16-inch. On Models

3005 and 3008-2, depress the clutch-brake pedal until brake is fully applied. Adjust clutch set collar so that spacer holds the clutch tension spring against the clutch rod guide. Spring should not be compressed when pedal is fully depressed.

Model 808

ADJUSTMENT. To adjust the clutch and brake linkage, refer to Fig. S6 and turn nut (A) on parking brake rod to end of threads. Engage parking brake and tighten nut (B) until the spring against it is fully compressed. Adjust nut (C) to obtain a distance of 1 to 1¼ inches between flat washer and brake band bracket as shown. Disengage parking brake and check to see that spring by nut (A) pushes brake band free of brake drum. If not, adjust nut (A) as required.

Clutching occurs when the pivot shaft assembly is moved to neutral position and belt tension is released from both the forward and reverse idler belts. Depressing the clutch-brake pedal or moving the direction control lever to NEUTRAL position will place the pivot shaft in neutral. Refer to Fig. S7 and loosen the set screw in directional control lever collar. Place directional control lever in NEUTRAL position. Rotate pivot shaft assembly to tighten forward idler belt, applying about five pounds pressure. Place a mark on directional control rod at front edge of rod guide.

Fig. S6 — Brake adjustment on Models 808 and 3008-3. Refer to text.

Fig. S7 — View of clutch and forward-reverse drive linkage on Model 808. Refer to text for adjustment procedure.

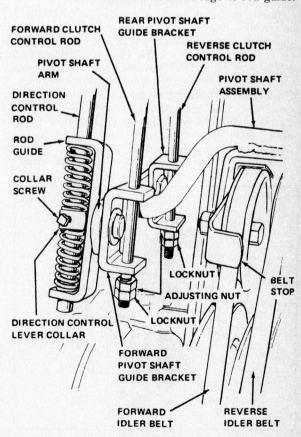

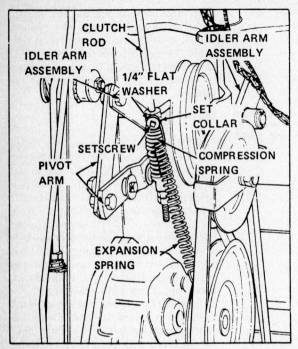

Fig. S8 — Clutch linkage on Model 3008-3. Refer to text for adjustment procedure.

Rotate pivot shaft in opposite direction to tighten reverse idler belt, once again applying about five pounds pressure. Place a second mark on directional control rod at front edge of rod guide. Distance between the two marks should be 11/16-inch. If this distance is incorrect, loosen the four cap screws securing pivot shaft in place. Move pivot shaft assembly forward or rearward as necessary to obtain the correct distance. Moving pivot shaft rearward will increase the distance. When the 11/16-inch distance is obtained, tighten the four cap screws to a torque of 15 ft.-lbs. Place a center mark on rod halfway between front and rear marks. Align front edge of rod guide with the center mark and tighten the collar set screw. Place the directional control lever in full forward position and pull pivot shaft downward. Loosen the locknut and turn adjusting nut until a clearance of ¼-inch exists between the adjusting nut and the forward pivot shaft guide bracket. Tighten the locknut. Move the directional control lever to full reverse position and push pivot shaft upward. Loosen the locknut and set the adjusting nut to a clearance of ¼-inch from the reverse pivot shaft guide bracket. Tighten the locknut.

Model 3008-3

ADJUSTMENT. To adjust the brake linkage, refer to Fig. S6 and turn nut (A) on parking brake rod to end of threads. Engage parking brake and tighten nut (B) until the spring against it is fully compressed. Then, adjust nut (C) on clutch-brake rod to obtain a distance of 1 to 1¼ inches between flat washer and brake band bracket as shown. Disengage parking brake and check to see that front spring by nut (A) pushes brake band free of brake drum. If not,

adjust nut (A) as required.

Clutching should occur when clutch-brake pedal is depressed approximately ½ of the pedal travel. If not, refer to Fig. S8 and loosen set screw in the set collar on clutch rod. Move set collar against or away from the compression spring as necessary, then retighten set screw. If clutching occurs too soon (before ½ of pedal travel), move set collar away from spring. If clutching occurs too late (pedal too far down) move set collar against the spring. Move set collar at ¼-inch increments until correct pedal travel is obtained.

Model 3008-FE3

ADJUSTMENT. To adjust the brake and clutch linkage, refer to Fig. S9 and proceed as follows: With clutch-brake pedal in up position, adjust elastic stop nut on front end of brake rod until nut is against the brake rod spring. Do no compress the brake rod spring. Fully depress the clutch-brake pedal. Pedal should stop 2½ inches from front edge of foot rest. If not, readjust the stop nut to obtain the 2½ inch distance.

With the clutch-brake pedal in up (clutch engaged) position, the distance between the clutch rod elastic stop nut and front end of rod guide should be ⅝-inch. Adjust the clutch rod stop nut to obtain the ⅝-inch distance. Fully depress clutch-brake pedal and adjust the set collar against rear of clutch rod spring so that spring is just free to rotate on the rod.

MAIN DRIVE BELTS

Models 305 – 315 – 355 – 3005 – 3008-2

REMOVE AND REINSTALL. To remove the main drive belt (1 – Fig.

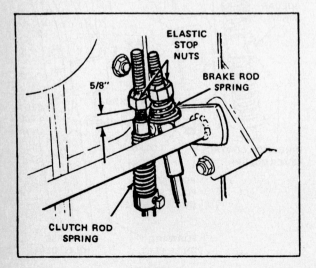

Fig. S9 — Brake and clutch linkage adjustment on Model 3008-FE3. Refer to text.

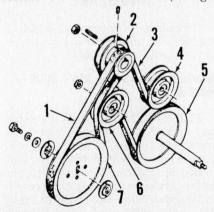

Fig. S10 — View showing main drive belt, pto drive belt and pulleys used on Models 305, 315, 355, 3005 and 3008-2.

1. Main drive belt	5. Pulley & jackshaft
2. Engine pulley	6. Main drive clutch idler pulley
3. Pto drive belt	7. Transmission input pulley
4. Pto (blade) clutch idler pulley	

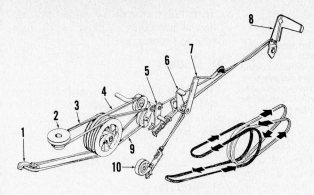

Fig. S11 – View showing clutch and brake linkage and drive belt arrangement on Model 808.

1. Primary drive belt idler
2. Engine pulley
3. Primary drive belt
4. Reverse drive belt
5. Pivot shaft assy.
6. Parking brake lever
7. Directional control lever
8. Clutch-brake pedal
9. Forward drive belt
10. Brake assy.

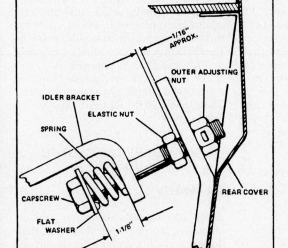

Fig. S12 – View showing primary drive belt idler adjustment on Model 808.

S10), unbolt and remove engine hood on all models so equipped. On electric start models, disconnect battery cables. On all models, disconnect spark plug wire, then unbolt and remove pulley and belt guard from left side. Remove the belt guide and place blade clutch lever in disengaged position. Remove the pto belt (3) from engine pulley. Depress the clutch-brake pedal and remove the main drive belt.

Install new belt by reversing the removal procedure. Adjust clutch and brake linkage as required.

Model 808

REMOVE AND REINSTALL. To remove the main drive belts (3, 4 and 9 – Fig. S11), disconnect spark plug wire, open the rear frame cover and on electric start models, remove the battery. On all models, push the primary drive belt idler (1) forward until primary belt (3) can be removed from engine pulley (2). Remove primary belt from left side of transmission pulley. Note position of belt stops on the forward and reverse control idler pulleys and loosen the mounting bolts. Remove reverse drive belt (4) and forward drive belt (9), then complete the removal of the primary belt.

Install new belts by reversing the removal procedure. Refer to Fig. S12 and

adjust outer adjusting nut until the length of the idler tension spring is 1 1/8 inches. Adjust the inner elastic stop nut to a clearance of 1/16-inch from frame bracket. Adjust clutch and brake linkage as outlined in previous paragraphs.

Model 3008-3

REMOVE AND REINSTALL. To remove the main drive belt (2 – Fig. S13), tie the clutch-brake pedal in fully depressed position. Disconnect spark plug wire and open rear frame cover. Working through rear opening, remove main drive belt from engine pulley. Remove the belt from transmission input pulley (5), flat idler pulley (4) and clutch idler pulley (3). If necessary, loosen clutch idler pulley mounting bolt to provide clearance to remove belt from under belt stop.

Install new belt by reversing the removal procedure. Adjust clutch and brake linkage as required.

Model 3008-FE3

REMOVE AND REINSTALL. To remove the main drive belt (4 – Fig. S14), disconnect spark plug wire and proceed as follows: Place the blade clutch control lever in disengaged position and remove mower drive belt from engine pulley. Tie the clutch-brake pedal in fully depressed position. Note position of belt stops at engine pulley, loosen cap screws securing belt stops to frame and move belt stops to the side. Remove main drive belt.

Install new belt by reversing the removal procedure. Make certain that belt stops are installed correctly. Adjust clutch and brake as necessary.

TRANSMISSION

All Models

The transmission used on Models 305, 315, 355, 3005 and 3008-2 are equipped with two forward gears and one reverse. The transmission used on Model 808 is equipped with two forward gears. No reverse gear is used in this transmission as the forward-reverse drive belt ar-

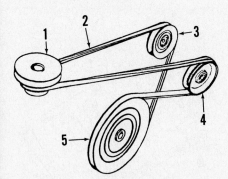

Fig. S13 – Main drive belt and pulleys used on Model 3008-3.

1. Engine pulley	4. Flat idler
2. Main drive belt	5. Transmission input
3. Clutch idler	pulley

Fig. S14 – Main drive belt and pulleys used on front engine Model 3008-FE3.

1. Engine pulley (pto drive)
2. Belt stops
3. Engine pulley (main drive)
4. Main drive belt
5. Idler pulley
6. Clutch idler arm
7. Transmission pulley

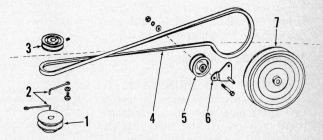

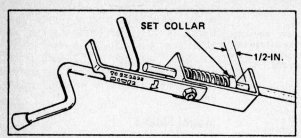

Fig. S15 — On Models 305, 315, 355, 3005 and 3008-2, when blade clutch control lever is in engaged position, clearance between set collar and rod bracket should be ½-inch.

rangement allows for forward or reverse operation in either gear. The transmissions used on Models 3008-3 and 3008-FE3 are equipped with three forward gears and one reverse. All units are of the transaxle type with the transmission gears and shafts, differential and axle shafts contained in one case.

NOTE: On models so equipped, make certain that safety interlock switches are connected and are in good operating condition before returning mower to service.

Models 305 — 315 — 355 — 3005 — 3008-2

REMOVE AND REINSTALL. The transmission shafts, shifter shafts and left axle shaft extend through the rear frame. The following procedures will outline the removal of components necessary for removal of the transmission and differential. Actual removal of the shafts, gears and differential is outlined in the Simplicity portion of the TRANSMISSION REPAIR section of this manual.

Unbolt and remove engine hood on all models so equipped. On electric start models, disconnect battery cables and remove battery. On all models, disconnect spark plug wire. Unbolt and remove the pulley and belt guard from left side, then remove the belt guide. Place blade clutch control lever in disengaged position and remove the pto drive belt from engine pulley. Depress the clutch-brake pedal and remove the main drive belt. Unbolt and remove transmission input pulley, then remove the rear hitch plate. Loosen the set screw and remove brake drum and key from brake shaft. Support rear of unit and remove rear wheel assemblies. Remove shift links, retaining rings and springs from shifter shafts and the transmission case. All gears, shafts and differential assembly can now be removed.

Reassemble by reversing the disassembly procedure. Lubricate with general purpose lithium grease.

Models 808 — 3008-3

REMOVE AND REINSTALL. To remove the transaxle assembly disconnect spark plug wire, open rear frame cover and on electric start models,

remove the battery. On Model 808, push primary drive belt idler forward until primary belt can be removed from engine pulley. Remove primary belt from left side of transmission pulley. Remove reverse drive and forward drive belts from transmission pulley. On Model 3008-3, tie the clutch-brake pedal in fully depressed position and remove main drive belt from engine pulley first, then from transmission input pulley. On all models, unbolt and remove brake band, loosen set screw and remove brake drum. Attach a hoist to rear frame, unbolt transmission and left axle housing from frame, then raise rear of unit to clear shift lever. Roll transaxle assembly from chassis.

Reinstall transaxle by reversing the removal procedure. Adjust clutch and brake as required.

Model 3008-FE3

REMOVE AND REINSTALL. To remove the transaxle assembly, disconnect spark plug wire and proceed as follows: Unbolt and remove the seat and rear fender assembly. Depress the clutch-brake pedal and remove main drive belt from transmission input pulley. Unbolt and remove the brake band. Support unit under main frame just ahead of pivot point. Loosen set screws and remove set collar from pivot shaft on side plate assembly. Roll transaxle assembly rearward. Unbolt and remove side plates. Unpin and remove rear wheel assemblies.

Fig. S16 — View of blade clutch (pto) linkage, pulleys and drive belts used on Models 305, 315, 355, 3005 and 3008-2. Belt (4) and pulleys (3 and 5) are used on Model 355 twin blade mower.

1. Mower pulley
2. Secondary mower belt
3. Mower pulleys (355)
4. Secondary mower belt (355)
5. Flat idlers (355)
6. Clutch rod
7. Tension spring
8. Set collar
9. Rod bracket
10. Blade clutch (pto) control lever
11. Spring
12. Pulley & jackshaft
13. Blade clutch idler pulley
14. Mower clutch belt
15. Engine pulley
16. Transmission input pulley

Reinstall transaxle by reversing the removal procedure. Adjust clutch and brake as necessary.

All Models

OVERHAUL. For transmission overhaul procedures, refer to the Simplicity paragraphs in the TRANSMISSION REPAIR section of this manual.

DIFFERENTIAL

All Models

R&R AND OVERHAUL. Transmission gears, shafts, differential and axle shafts are contained in one case. To remove the transaxle assembly, refer to the R&R procedures outlined previously in TRANSMISSION paragraphs.

For differential overhaul procedures, refer to the Simplicity paragraphs in the TRANSMISSION REPAIR section of this manual.

MOWER BLADE CLUTCH AND BELT

All Models

Belt idler type blade clutch is used on all models. If the mower drive belt slips during normal operation, check and adjust the clutch tension spring. If belt cannot be adjusted, due to excessive wear or stretching, renew the belt. On models equipped with two belts, R&R procedures will be given for both belts.

Models 305 — 315 — 355 — 3005 — 3008-2

To adjust the blade clutch, move the control lever to fully engaged position. Clearance between the set collar and bracket should be ½-inch as shown in Fig. S15. If clearance is incorrect, disengage blade clutch, loosen set screw and

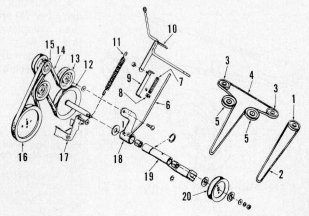

17. Pto brake (3005 & 3008-2)
18. Clutch idler arm

19. Jackshaft pulley bracket
20. Jackshaft pulley

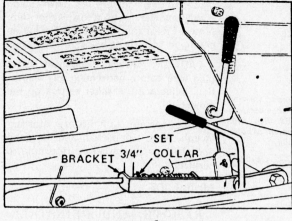

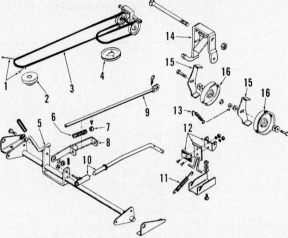

Fig. S17—On Models 808 and 3008-3, when blade clutch control lever is in engaged position, clearance between set collar and rod bracket should be 3/4-inch.

Fig. S18—View of blade clutch linkage, pulleys and mower drive belt used on Models 808 and 3008-3.

1. Belt stops
2. Engine pulley
3. Mower drive belt
4. Mower pulley
5. Blade clutch control lever
6. Tension spring
7. Set collar
8. Rod bracket
9. Clutch rod
10. Blade brake rod
11. Brake spring
12. Blade brake
13. Spring
14. Blade clutch idler arm
15. Belt guides
16. Clutch idler pulleys

On Model 808, refer to Fig. S11 and push primary drive belt idler (1) forward until primary (transmission drive) belt (3) can be removed from engine pulley (2). Loosen mounting bolts and move belt stops (1–Fig. S18) away from engine pulley. Remove mower clutch belt (3) from engine pulley.

On Model 3008-3, tie the clutch-brake pedal in fully depressed position. Working through rear frame opening, remove transmission main drive belt (2–Fig. S13) from engine pulley (1). Loosen mounting bolts and move belt stops (1–Fig. S18) away from engine pulley. Remove mower clutch belt (3) from engine pulley.

On all models, note position of belt guides (15–Fig. S18) on idler pulleys (16) and loosen the mounting bolts. Move belt stops as required to remove the mower clutch belt (3).

Install new mower belt by reversing the removal procedure. Make certain that belt guides and belt stops are properly installed. Adjust blade clutch as necessary.

Model 3008-FE3

To adjust the blade clutch, place the control lever in engaged position. Clearance between the set collar and rod bracket should be 5/8-inch measured as shown in Fig. S19. If clearance is not correct, disengage blade clutch, loosen set screw and reposition set collar on rod end. Engage clutch and recheck clearance.

To remove the blade clutch belt (2–Fig. S21) and blade belt (6), disconnect spark plug wire and proceed as follows: Place the blade clutch control lever in disengaged position and remove the belt from engine pulley (1) and mower drive pulley (3). Disconnect rod bracket (5–Fig. S20) from clutch idler arm (6). Disconnect mower lift link, unpin front of mower and slide mower unit from under left side. Note position of belt

reposition set collar. Engage clutch and recheck clearance.

To remove the mower clutch belt (14–Fig. S16), unbolt and remove engine hood on models so equipped. Disconnect spark plug wire, then unbolt and remove pulley and belt guard from left side. Remove belt guide and place blade clutch lever (10) in disengaged position. On models so equipped, pull the pto brake (17) away from belt. Remove the mower drive clutch belt. To remove the secondary mower belt (2 or 5), unbolt and remove pulley and belt cover from mower housing. Pull the jackshaft pulley bracket (19) forward and remove the belt from pulley (20). Remove belt from mower.

Install new belts by reversing the removal procedure and adjust blade clutch as required.

Models 808 – 3008-3

To adjust the blade clutch, move the control lever to fully engaged position. Clearance between the set collar and rod bracket should be 3/4-inch measured as shown in Fig. S17. If clearance is not correct, disengage blade clutch, loosen set screw and reposition set collar on rod. Engage clutch and recheck clearance.

To remove the mower drive belt (3–Fig. S18), first remove mower unit as follows: Disconnect spark plug wire and on electric start models, open rear frame cover and remove battery. Place blade clutch lever in disengaged position. Disconnect the mower lift link and blade clutch rod, then unpin front of mower from frame. Unbolt pulley and belt cover from mower housing. Remove mower drive belt from mower pulley and slide mower unit out from under left side.

Fig. S19—On Model 3008-FE3, when blade clutch control lever is in engaged position, clearance between set collar and rod bracket should be 5/8-inch.

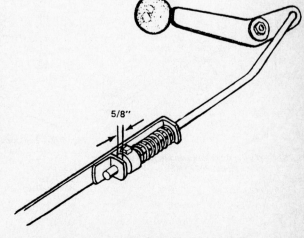

5/8"

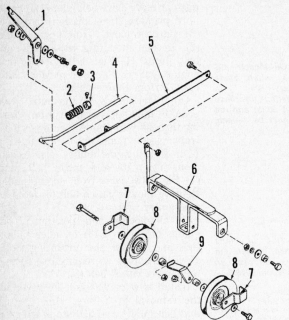

blade rotary mowers. A twin blade rotary mower is used on Model 355. Mower used on Model 3008-FE3 is equipped with three blades.

CAUTION: Always disconnect spark plug wire before performing any inspection, adjustment or other service on the mower.

Make certain that safety starting switches are connected and in good operating condition before returning to service.

Model 305

REMOVE AND REINSTALL. Remove the mower unit as follows: Un-

Fig. S20—Exploded view of blade clutch linkage used on Model 3008-FE3. See Fig. S21 for arrangement of belts and pulleys.

1. Blade clutch control lever
2. Tension spring
3. Set collar
4. Clutch rod
5. Rod bracket
6. Blade clutch idler arm
7. Belt stops
8. Clutch idler pulleys
9. Stop plate

stops (7) on clutch idler pulleys (8). Loosen pulley mounting bolts and remove clutch belt (2–Fig. S21). Unbolt and remove the 2-piece cover from mower housing. Unhook spring (9), move belt idler to the left and remove blade belt (6).

Install new belts by reversing the removal procedure. Make certain that belt stops are properly installed. Adjust blade clutch as necessary.

MOWER BLADES AND SPINDLES

All Models

Models 305, 315, 3005, 808, 3008-2 and 3008-3 are equipped with single

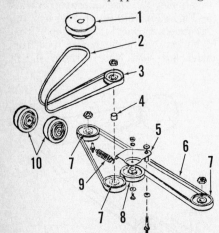

Fig. S21—Mower drive belts and pulleys used on Model 3008-FE3.

1. Engine pulley
2. Mower clutch belt
3. Mower drive pulley
4. Spacer
5. Idler arm
6. Blade belt
7. Mower pulleys
8. Flat idler
9. Spring
10. Clutch idler pulleys

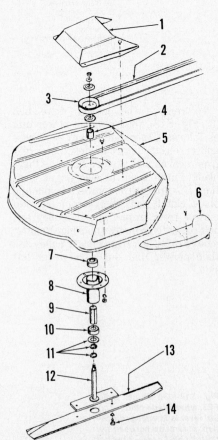

Fig. S22—Exploded view of the 24" mower unit used on Model 305.

1. Cover
2. Mower drive belt
3. Mower pulley
4. Spacer
5. Mower housing
6. Deflector
7. Bearing
8. Bearing housing
9. Spacer
10. Bearing
11. Washers
12. Spindle
13. Blade
14. Cap screw

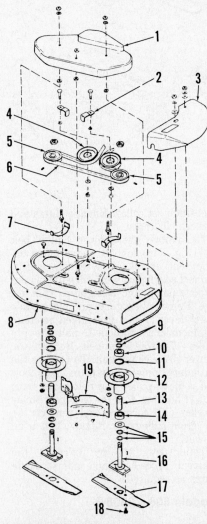

Fig. S23—Exploded view of the 28" twin blade mower unit used on Model 355.

1. Cover
2. Belt stops
3. Deflector
4. Flat idler pulleys
5. Mower pulleys
6. Mower drive belt
7. Belt guides
8. Mower housing
9. Washers
10. Bearing
11. Backing ring
12. Bearing housing
13. Spacer
14. Bearing
15. Washers
16. Spindle
17. Blade
18. Cap screw
19. Baffle plate

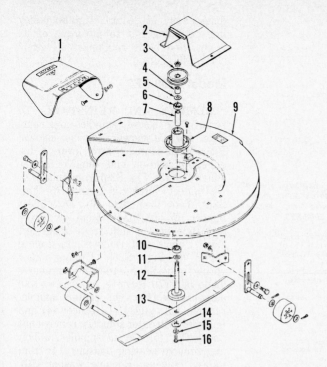

Fig. S24—Exploded view of the 26" mower unit used on Models 315 and 3005.

1. Deflector
2. Cover
3. Mower pulley
4. Spacer
5. Washer
6. Bearing
7. Spacer
8. Bearing housing
9. Mower housing
10. Bearing
11. Washer
12. Spindle
13. Blade
14. Belleville washer
15. Shoulder washer
16. Cap screw (LH thread)

pulley retaining nuts to a torque of 45 ft.-lbs. Tighten blade cap screws (18) to a torque of 30 ft.-lbs.

Models 315—3005

REMOVE AND REINSTALL. To remove the blade, spindle and bearings, remove mower unit as follows: Place blade clutch control lever in disengaged position. Pull jackshaft pulley housing (19—Fig. S16) forward and remove mower belt (2) from pulley (20). Unhook rear of mower and unpin front of mower, then slide mower unit from under left side. Unbolt cover (2—Fig. S24) and remove mower belt. Remove left hand thread cap screw (16), washers (14 and 15) and blade (13). Remove nut, pulley (3), spacer (4) and washer (5) from above and spindle (12) and washer (11) from below. Unbolt bearing housing (8) and separate bearings (6 and 10) and spacer (7) from the housing.

Clean and inspect all parts and renew any showing excessive wear or other damage. Reassemble by reversing the disassembly procedure. Tighten mower pulley retaining nut to a torque of 70 ft.-lbs. and left hand thread cap screw (16) to a torque of 80 ft.-lbs.

Models 808—3008-2—3008-3

REMOVE AND REINSTALL. To remove the blade, spindle and bearings,

bolt and remove cover (1—Fig. S22). Place blade clutch control lever in disengaged position. Pull jackshaft pulley bracket forward and remove mower drive belt from mower pulley (3). Unhook rear of mower and unpin front of mower, then slide mower unit from under left side. Remove the two cap screws (14) and remove blade (13). Remove nut, washers, pulley (3) and spacer (4) from above mower housing. Unbolt and remove bearing housing (8), then separate bearings (7 and 10), spacer (9), washers (11) and spindle (12) from bearing housing.

Clean and inspect all parts and renew any showing excessive wear or other damage. Reassemble by reversing the disassembly procedure. Tighten mower pulley retaining nut to a torque of 25 ft.-lbs. Tighten blade cap screws (14) to a torque of 30 ft.-lbs.

Model 355

REMOVE AND REINSTALL. To remove the blades, spindles and bearings, first remove mower unit as follows: Place blade clutch lever in disengaged position. Pull jackshaft pulley bracket (19—Fig. S16) forward and remove mower drive belt (4) from pulley (20). Unhook rear of mower and unpin front of mower, then slide mower from under left side. Refer to Fig. S23 and unbolt and remove cover (1). Remove mower drive belt (6). Remove cap screws (18) and remove blades (17). Remove nuts, mower pulleys (5) and washers (9) from above mower housing (8). Unbolt bearing housings (12), then separate bearings (10 and 14), backing

rings (11), spacers (13), washers (15) and spindles (16) from bearing housings. Unbolt and remove flat idler pulleys, if necessary.

Clean and inspect all parts and renew any showing excessive wear or other damage. Reassemble by reversing the disassembly procedure. Tighten mower

Fig. S25—Exploded view of the 30" mower unit used on Models 808, 3008-2 and 3008-3.

1. Covers
2. Deflector
3. Rock guard
4. Mower pulley
5. Hub
6. Mower housing
7. Spacer
8. Bearing
9. Spacer
10. Bearing housing
11. Bearing
12. Washers
13. Wave washer
14. Washer
15. Spindle
16. Blade
17. Cap screw

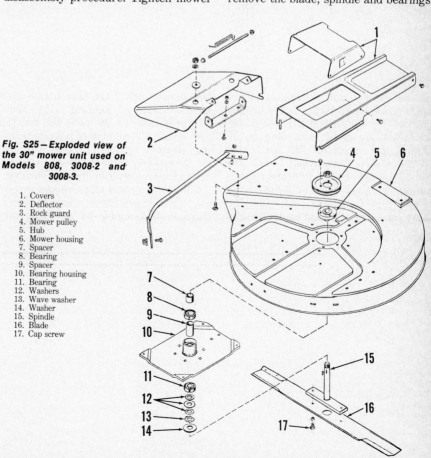

disassembly procedure. Tighten pulley hub retaining nut to a torque of 95 ft.-lbs. and blade retaining cap screw (17) to 45 ft.-lbs.

Model 3008-FE3

REMOVE AND REINSTALL. To remove the blades, spindles and bearings, remove mower unit as follows: Place blade clutch control lever in disengaged position and remove mower clutch belt (3 – Fig. S26) from pulley (1). Disconnect clutch rod bracket (5 – Fig. S20) from idler arm (6). Disconnect mower lift link, unpin front of mower and slide mower unit from under left side. Unbolt and remove the 2-piece cover (2 – Fig. S26). Unhook idler spring, move idler to the left and remove blade belt (7). Remove cap screws (20) and blades (19). On center spindle, remove nut, pulley (1), spacer (4) and pulley (5). On outer spindles, remove nut and pulley (5). On all spindles, unbolt and remove bearing housing (14) from mower housing. Remove washer (10), spacer (11), bearing (12) and backing ring (13) from top of bearing housing, then remove spindle (18), washers (17), bearing (16) and spacer (15) from bottom of bearing housing. Remove flat idler pulley (6), if necessary.

Clean and inspect all parts and renew any showing excessive wear or other damage. Reassemble by reversing the disassembly procedure. Tighten pulley retaining nuts on spindles to a torque of 95 ft.-lbs. and blade retaining cap screws (20) to 45 ft.-lbs.

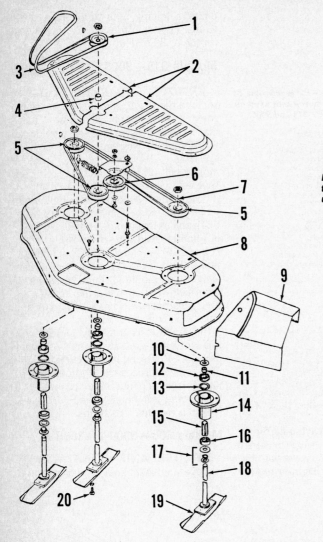

Fig. S26 – Exploded view of the 36" three blade mower unit used on Model 3008-FE3.

1. Mower drive pulley
2. Covers
3. Mower clutch belt
4. Spacer
5. Mower pulleys
6. Flat idler pulley
7. Blade belt
8. Mower housing
9. Deflector
10. Washer
11. Spacer
12. Bearing
13. Backing ring
14. Bearing housing
15. Spacer
16. Bearing
17. Washers
18. Spindle
19. Blade
20. Cap screw

LUBRICATION

All Models

Lubricate all linkage pivot points with SAE 30 oil. Use multi-purpose lithium grease on all models equipped with lubrication fittings on bearing housings. Others are equipped with sealed bearings and require no additional lubrication.

remove mower unit as follows: Place blade clutch control lever in disengaged position. On Models 808 and 3008-3, disconnect blade clutch rod, then on all models, disconnect lift link and unpin front of mower from frame. Unbolt and remove covers (1 – Fig. S25) and remove mower drive belt from pulley (4). Slide mower unit from under left side. Remove cap screws (17) and blade (16). Unbolt mower pulley (4), then unbolt and remove bearing housing (10) from mower housing (6). Remove nut, hub (5) and spacer (7) from top of bearing housing. Remove spindle (15) and washers (12, 13 and 14). Separate bearings (8 and 11) and spacer (9) from bearing housing.

Clean and inspect all parts and renew any showing excessive wear or other damage. Reassemble by reversing the

SNAPPER

McDONOUGH POWER EQUIPMENT CO.
P.O. Box 777
McDonough, GA 30253

Model	Make	Engine Model	Horsepower	Cutting Width, In.
265X/XS	Tec.	V50	5	26
268X/XS	B&S	190000	8	26
308X/XS	B&S	190000	8	30
267X/XS	B&S	170000	7	26
305X/XS	Tec.	V50	5	30
307X/XS	B&S	170000	7	30
266X/XS	B&S	140000	6	26
306X/XS	Tec.	VH60	6	30
417X/XS	B&S	170000	7	41
418X/XS	Wisc.	HS8D	8	41
2550/S	B&S/Tec.	130000/V50	5	25
2652/S	B&S/Tec.	130000/V50	5	26
2651/S	Tec.	V50	5	26
2681/S/W/WS	B&S/Tec.	190000/VM80	8	26
3081/S/W/WS	B&S/Tec.	190000/VM80	8	30
2650/S	Tec.	V50	5	26
2680/S	B&S/Tec.	190000/VM80	8	26
2681/S	B&S/Tec.	190000/VM80	8	26
2810/W/WS	B&S	250000	11	28
2880/W/WS	B&S/Tec.	190000/VM80	8	28
3010/WS	B&S/Tec.	250000/VM110	11	30
3080/S	B&S/Tec.	190000/VM80	8	30
3081/W/WS	B&S/Tec.	190000/VM80	8	30
4210/W/WS	B&S	250000	11	42
25063/S	B&S/Tec.	140000/VM60	6	25
25083/S	B&S/Tec.	190000/VM80	8	25
26062/S	B&S/Tec.	140000/VM60	6	26
26063/S	B&S/Tec.	190000/VM80	8	26
26083/S	B&S/Tec.	190000/VM80	8	26
28083/S	B&S/Tec.	190000/VM80	8	28
28113/S	B&S	250000	11	28
30083/S	B&S/Tec.	190000/VM80	8	30
30113S	B&S	250000	11	30
33113S	B&S	250000	11	33
42113S	B&S	250000	11	42

FRONT AXLE

All Models

All steering components are mounted in a forward support deck which is fitted to a single-rail tubular chassis member which extends from engine and drive deck at rear of unit. Steering is non-adjustable, except on some early models, tie rods are fitted with adjustable ball joints threaded to inner ends of tie rods. Steering gear in all subsequent production is as shown in Fig. SN1.

All models are fitted with a pair of vertical bumpers bolted to rear of engine platform which will serve as a rear stand for convenience when service is performed.

CAUTION: Frequently, the performance of maintenance, adjustment or repair operations on a riding mower is more convenient if mower is standing on end. This procedure can be considered a recommended practice providing the following safety recommendations are performed:

1. Drain fuel tank or make certain that fuel level is low enough so that fuel will not drain out.
2. Close fuel shut-off valve if so equipped.
3. Remove battery on models so equipped.
4. Disconnect spark plug wire and tie out of way.
5. Although not absolutely essential, it is recommended that crankcase oil be drained to avoid flooding the combustion chamber with oil when engine is tilted.
6. Secure mower from tipping by lashing unit to a nearby post or overhead beam.

R&R AND OVERHAUL. Set unit up on rear bumper bars. Remove hub caps (15–Fig. SN1), cotter pins (14) and wheels. Take care not to drop washers (13) or inner dust caps (11). Clean all hub parts, especially bearings (37) and felt washers (16). Renew hub parts which are worn or defective. Remove cotter pins (10) at each end of tie rods (18 and 19). Remove tie rods, then clean and evaluate tie rod bearing parts (7 and 29) and renew if necessary. Note that tie rods, though very similar, are not interchangeable. Use an "external" snap ring tool, remove king pin retainer (4), then lower king pins (8) out of platform assembly. Clean and inspect shoulder bushings (5) and friction surfaces of king pin and spindle assemblies. It should also be noted that king pins do not interchange from side to side. Steering shaft (17) can be lowered out of its support after removing bolts which secure steering wheel (27) to upper end.

NOTE: All bearings used in steering linkage, king pins and for steering shaft are shouldered nylon type. Manufacturer's lubrication instructions specify manual oiling with SAE 20 oil at 10 operating hour intervals. Liberal oiling will help to clear out abrasive dirt attracted by oily surfaces. Wipe away excess. Front wheel bearings are lubricated by use of pressure gun at fitting on inner wheel hub. Use Snapper No. 00 grease, also at 10 hour minimum intervals. See Fig. SN3.

Reassemble steering system in reverse of disassembly order.

STEERING WHEEL KIT. Fig. SN2 shows an exploded view of optional steering wheel assembly used on some models. As with other steering system components, there is no adjustment procedure to compensate for wear or defective parts.

For inspection, service and lubrication of reduction gears (4, 6 and 7) remove acorn nut (13) and slide gear cover (12) upward on shaft tube.

ENGINE

All Models

For overhaul and repair procedures on engines listed in Specification Table refer to Small or Large Air Cooled Engines Service Manual.

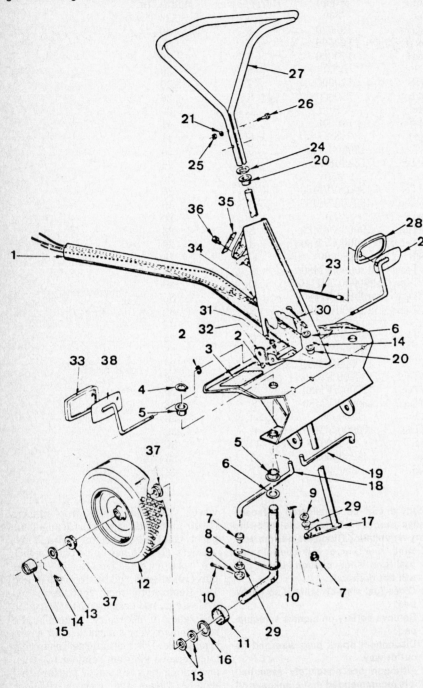

Fig. SN1—Exploded view of a typical front support, steering system with clutch and brake pedal mounts.

1. Main frame tube	11. Dust cap	21. Lockwasher	31. Cap screw (5/16-18)
2. Pedal snap ring	12. Wheel & tire	22. Clutch pedal	32. Pedal shaft mount
3. Pad	13. Hub washer	23. Clutch cable	33. Pedal pad
4. King pin snap ring	14. Cotter pin (5/32)	24. Shim	34. Throttle control
5. King pin bearing	15. Hub cap	25. ¼-28 nut	35. Throttle plate
6. Washer	16. Felt washer	26. ¼-inch bolt	36. WF cap screw, 10-24 (2)
7. Tie rod bearing	17. Steering shaft	27. Steering wheel	37. Ball bearing (4)
8. King pin/spindle	18. RH tie rod	28. Pedal pad	38. Brake pedal
9. Tie rod washer	19. LH tie rod	29. Tie rod bearing	
10. Cotter pin	20. Nylon bushing	30. Brake cable	

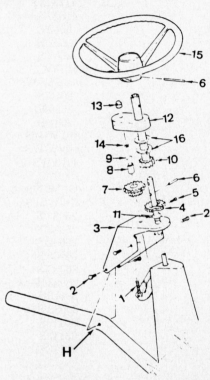

Fig. SN2—Exploded view of optional steering wheel kit furnished as a modification on some models. Diameter of hole (H) is 7/32-inch.

1. Cap screw, ⅜X1¾	8. Split spacer
2. ¼-20X⅝ ST screw (4)	9. Washer
3. Control panel	10. Steering tube
4. Steering shaft gear	11. Bushing
5. Cap screw (socket head)	12. Gear cover
6. Roll pin, ¼X1½	13. Acorn nut, ⅜-24
7. Compound gear	14. Jam nut, ⅜-24
	15. Steering wheel
	16. Fiber washer

Fig. SN3—Lubrication of front system is convenient when unit is placed on its rear stand. Use SAE 30 oil on linkages and Snapper 00 gun grease (2 shots) on wheel bearings. See text.

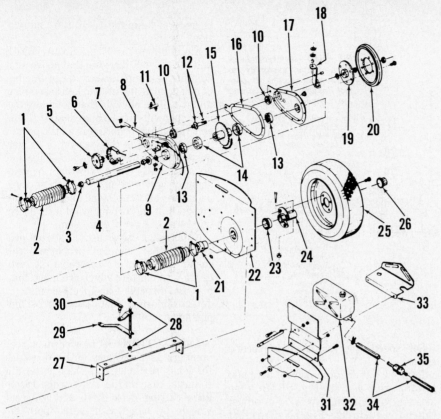

1. Boot seal ring (4)
2. Drive tube boot (2)
3. Tube bearing (2)
4. Hex drive tube
5. Brake drum
6. Brake band
7. Roller
8. Extension rod
9. Chain case half, RH
10. Bearing (2)
11. Shift link
12. Sprocket shaft & key
13. Needle bearings (2)
14. Nylon thrust washer (2)
15. Sprocket and hub
16. Case gasket
17. Chain case half, LH
18. Brake lever assy.
19. Input hub
20. Driven disc
21. Bearing
22. Fender, LH
23. Seal cap
24. Wheel hub, LH
25. Wheel & tire
26. Hub cap
27. Cross brace
28. Nylon bearing (2)
29. Shift crank
30. Shift link
31. Tank bracket
32. Fuel tank
33. Tank support clamp
34. Fuel line
35. Fuel filter

Fig. SN4A — Exploded view of left side portion of typical variable speed friction drive system. Note hex drive tube (4) and sliding chain case (7 through 17). Drive tube boots (2) do not interchange though appearance is similar.

REMOVE AND REINSTALL. To remove engine, disconnect spark plug lead and if unit is equipped for electric start, disconnect cable from starting motor. Shift mower drive control lever to "OUT" position then unbolt and remove spindle cover from mower deck. Relieve tension from idler pulley, then slip belt off mower spindle pulley (or pulleys). Working beneath engine platform, slip belt from its sheave on driving disc hub and pull clear.

On units with separate fuel tanks, disconnect fuel hose at carburetor end and plug hose to prevent drainage.

Uncouple throttle cable at carburetor, then disconnect red wire lead from ignition switch (all models) and white wire lead from rectifier on electric start models. Check carefully for other items which might impede engine removal. Now, unbolt engine from deck and lift away for other service.

Reinstall engine by reversal of these steps.

TRACTION DRIVE CLUTCH

All Models

All models are equipped with a cable-operated clutch. Cable extends from a keyhole slot in left hand pedal through main frame tube and is attached to lever arm portion of lift yoke (5 – Fig. SN4C). When clutch pedal is depressed, cable tension pulls lift yoke forward so that extension rod (8 – Fig. SN4A) of sliding

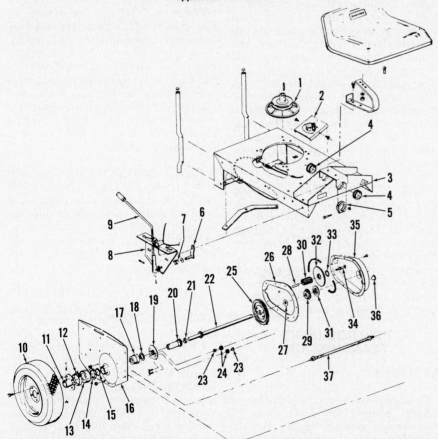

1. Drive disc
2. Switch plate
3. Main frame
4. Tube grommet (2)
5. Tube clamp
6. Shift arm
7. Lever spring
8. Detent & neutral switch
9. Shift lever
10. Wheel & tire
11. Wheel hub, RH
12. Boot clamp
13. Seal cap
14. Oil seal
15. Clamp ring
16. Fender, RH
17. Bearing
18. Thrust washer
19. Differential plate
20. Short axle assy.
21. Spacer
22. Long axle assy.
23. Pinion spacers (4)
24. 12-tooth pinions (4)
25. 63-tooth reduction gear
26. Gasket
27. Flange washer
28. Spacer
29. Sprocket
30. 11-tooth gear
31. Washer, special
32. Sprocket
33. Retainer
34. Idler bolt
35. Case
36. Plastic lube plug
37. Tie bolt

Fig. SN4B — Exploded view of right hand portion of friction drive axle. Reduction gear set sprockets (29 through 32) and differential parts (9 through 25) are shown. Long axle (22) extends through drive tube (4 – Fig. SN4A) to left wheel hub.

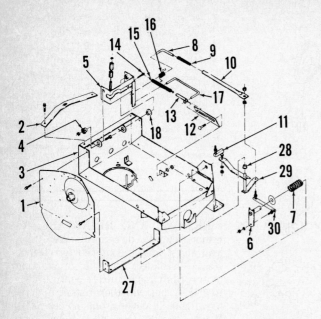

is generally limited to routine lubrication.

Examination of Figs. SN4A, SN4B and SN4C will disclose that drive unit contains a comparatively large number of small parts and interlinked assemblies and it will be obvious that complete disassembly will entail a high degree of concentration and orderliness on the part of a mechanic. For this reason, it is suggested that if a malfunction in drive unit is suspected, the following trouble-shooting guide be used to isolate cause or causes of trouble and that disassembly be limited to only what will be required to correct the problem.

Note that in using these check-out procedures that simpler problems with easier remedies are offered first, proceeding to matters which are more complex and difficult. Careful observation to detect abnormal conditions is most important.

INSPECTION. Set mower up on rear bumpers as previously outlined. Check for leaks at flanges of chain case, differential case and at shaft seals. Disengage cutting deck drive and roll rear wheels by hand with clutch both depressed and released and with speed selector in neutral and also when shifted into drive. Loosen clamps on drive tube boots and pull back so that rotation of hex drive tube can be observed. Check for normal action of differential pinions. Jamming, locking or noise in differential case can be caused by damaged differential parts or broken drive chain. Undue noise or abnormal operation in differential case or in sliding chain case will determine which assemblies should be removed for inspection and evaluation of damage or problem.

R&R AND OVERHAUL. When decision is made as to extent of disassembly needed to make necessary repairs and whether sliding chain case or differential case is the trouble point and will require disassembly, use whichever of the following procedures is considered applicable.

LEFT SIDE. (SLIDING CHAIN CASE). Place mower on rear stand as previously outlined then proceed as follows:

Unbolt left rear wheel from axle hub and lift off wheel and tire (25 – Fig. SN4A). Remove hub cap (26), axle bolt, hub (24) and seal cap (23). Now, carefully unbolt left hand side plate/fender (22). Wipe axle end clean, then loosen boot clamp (1), pull drive tube boot (2) from inner shoulder of bearing (21) and lift off plate/fender (22) to expose sliding chain case. On models so equipped, remove separate fuel tank (32).

---

chain case, which is engaged in horizontal slot of lift yoke, will be depressed to lower driven disc (20) out of contact with drive disc (1 – Fig. SN4B).

ADJUSTMENT. Linkage design has a great deal of built-in tolerance for wear, cable stretch and distortion of parts. Allowance is made for normal wear of driven disc (20 – Fig. SN4A). When disc rubber facing is worn down to 1/16 to 1/32-inch, disc must be renewed, and no adjustment will be required for linkages after new disc is installed. If, however, linkage parts have been altered due to extreme wear, damage or tampering, adjustment will be required before installation of a new driven disc, as follows:

Modify a new driven disc by cutting or grinding away circumference rubber on one side so that radius of one-half of disc is reduced to 2¾ inches while radius of opposite circumference section remains at a full three inches. If this modification of another new disc is a problem due to parts scarcity, a measuring template can be made up from 10-11 gage (approx. ⅛-inch) aluminium or sheet iron or from a scrap of ¼-inch masonite hard board. Use a fly-cutter in a drill press to score or cut concentric circles, of exactly three and exactly six inches diameter in material selected. Result should be a six inch disc with a three inch circle cut out at its center. Mounting bolt holes will not be needed. Cut away ¼-inch from outer circumference on one-half of disc and template is ready for use. Mower should be set up on rear stand as previously outlined.

Now, install modified disc or made-up template on hub (19). If substitute template is used, it will also be necessary to first remove brake lever assembly (18).

Set speed selector lever in second speed.

Rotate 2¾ inch radius side of template toward drive disc (1 – Fig. SN4B), then loosen its mounting bolt and adjust clutch rod guide (12 – Fig. SN4C) so that low (2¾ inch) side of template clears driving disc by 0.005-0.010 inch. Retighten mounting bolt. Now, rotate template half way and depress clutch pedal all the way to bottom in brake applied position, clamp or lash pedal down, and measure between three inch radius side of disc template and driving disc surface. Clearance should be 1/32-inch (0.031) minimum with sliding chain case in light contact with lift yoke. When so adjusted, driven disc will be pulled clear of surface of driving disc when clutch pedal is depressed during regular operation.

R&R AND OVERHAUL. Detailed disassembly and repair to clutch are included under FRICTION DRIVE TRANSMISSION paragraphs. Springs and minor linkage parts including operating cable can be readily renewed upon inspection with minimal disassembly. Worn or defective driven disc is easily unbolted for removal and renewal as covered in preceding section.

FRICTION DRIVE TRANSMISSION

All Models

On all models, the friction drive transmission incorporates all functions of clutch-brake, auxiliary brake, variable speed drive, reduction gearing, differential and divided axle. There are no adjustments, except as covered in CLUTCH and BRAKE sections. Service

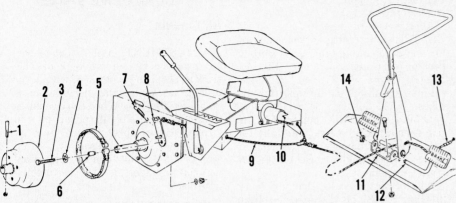

Fig. SN5 — Exploded view of band-drum brake kit offered as an option for 1971 through 1974 models. Refer to text.

1. Axle bolt
2. Special hub/drum (2)
3. Anchor-pivot bolt (2)
4. Spring washer (2)
5. Brake band (2)
6. Pivot bushing (2)
7. Return spring (2)
8. Brake arm (2)
9. Band cable
10. Clutch-brake cable
11. Pedal mount
12. Pedal
13. Cable stop adjusters
14. Pedal retainer

Disconnect shift link (11) from lug on inboard side of chain case and release brake cable from its guide on chain case. Unbolt anchored end of brake band (6) and slip off brake drum (5). Uncouple clutch cable from arm portion of lift yoke (5 – Fig. SN4C), then disconnect springs and other linkage elements and swing yoke clear of extension rod (8 – Fig. SN4A). Slide chain case outward off hex drive tube (4).

NOTE: At this point, it is important that small parts such as boot seal rings, spacer washers and bearings from drive tube not be dropped and damaged or lost. Maintain assembly order by laying removed parts out in sequence.

Use a strap wrench to hold driven disc (20) and remove the cap screw which holds brake drum (5) on shaft (12) and the nut which holds input hub (19). Unbolt disc (20) from hub (19) and use a puller if necessary to draw hub off sprocket shaft (12). Do not lose Woodruff key. Now, unbolt case halves (9 and 17), taking care not to drop or lose split spacer which serves as pivot for brake lever (18). When case halves are separated, sprockets (12 and 15) and chain can be removed for thorough cleaning. Pay special attention to internal machined surfaces of hex hub portion of large sprocket (15). Needle bearings (13), nylon thrust washers (14) and input shaft bearings (10) must be carefully evaluated for possible renewal.

Remove all clamps from drive tube boots (2) and remove boots for thorough cleaning, inside and out. Seals at collar ends of boots should be renewed. Thoroughly clean and lightly polish surface of hex drive tube (4). Service to bearings inside tube which support long axle is not performed unless right side of drive and differential case are removed as covered in following section.

Reassemble in reverse of disassembly order and when fender/side plate (22) is bolted in place, check for end play of hex drive tube (4). If end play exceeds 1/16-inch, remove side plate and insert nylon thrust washers (SNAPPER Part No. 1-1071) as needed between end of drive tube and bearing (21) and recheck end play after reinstalling side plate. Normal end play is 1/32 to 1/16-inch. Lubricate with two ounces of SNAPPER 00 grease and install a new plastic plug.

RIGHT SIDE. (DIFFERENTIAL CASE). To disassemble only differential case, final drive, axle and differential, set unit up on rear stand as previously outlined.

Remove axle cross bolt from hub (24 – Fig. SN4A) of left rear wheel and remove wheel and hub from axle. Unbolt right side fender (16 – Fig. SN4B) from main frame (3) and from tie bolt (37). Loosen seal clamps (1 – Fig. SN4A).

Grasp wheel (10 – Fig. SN4B) and pull fender (16), drive case (35) and axle (22) out of drive tube and its boot and set up with axle (22) pointed upward and with entire assembly resting on wheel (10).

Unbolt and lift off case (35) from inside of fender (16). Note that some binding may impede removal of idler bolt (34) from bore of gear (30). Watch for spacer (28) during removal. When case (35) is lifted off, drive chain will go slack and sprocket (32) and gear (30) can be lifted out as an assembly. If sprocket or gear are damaged so as to require renewal, remove retainer (33) to separate. Thoroughly clean all parts so far removed for inspection and evaluation and lay them out in assembly sequence.

IMPORTANT NOTE: On recent models (2651, 2652, 2681, 3081) hex drive tube is integral with small sprocket (29) and there are no bearings inside drive tube to sup- port long axle (22). These minor design changes do not affect overhaul procedures.

Remove axle bolt from hub (11) and pull wheel and tire (10) and hub (11) off short axle (20). At this point, complete driving axle (parts 19 through 25) should be pulled out of axle bearing (17).

When pinion shaft bolts are backed out of large gear (25), differential plate (19) can be removed for separation of short axle (20) and spacer (21) from long axle (22). All four differential pinions (24), pinion spacers (23) and axle spacer (21) should be cleaned and evaluated for re-use or renewal.

Be sure to clean bore of short axle shaft (20) and carefully inspect teeth of its sun gear and the sun gear on long axle (22). Check condition of small sprocket (29) and special washer (31). Renew if necessary.

Use a small amount of SNAPPER 00 grease on differential pinions and spacers during reassembly and be sure that pinions and spacers are correctly installed on pinion shaft bolts after bolts are inserted through differential plate (19). Spacers (23) must be alternated so that two opposite pinions (24) will mesh with sun gear portion of short axle (20) and alternating set of pinions engaging sun gear of long axle (22). Be sure to install center spacer (21). When differential parts are installed and aligned and with differential plate (19) bolted back in place against gear (25), manually rotate long and short axles in opposite directions to be sure that assembly was correctly performed. If axles cannot be so turned against one another, unbolt and check assembly.

Reinstall hex drive tube over long axle with drive chain over small sprocket (29) and special washer (31) in place, then install large sprocket (32) with pinion gear (30) and spacer (28) on idler bolt (34). Be sure that retainer (33) is in place and do not overlook "O" rings and thrust washers. Use No. 3 Permatex, "Hightack" or other gasket adhesive to hold new gasket (26) in place during reassembly. Roll chain into place on sprockets, then slip short axle (20) portion through thrust washer (18) and bearing (17) in fender (16). Bolt case (35) loosely in position and manually test shafts and gears for freedom of rotation without humping or binding. Tighten case bolts evenly in an alternating sequence around case flange. Lubricate differential case by adding 12 ounces of SNAPPER 00 grease and insert a new plastic plug (36).

Fit open end of axle and hex drive into pre-assembled boot, seal and clamps and through sliding chain case. Be sure that surface of hex drive tube is lightly lubricated to prevent galling. End of axle (22) should extend through left fender

(22 – Fig. SN4A). Do not lose seal cap (23).

Retighten boot clamps and fit right side fender (16 – Fig. SN4B) over tie bolt (37) and cross brace for re-bolting back in position on side of frame. Recheck alignment and freedom of movement and reconnect control linkages for clutch, brake and speed selector. If clutch requires adjustment, refer to CLUTCH paragraphs.

CONTROL LINKAGE. Speed selector shift lever (9 – Fig. SN4B) is not adjustable, except that bracket (8) has slotted mounting holes so that neutral adjustment can be made, as follows: Manually slip sliding chain case along hex drive tube into neutral position and if shift lever does not fit into its neutral notch, loosen mounting bolts and shift detent asembly bracket to correct position. On 1975 and 1976 models, this adjustment, if neglected, may cause a problem with neutral-start interlock switch (Part No. 5-0681). This microswitch is not adjustable and proper func-

tion depends upon correct position of shifter bracket.

Adjustment of clutch, covered in CLUTCH paragraphs, is performed by loosening mounting bolt and shifting position of clutch rod guide (12 – Fig. SN4C).

Linkage ball joints, springs, rod and tubes should be kept clean. Use very light lubrication at wear points. If operation seems faulty, inspect carefully for defective parts damaged by accident or neglect which might cause binding or limited movement.

It is not recommended that shift lever (9 – Fig. SN4B) be disconnected from shaft portion of shift arm (6). Proper lubrication at grease fitting should prevent undue wear. If, however, it becomes necessary to remove pivot pin from lever (9), for any reason, it is suggested that a 6 or 8 inch "C" clamp be used to aid in reassembly for lining up pin holes. Lever spring (7) is quite strong. A large nut or socket can be set over grease fitting for protection when using a "C"-clamp to compress spring.

GROUND DRIVE BRAKE

All Models

ADJUSTMENT. Adjustment of service brakes is limited to take-up of cable slack at pedal and renewal of worn lever or band assemblies when performance is no longer effective. When cable requires tightening – ususally if pedal at full stroke reaches about ½-inch from its bottom limit – take up cable by shifting stop billets on cable end through pedal key hole to underside of pedal to eliminate slack.

IMPORTANT NOTE: It may at times become necessary, during other service operations, to adjust chassis tube to take up looseness in drive belt to cutting deck. Whenever clamp on chassis tube is loosened for adjustment of its length, for whatever reason, always check brake (and clutch) cables for correct operating length and adjust accordingly. Normal free pedal travel should be ½ to ¾-inch.

BRAKE TYPES. CLUTCH-BRAKE WITH AUXILIARY BRAKE. All models since 1973 have been equipped with a combination clutch-brake pedal on operator's left and a single-function auxiliary brake pedal at right. Sufficient force is developed at either pedal to provide adequate service braking. Clutch-brake is applied when pedal is pressed down to its limit. At this point, brake lever (18 – Fig. SN4A) rotates on its center pivot so that it's lined shoe portion is in solid contact with inner rim of driven disc (20). Pressure on upper end of brake lever is exerted by lift yoke (5 – Fig. SN4C) when actuated by tension on clutch cable.

Auxiliary band/drum type brake is mounted on opposite end of sprocket shaft (12 – Fig. SN4A) from driven disc (20) so that when brake pedal is depressed in addition to clutch-brake pedal then all braking force is concentrated on sprocket shaft (12) which is the input shaft of transmission chain case.

CLUTCH-BRAKE. Prior to 1973, all models were equipped with clutch-brake only operated by single left hand pedal as previously described. These models were: 265X, 266X, 267X, 268X, 305X, 306X, 307X, 308X, 417X and 418X.

PARKING BRAKE. Only Models 265X, 268X and 308X (1972 production) were fitted with a parking brake. This was a lever-operated cable clamping device mounted on chassis tube behind clutch-brake pedal to lock cable with pedal fully depressed.

REAR WHEEL DRUM BRAKES. An optional brake kit was offered as an

Trouble Shooting Guide

Problem	Possible Cause	Corrective Action
No drive – mower will not move in either direction.	Oil or grease on drive or driven disc.	Clean with solvent and wipe dry.
	Excessive clearance between drive and driven discs.	Check clutch adjustment. Check speed selector lever linkage and lift yoke for jamming or damage. Repair or renew.
	Rubber trend of driven disc damaged or worn out.	Renew driven disc.
	Breakage in chain case or differential case.	Use check procedure to locate trouble. Disassemble and repair.
Selector hard to move when shifting through speed range.	Hex shaft dry-galled or burred.	Remove burrs, polish hex surface lightly, check shifting action.
	Jammed or damaged control linkage.	Disassemble linkage only, clean, lube and renew parts as needed.
Noisy drive.	Damaged drive chain or bearings.	Use check procedure to locate problem. Repair as needed.
Overheating of differential case or chain case.	Insufficient lubrication.	Lubricate as required.
	Lube leaks from case.	Disassemble leaky case for renewal of gasket. Check for other possible damage.

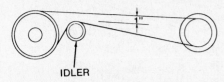

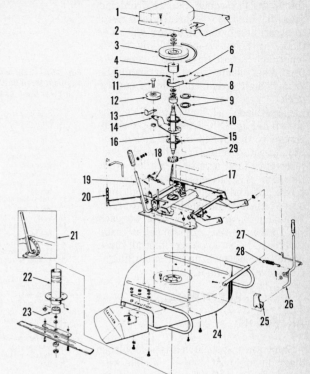

Fig. SN6—Mower blade drive belt clearance should be one inch at idler pulley as shown. See procedure in text.

Fig. SN7—Exploded view of a typical mower deck assembly.

1. Spindle cover
2. Locknut
3. Driven pulley
4. Brake drum
5. Cotter pin
6. Return spring
7. Wire link
8. Brake assy.
9. Retainer rings
10. Upper spindle bearing
11. Idler shaft bolt
12. Idler pulley
13. Keeper
14. Idler arm
15. Retainer rings
16. Spindle shaft
17. Deck rail assy.
18. Timing link & rod
19. Lift handle
20. Suspension chain (2)
21. Height indicator
22. Spindle housing
23. Lower spindle bearing
24. Cutting deck
25. Handle mount
26. Blade control handle
27. Idler control rod
28. Idler spring
29. Tolerance ring

"aftermarket" item for 1971 through 1974 models. Installation of this modification calls for removal of original equipment wheel hubs and substitution of special hubs with drums (2 – Fig. SN5). In factory production, frames and fender assemblies were pre-drilled to accommodate this kit. An instruction sheet with specifications is furnished for installation of this option. Brake kit is SNAPPER part number 6-0170. Service is limited to parts renewal only, and adjustment is confined to shifting of cable stop adjusters (13).

INDIVIDUAL WHEEL BRAKE KIT. All SNAPPER Comet models, 26 through 41 inch, having an "X" suffix to model number (1972 and prior production) offered an option kit (part number 6-0113) to provide a single shoe-drum brake for each rear wheel. These hand-lever operated individual wheel brakes will function as a steering brake on either rear wheel for assistance in making tight turns and can be locked in applied position to serve as parking brakes. Slotted mounting bolt holes for handle guide brackets provide the only adjustment. Adjust forward or back to set "release" position of each brake handle.

BLADE DRIVE BRAKE

Models So Equipped

On models equipped with spindle brake, when mower control handle is moved to "OUT" position, blade rotation should stop automatically in 3 to 5 seconds. To check, remove spindle cover (1 – Fig. SN7), start engine, engage mower control and allow speed to build up to normal, then move handle to "OUT" and note time required for rotation to halt. If not satisfactory, adjust brake as follows:

Disconnect wire link (7) from spindle brake band (8) and reconnect link in another, closer hole to increase brake band tension.

Retest. If moving wire link to closer hole does not increase tension sufficiently, it may be necessary to renew brake parts.

MOWER CLUTCH AND DRIVE BELT

All Models

Mowing unit clutch is a hand lever engaged, spring-loaded belt idler type. Clutch mechanism is not adjustable; however, if belt stretch and general wear cause belt sections to rub together at idler behind spindle pulley, adjust frame to gain minimum one-inch clearance at idler as shown in Fig. SN6. Proceed as follows:

Disconnect engine spark plug and remove spindle cover (1 – Fig. SN7) from mower deck. With mower drive control handle (26) in engaged position, loosen tube clamp under operator's seat and pull main chassis tube forward until drive belt is drawn up enough to provide required one inch clearance on inner surface of drive belt. Retighten tube clamp.

When belt slack is removed, be sure to check clutch-brake pedal and auxiliary brake pedal, if so equipped, for correct free travel and proper operation. Correct as necessary, making reference to CLUTCH and BRAKE paragraphs.

MOWER DECK SPINDLE

All Models

Mower spindles are fitted with sealed ball bearings. Lubricate at grease fitting with a single shot of SNAPPER 00 grease from a hand grease gun once per mowing season. If spindles must be disassembled for overhaul due to neglect or accidental damage, observe parts assembly sequences shown in Figs. SN7 and SN8.

IMPORTANT: Spindle housings (16 and 27 – Fig. SN8) must be reinstalled exactly plumb to deck.

MOWER CUTTING HEIGHT

All Models

As delivered from manufacturer's plant, these mowers are set to make a low cut at one inch and the cut is adjustable upward (8 positions) to four inches. When adjustment is required, mainly due to wear in linkage, proceed in this order:

Park unit on level surface, set mower lift handle (19 – Fig. SN7) in low position and remove spindle cover (1). Place mower drive clutch in "OUT" position.

Check straightness of blade by measuring height of each blade tip above the same point on floor at forward edge of mower deck. If tip heights are identical, blade is straight.

Adjust blade tip height to $1\frac{5}{8}$ inches at rear by bending support arms for suspension chains (20). Equalize bend on each side so that blade remains level, measured side-to-side.

With blade tip at front of mower deck, adjust tip height to one inch by use of timing link (18) which connects front and rear lift arms. Shortening (turning clevis to right) link by one-half turn will raise blade tip about ⅛-inch. If link rod is lengthened, blade tip will be lowered. As finally set, blade tip at rear should be ½-inch higher than at front of cutting deck.

Three-blade, 41 inch mower deck is shown in Fig. SN8. Service is essentially similar to that for 26 and 30 inch decks. Side-to-side leveling is performed by adjustment of set screw (12) on left front lift arm (17). Release locknut on set screw (12) and turn screw in or out until outer tips of outboard blades measure at same height above level floor on each side. Front-to-rear adjustment is made on link rod (18). Leading tip of all three blades should measure at same height from level shop floor.

ELECTRICAL

All Models

Shown in Fig. SN9 is a typical wiring diagram for most models. Models with recoil or push button start will be similar, but will differ with respect to their starting system.

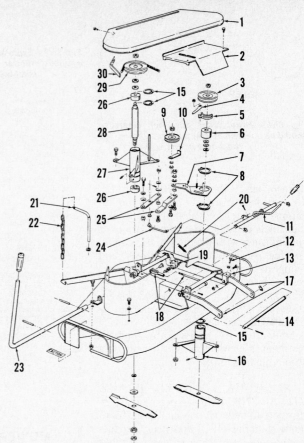

Fig. SN8 — Exploded view of a typical three-blade, 41 inch mower deck assembly.

1. Spindle cover
2. Front cover
3. Spindle pulley (2)
4. Link
5. Spindle brake
6. Brake drum
7. Idler arm
8. Retainer rings
9. Idler pulley
10. Keeper
11. Idler control handle
12. Set screw
13. Level adjuster
14. Suspension bar
15. Retaining ring (2)
16. Spindle housing (2)
17. Front lift arms
18. Timing rod & clevis
19. Lift cams
20. Cam spring
21. Chain arm
22. Deck chain
23. Lift handle
24. Idler link rod
25. Idler link arms
26. Spindle bearings
27. Center spindle housing
28. Spindle shaft
29. Spindle pulley
30. Belt guard

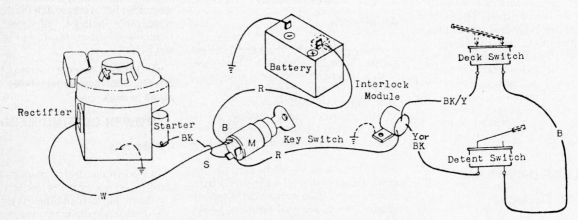

Fig. SN9 — View of a typical key switch electric start model.

1. BK-Black	3. B-Brown	5. W-White
2. B/Y-Black and Yellow	4. R-Red	6. Y-Yellow

WARDS

MONTGOMERY WARD
Montgomery Ward Plaza
Chicago, IL 60671

Model	Make	Engine Model	Horsepower	Cutting Width, In.
33857A	B&S	250000	11	36
33857B	B&S	250000	11	36
33867A	B&S	250000	11	36
33877A	B&S	250000	11	38
33887A	B&S	250000	11	38
33889A	B&S	250000	11	38

For service information and procedures, use the following model cross reference and refer to the GILSON section of this manual. Parts are not necessarily interchangeable and should be obtained from Montgomery Ward Co.

Wards Models	Gilson Models
33857A	52051C
33857B	52051C
33867A	52061
33877A	52060
33887A	52066
33889A	52074

WHEEL HORSE

WHEEL HORSE PRODUCTS, INC.
515 West Ireland Road
P.O. Box 2649
South Bend, IN 46680

Model	Make	Engine Model	Horsepower	Cutting Width, In.
R-26	Tecumseh	V70	7	26
A-50	Tecumseh	V70	7	26
A-51	B&S	130000	5	32
A-60	B&S	130000	5	26
A-70	B&S	191000	8	32
A-81	B&S	191000	8	32/36
A-111	B&S	252000	11	32/36
RR-532	B&S	130000	5	32
RR-832	B&S	191000	8	32

FRONT AXLE AND STEERING SYSTEM

Models R-26 and A-50

These mowers are fitted with a non-adjustable, center control, gear and sector steering system. There is no front axle as such and steering spindles are set into bushed holes in front body support.

R&R AND OVERHAUL. Raise front of unit, block up under body frame to support front wheels clear of surface and lock parking brake. Remove hub cap (22–Fig. WH1), "E" ring retainer (20) and wheel and tire assemblies (21). Remove tie rods (18), disengage "E" ring from slot at top of spindles (19) and lower each spindle out of bushings (7). Thoroughly clean spindles (19), bushings (7), and wheel hub bores. Carefully inspect all parts for excessive wear or any other damage and renew as needed.

Drive out roll pin (5) at bottom of steering shaft assembly tube (6) and separate from lower shaft and pinion assembly (12). Unbolt and remove two plates which make up steering support (10), then perform a thorough cleaning and inspection of shaft and pinion (12), all bearings (7), washers (11 and 13), retainer (9) and spacers (16). Renew all defective parts and reassemble, observing parts arrangement in Fig. WH1. Lubricate, using a good grade of chassis grease, by hand on spindles (19) and bearings (7) and with a pressure gun on each front wheel hub and at fitting (15).

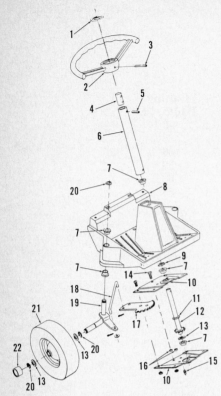

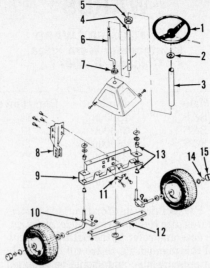

Fig. WH1 – Exploded view of front system and steering control parts used on Models R-26 and A-50.

1. Logo insert
2. Steering wheel
3. Spirol pin (¼X2)
4. Adapter
5. Roll pin (¼X1½) (2)
6. Assembly tube
7. Sintered iron bearing
8. Body support
9. Retainer
10. Steering support (2 pc.)
11. Washer
12. Lower shaft & pinion
13. Washer
14. Pivot bolt
15. Lube fitting
16. Spacers (3)
17. Steering sector
18. Tie rod (2)
19. Steering spindle
20. "E" ring
21. Wheel & tire
22. Hub cap

Model A-60

R&R AND OVERHAUL. To remove and disassemble axle main member (9– Fig. WH2), remove pull pin to disconnect front of mower from hanger bracket (8). Raise and block front of unit. Remove hub cap (15), "E" ring retainer (14) and front wheel assemblies. Unbolt and remove steering link (12) from spindles (10).

Remove "E" rings and washers, then lower spindles from axle and remove spindle bushings (13) from axle ends. Unbolt "U" bracket (11) securing steering shaft (6) to axle. Unbolt clutch and brake pedal brackets from axle. Unbolt and remove axle from frame. Reassemble by reversing removal procedure. Lubricate bushings, front wheel hubs and steering pivot points with SAE 30 oil.

To remove steering shaft (6–Fig. WH2), drive out roll pin and remove steering wheel from steering tube (4), then pull steering tube housing (3) out of

Fig. WH2 – Exploded view of front axle and steering system used on Model A-60.

1. Steering wheel
2. Plastic washer
3. Steering tube housing
4. Steering tube
5. Speed nut
6. Steering shaft
7. Flange bushing
8. Deck hanger bracket
9. Front axle
10. Spindle
11. "U" bracket
12. Steering link
13. Spindle bushings
14. "E" ring
15. Hub cap

flanged bushing (7). Drive out bottom roll pins and remove steering tube. Unbolt steering link (12) from spindles (10), then raise front of unit and withdraw steering shaft. Reinstall by reversing removal procedure.

Model A-70

R&R AND OVERHAUL. To remove axle main member (24–Fig. WH3), place mower unit in lowest position and unpin mower from mower hanger. Support front of unit and remove front wheels. Disconnect drag link end (20) from steering arm (22) and tie rod ends (27) from spindles. Drive out roll pin,

Fig. WH3 – Exploded view of front axle and steering system used on Model A-70.

1. Steering wheel
2. Cover
3. Steering shaft
4. Bushings
5. "E" ring
6. Pinion gear
7. Roll pin
8. Sector gear
9. "E" ring
10. Bushing
11. Roll pin
12. Steering support
13. Bushing
14. Sector shaft
15. "E" ring
16. Spindle bushings
17. Spindle R.H.
18. Sector arm
19. Adjusting nut
20. Drag link ends
21. Drag link
22. Spindle arm
23. Spacer
24. Axle
25. Pivot bolt
26. Spindle L.H.
27. Tie rod ends
28. Tie rod
29. "U" bracket
30. "E" ring

remove steering arm (22) and remove spindle (26) from axle. Remove "E" ring (15) and remove spindle (17). Remove pivot bolt (25), then lower axle main member from frame. Inspect spindle bushings (16) for excessive wear and renew as necessary. Reassemble by reversing disassembly procedure. Check front wheel toe-in and adjust ends (27) on tie rod (28) to obtain a toe-in of ⅛-inch.

To remove steering gears and shafts, drive out roll pin and remove steering wheel (1) and cover (2). Unbolt and remove console from around steering unit. Disconnect drag link end (20) from quadrant arm (18) and remove cotter pin, adjusting nut (19), quadrant arm and washers. Drive out roll pin (11) and remove "E" ring (9), then remove shaft (14) from front and lift out sector gear (8). Drive out roll pin (7) and remove "E" ring (5). Withdraw steering shaft (3) and remove pinion gear (6). Bushings (4, 10 and 13) can now be removed. Clean and inspect all parts for wear or other damage and renew as necessary. When reassembling, adjust steering gear free play as follows: Turn adjusting nut (19) clockwise to remove excessive play, then install cotter pin. Free play should be adjusted to a minimum but gears should not bind.

Lubricate all bushings and pivot points with SAE 30 oil. Apply a light coat of lithium grease to pinion gear and quadrant gear teeth.

Models A-51, A-81, A-111, RR-532 and RR-832

R&R AND OVERHAUL. To remove axle main member (19–Fig. WH4), place mower unit in lowest position and unpin mower from front hanger. Raise and block front of unit and remove front

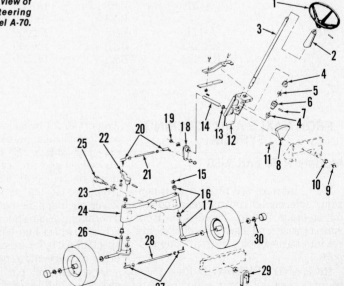

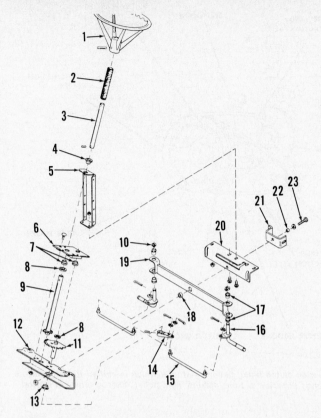

Fig. WH4 — Exploded view of front axle and steering system used on Models A-51, A-81, A-111, RR-532 and RR-832.

1. Steering wheel
2. Cover
3. Steering shaft extension
4. Bushing
5. Steering support
6. Reinforcing plate
7. Bushings
8. Shim washer
9. Steering shaft
10. Retaining ring
11. Steering sector
12. Axle support
13. Bushing
14. Steering bar
15. Tie rod
16. Spindle
17. Bushings
18. Bushing
19. Axle
20. Rear support
21. Pivot bracket
22. Bushing
23. Pivot bolt

wheel assemblies. Disconnect tie rods (15) from spindles (16). Remove retaining rings (10) and withdraw spindles and bushings (17) from axle ends. Remove pivot bolt (23), then lower axle from frame. Inspect bushings for excessive wear and renew as necessary. Reassemble by reversing disassembly procedure.

To remove steering gears and shafts, drive out roll pin and remove steering wheel (1 – Fig. WH4) and steering shaft cover (2). Drive out roll pin and remove steering shaft extension (3) and flanged bushing (4). Remove steering column shroud, then unbolt and remove reinforcing plate (6), bushings (7) and shim washers (8). Raise front of unit and withdraw steering shaft (9). Disconnect tie rods from steering bar (14). Drive out roll pin and remove steering bar, steering sector (11) and bushing (13). Clean and inspect all parts and renew as neces-

sary. Reassemble by reversing disassembly procedure. Use shim washers (8) as required to adjust end play of steering shaft and sector. Apply light coat of lithium base grease to gear teeth. Lubricate bushings and pivot points with SAE 30 oil.

ENGINE

All Models

For overhaul and repair procedures on engines listed in Specification Table, refer to Small Air Cooled Engines or Large Air Cooled Engines Service Manual.

Models R-26 and A-50

REMOVE AND REINSTALL. Remove mower as outlined in MOWER DECK paragraphs. Disconnect throttle

control cable at carburetor. Disconnect switch lead and solenoid cable from terminals on engine ends, then disconnect solenoid cable from post on starter motor. Set transmission shift lever in neutral and place traction clutch lever in its neutral notch. Slip forward traction belt from lower sheave of transmission pulley and work it up against transmission input shaft for slack, then remove from engine pulley and through belt guide. Release traction clutch lever from neutral notch and allow it to move forward, then, slip reverse traction belt from its sheave in transmission pulley and off reverse engine pulley (camshaft). Work both belts up through access hole in deck. Unbolt engine mounts and lift engine clear of frame. Reverse this procedure to reinstall engine.

Models A-60 and A-70

REMOVE AND REINSTALL. Loosen engine pulley belt guides, then remove mower as outlined in MOWER DECK paragraphs. On Model A-70, unbolt and remove engine cover and air intake duct. Shut off fuel at tank and disconnect fuel hose. On all models disconnect throttle control cable and ignition wires. On Model A-70 disconnect starter cable, alternator wire and unbolt muffler from engine. On all models, remove main clutch idler tension spring, then remove drive belt from engine pulley. Remove engine mounting bolts and lift off engine. Reinstall engine by reversing removal procedure.

Models A-51, A-81, A-111, RR-532 and RR-832

REMOVE AND REINSTALL. Unbolt rear of body, then raise body and support with support rod. On electric start models, disconnect battery cable and starter and alternator wires. On all models disconnect ignition wires and throttle control cable. On Model A-111 unbolt muffler. Remove mower as outlined in MOWER DECK paragraphs. Remove lower nuts, washer, spacer and clevis pin securing pto clutch-brake plate, then remove pto pulley (Fig. WH5). Unhook traction clutch idler spring and remove belt from transmission pulley and engine pulley. Unbolt and remove engine pulley. Remove engine mounting bolts and lift engine from frame. Reinstall engine by reversing removal procedure.

TRACTION DRIVE CLUTCH AND BELT

All Models

All models use belt idler type clutch. On Models R-26 and A-50, idler shifts

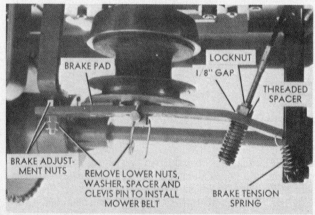

Fig. WH5 — Underside view of pto pulley and clutch-brake plate.

between forward drive belt (engine crankshaft pulley to transmission pulley) and reverse drive belt (engine camshaft pulley to transmission pulley). See Fig. WH6. All other models use a single belt and idler arrangement.

Models R-26 and A-50

REMOVE AND REINSTALL. To remove drive belts, disengage mower clutch and traction clutch. Remove mower drive belt from crankshaft pulley, then slip traction drive belts (Fig. WH6) off transaxle pulley and engine pulleys. Renew belts by reversing removal procedure.

IMPORTANT: Brake has a clutch override feature for automatic application. See BRAKE paragraphs for adjustment procedure.

All Other Models

R&R AND ADJUST. Remove mower as outlined in MOWER DECK paragraphs. Disconnect spring from idler bracket and remove idler pulley. On Model A-60, disconnect clutch rod from idler bracket. Loosen belt guides as needed and slip belt out of pulleys. Install new belt by reversing removal procedure. Belt guides should be 1/16 to ⅛-inch from belt with belt under tension.

TRANSAXLE

Models R-26 and A-50

REMOVE AND REINSTALL. Disconnect spark plug lead and positive cable from battery if unit is equipped for electric start. Remove mower deck as outlined in MOWER DECK paragraphs, and block front wheels securely. Raise and block under chassis and remove rear wheels. Disconnect transaxle shift lever from fork (25–Fig. WH7). For convenience, also unbolt shift lever hanger and grommet at forward end and move lever forward out of way. Set traction clutch lever in neutral and remove traction drive belts from drive pulleys on engine. Now, unbolt flanged bearing (29) from axle support on frame, place a temporary support under axle and remove cap screws which hold gear case to right hand axle support. Lower and remove entire transaxle assembly from frame. Reverse these steps to reinstall transaxle.

OVERHAUL. With transaxle removed and thoroughly cleaned, remove cap screw (1–Fig. WH7) and withdraw hub and flange (4). Remove cap screw (40) and, using a puller, remove input pulley (39) from input shaft (38). Set assembly up securely in shop vise with

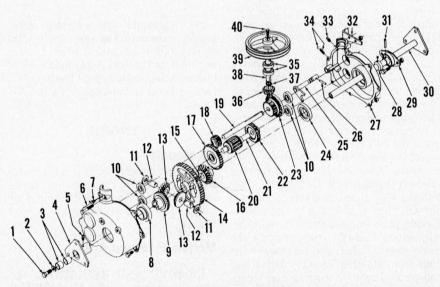

Fig. WH6 — View to show traction clutch belts, pulleys and idler which is shifted from forward to reverse traction belts to change direction of transmission input pulley. Idler is shown in neutral.

Fig. WH7 — Exploded view of Model 5056 Wheel Horse transaxle. Note double-sheave driven pulley (39) for forward or reverse operation.

1. Cap screw	11. Thrust washer (2)	21. Sliding gear shaft
2. Washer	12. Differential pin (2)	22. High gear
3. Bronze bushings (2)	13. Differential pinion (2)	23. Combination gear
4. Hub & flange	14. Final drive gear	24. Ball bearing LH
5. Grease fitting	15. Woodruff key	25. Shift fork & shaft
6. Gear case RH	16. Side gear LH	26. Roll pin
7. Grease fitting	17. Low gear	27. Gear case LH
8. Ball bearing RH	18. Low gear pinion	28. Collar & set screw
9. Side gear RH	19. Pinion shaft	29. Flanged ball bearing
10. Ball bearing (4)	20. Sliding gear	30. Rear axle
		31. Roll pin

32. Shift rod bracket	
33. Grease fitting	
34. Detent ball & spring	
35. Ball bearing (2)	
36. Input pinion	
37. Snap ring	
38. Input shaft	
39. Transmission pulley	
40. Bolt, nylok	

axle (30) pointed down. Remove the eight bolts which hold case halves together, then separate case halves part way and remove input shaft (38) along with bearings (35) and input pinion (36). Take care shift fork and shaft (25) do not slip from bore in left case (27) as detent

ball and spring (34) may be dropped. Continue to carefully separate case halves and lift off case (6), exposing all internal gears. Bearings (8 and 10) should remain in bores in case. Differential side gear (9) may lift out of place. Lift off side gear (9) and final drive gear

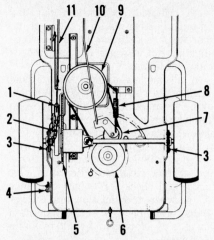

Fig. WH8—Underside view of traction drive clutch, brake linkage and differential assembly used on Model A-60.

1. Brake rod clevis	6. Engine pulley
2. Return spring	7. Clutch idler pulley
3. Axle bearing plates	8. Tension spring
4. Chain adjustment bracket	9. Transmission pulley
5. Chain	10. Clutch rod
	11. Brake rod

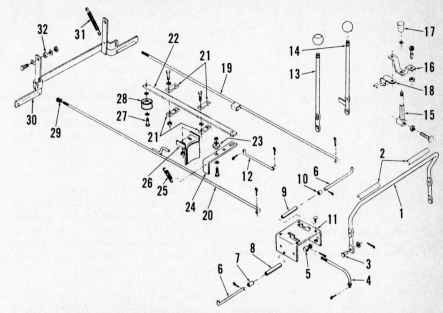

Fig. WH9—Exploded view of typical clutch, brake and control systems. Model A-50 is shown. Some early models have coil springs compressed on control bracket rods (6) as well as spacers shown.

1. Brake pedal	10. Spacer, 0.46X0.62X0.70	18. Ignition interlock module
2. Surface material	11. Control bracket	19. Brake rod, LH
3. Clevis pin	12. Traction clutch link	20. Brake rod, RH
4. Mower clutch link	13. Traction clutch lever	21. Guide blocks (4)
5. Trunnion stud	14. Mower clutch lever	22. Traction clutch bar
6. Control bracket rod (2)	15. Parking brake stem	23. Bushing
7. Spacer, 1.82-inch	16. Parking brake bracket	24. Clutch-brake pivot bar
8. Spacer, 4.32-inch	17. Knob	25. Traction clutch spring
9. Spacer, 5.46-inch		26. Blade brake/lining
		27. Hex bolt
		28. Forward-reverse idler
		29. Elastic stop nut
		30. Brake arm assembly
		31. Return spring
		32. Spacer

(14) with differential pinions (13) and their pins (12). Left hand side gear (16) is keyed to axle (30) by Woodruff key (15) and its removal may call for use of a small gear puller. After side gear (16) is removed, lift gear case (27) from axle, then lift off low gear pinion (18), pull shaft (19) out of bearing (10) and remove combination gear (23). Lift low gear (17) from sliding gear shaft (21), then pull shaft out of sliding gear (20) and disengage shift collar on sliding gear from shift fork (25). High gear (22) will be released at the same time.

If there is no need to remove shift fork and shaft (25), leave it in place in its bore in gear case (27); however, if it must be removed, take care not to lose detent ball and spring (34).

Remove ball bearings (8, 10 and 24) with care so as not to damage case halves. Install new bearings and reassemble gears and shafts in reverse order in which they were removed. Lubricate wear surfaces of shafts and bores of gears during reassembly, then coat gear teeth by hand with a good grade of lithium base chassis grease. Torque case bolts to 18-20 ft.-lbs. Use a grease gun at fittings (7 and 33) before reinstalling transaxle assembly.

TRANSMISSION

All Models

Models A-60 and A-70 use Peerless 500 series tranmissions. Models A-51, A-81, A-111, RR-532 and RR-832 are equipped with Peerless 700 series transmissions. For overhaul procedures, refer to Peerless paragraphs in TRANS-

MISSION REPAIR section of this manual.

REMOVE AND REINSTALL. Unbolt and raise or remove body components as needed for access to transmission. Disconnect wiring to transmission safety switch. Remove master link from drive chain and disconnect chain. Remove mower as outlined in MOWER DECK paragraphs. Slip drive belt off transmission pulley. Remove retaining ring from transmission input shaft and pull transmission pulley off shaft. Unbolt transmission and withdraw from chassis.

Reinstall by reversing removal procedure. Adjust drive belt and drive chain as needed.

DRIVE CHAIN

Models A-60 and A-70

R&R AND ADJUST. To remove chain (5–Fig. WH8), rotate differential sprocket to locate master link. Disconnect master link and remove chain. Inspect chain and sprockets for excessive wear and renew as needed. Lubricate chain with light coat of SAE 30 oil.

To adjust chain, raise and block rear of unit. Loosen bolts in rear axle bearing mounting brackets (3–Fig. WH8). On Model A-60, loosen brake bracket bolts, disconnect brake return spring (2) and

disconnect clevis (1) from parking brake arm. On both models, tighten chain by tightening bolt in chain adjustment bracket (4). Move other end of axle equal amount to maintain axle to frame alignment. When adjustment is correct, chain deflects approximately ½ to ¾-inch with light finger pressure, retighten axle mounting bolts. On Model A-60, reconnect brake linkage components and check brake operation.

Models A-51, A-81, A-111, RR-532 and RR-832

R&R AND ADJUST. To remove chain, rotate differential sprocket to locate master link. Disconnect master link and remove chain. Inspect chain and sprockets for excessive wear and renew as needed. Lubricate chain with light coat of SAE 30 oil.

Adjust chain tension, by adjusting position of rear idler sprocket (idler mounting plate is slotted for this purpose).

DIFFERENTIAL

All Models

Models A-60, A-70 and early production A-51, A-81, A-111, RR-532 and RR-832 are equipped with Peerless Series 100 differential. Late production Models A-51, A-81, A-111, RR-532 and

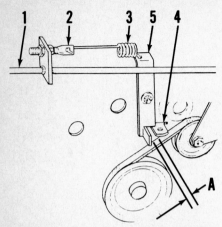

Fig. WH10—Underside view of brake linkage. Gap "A" should be ⅛-inch. Hold pivot lever to front and brake lever to rear when checking gap.

1. Clutch-brake rod
2. Adjustment bolt
3. Brake spring
4. Brake pivot lever
5. Transmission brake lever

RR-832 are equipped with Stewart Model 9500 differential. For overhaul procedures, refer to DIFFERENTIAL REPAIR section of this manual.

Models A-60 and A-70

REMOVE AND REINSTALL. Raise and block rear of unit. Remove hub cap, retaining ring and rear wheel assemblies. On Model A-60, unhook brake return spring (2-WH8) and disconnect brake rod clevis (1) from brake lever. On all models, disconnect drive chain at master link. Unbolt bearing plates (3) and remove differential assembly. Clean rust, paint or burrs from axle shaft and remove bearings. On Model A-60, remove nuts from inside brake drum and remove drum and sprocket.

Reinstall differential assembly by reversing removal procedure. Lubricate axle bearings with SAE 30 oil.

Models A-51, A-81, A-111, RR-532 and RR-832

REMOVE AND REINSTALL. Raise and block rear of unit. Remove hub cap and retaining ring and remove wheel assemblies. Disconnect drive chain at master link. Loosen set screw in axle lock collar and unbolt bearings from axle brackets. Remove differential and axle assembly from chassis (note location of spacers and washers).

Reinstall by reversing removal procedure.

GROUND DRIVE BRAKE

Models R-26 and A-50

ADJUSTMENT. Adjustment is performed on elastic stop nuts (29–Fig. WH9) at threaded rear end of brake rods (19 and 20). With brake relaxed (brake bars not contacting rear tires), nuts should be positioned so slack is equal on each rod.

To adjust so traction clutch will disengage automatically when brakes are applied, set traction clutch lever in **FORWARD** position and depress brake pedal, observing if traction clutch lever (13) moves to about ¾-inch from neutral notch just before brake bars contact rear tires. Turn nuts (29) an equal amount on each rod so braking will be balanced. Start engine and put mower in motion; then, observe action of traction clutch lever as brake is applied. If lever does not move to neutral notch, readjust nuts (29). If no declutching action occurs, check under chassis for condition of brake rod (20), especially hook portion welded to middle section of rod, spring (25) and pivot bar (24). While observing beneath deck, reach up and operate lever (13) and check for proper action of link (12), pivot bar (24) and its bushing (23) and traction clutch bar (22). Take steps to correct any binding or blockage found, and if parts are broken or defective, renew as necessary.

ADJUST PARKING LOCK. If parking brake does not hold when lock is engaged after brake pedal is applied, adjust as follows: Increase length of carriage-type bolt by backing out of parking brake stem (15–Fig. WH9) until it contacts welded block at middle of left hand brake rod (19) and holds brake arm (30) securely against rear tires. Test for holding by pushing mower by hand. Retighten locknut on adjuster.

Model A-60

ADJUSTMENT. To adjust brake, raise and block rear of unit and remove right rear wheel. Unhook brake return spring (2–Fig. WH8) and disconnect brake rod clevis (1) from brake lever. Loosen jam nut and turn clevis clockwise to tighten brake adjustment. Reconnect and check adjustment.

R&R AND OVERHAUL. Normal overhaul consists of renewing brake band. To remove brake assembly, raise and block rear of unit and remove right rear wheel. Unhook brake return spring and disconnect brake rod clevis from brake lever. Unbolt brake cam bracket and remove brake band roll pins, then remove brake band and cam bracket. Inspect for excessive wear and renew as needed. Reassemble by reversing removal procedure.

Models A-70, A-51, A-81, A-111, RR-532 and RR-832

ADJUSTMENT. To adjust disc brake, first unbolt and raise body. With brake released, loosen jam nut and tighten adjusting nut on brake lever until brake pads just clear brake disc. Lock adjusting nut with jam nut and recheck brake operation.

On all models except A-70, adjust clutch-brake rod (1–Fig. WH10) as follows: Disconnect idler plate spring. With clutch-brake pedal fully forward, idler plate should clear engine pulley by 1/16 to ⅛-inch. Adjust length of rod with threaded trunnion at front of rod to obtain correct clearance. Reinstall idler spring and check for ⅛-inch clearance "A" between transmission brake lever (4) and brake pivot lever (5). Reposition jam nuts on adjustment bolt (2) to obtain clearance.

R&R AND OVERHAUL. Normal overhaul consists of renewing brake pads. To remove brake assembly, unbolt and raise body. On Model A-70, disconnect brake rod from disc brake lever. Unbolt brake pad holder from transmission and remove holder and outer brake pad, then remove brake disc and inner brake pad. Inspect parts for excessive wear and renew as needed. Reassemble in reverse order of removal and adjust brake as previously outlined.

PTO DRIVE BRAKE

Models A-51, A-81, A-111, RR-532 and RR-832

ADJUSTMENT. Check and adjust pto clutch and brake as follows: With pto engaged, there should be .010 inch clearance between brake pad (Fig. WH5) and pulley and ⅛-inch clearance between hex head of threaded spacer and clutch-brake plate. To adjust brake, adjust position of adjustment nuts on clutch bracket stud. To adjust pto cable, loosen locknut and turn threaded spacer as required, then tighten locknut.

MOWER DRIVE BELT

Models R-26 and A-50

ADJUSTMENT. To adjust mower drive belt tension, disconnect trunnion stud (5–Fig. WH9) from mower clutch lever (14), then swing mower clutch link (4) to one side and back off threaded trunnion to increase length of clutch link. Check adjustment by reconnecting trunnion stud to clutch lever and move lever to engaged position. Tension of drive belt should be felt as lever reaches locking notch and a slight extra effort should be needed to set lever into engaged position.

To adjust blade brake (26), loosen brake adjusting nuts on floor under operator's seat and push mower clutch

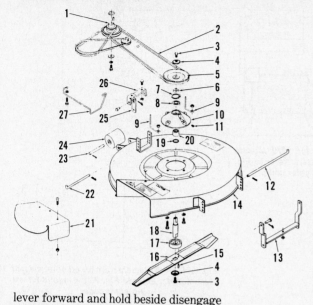

Fig. WH11—Exploded view of typical single spindle mower deck to show parts identification and arrangement.

1. Engine pulley
2. Mower drive belt
3. Nylok bolt
4. Dome washer (2)
5. Blade drive pulley
6. Spacer
7. Snap ring
8. Ball bearing
9. Belt guide (2)
10. Spindle housing
11. Grease fitting
12. Front support rod
13. Front hanger
14. Cutting deck
15. Roll pin
16. Blade
17. Blade driver
18. Spindle
19. Seal
20. Needle bearing
21. Deflector
22. Rear support rod
23. Roller shaft
24. Roller
25. Rear lift link (2)
26. Rear lift bracket
27. Belt retainer

Fig. WH13—Exploded view of typical triple spindle mower to show parts location and arrangement. Two different styles of spindles are shown.

1A. Blade
2A. Spindle cup
3A. Spindle shaft & bearing
4A. Spindle housing
1. Blade
2. Spindle shaft
3. Spindle cup
4. Spacer
5. Snap ring
6. Bearing
7. Spacer
8. Spindle housing
9. Bearing
10. Spacer
11. Seal
12. Deck
13. Bushing
14. Idler support bar
15. Belt guard
16. Pivot bracket
17. Bushings
18. Clutch lever
19. Tension spring
20. Blade pulley
21. Center pulley
22. Belt tension idler
23. Spindle belt
24. Blade pulley
25. Bushing
26. Clutch idler pulley
27. Bushing
28. Idler arm
29. Belt retainer
30. Return spring
31. Trunnion
32. Brake rod
33. Brake lever
34. Bushing

lever forward and hold beside disengage hooking notch of control panel; then, slide blade brake (26) forward so lining is in contact with edge of pulley and tighten adjusting nuts. Now, when lever is pushed and set firmly into blade disengage notch, positive stop of blade rotation should result. If blade brake lining is worn to the point of metal-to-metal contact, renew parts to prevent damage to mower drive pulley.

Models A-60 and A-70

ADJUSTMENT. Remove mower front hitch pull pin. On Model A-60 loosen front hitch jam nut and turn hitch clockwise to tighten belt. On Model A-70, turn leveling rod clevis clockwise to tighten belt. Check adjustment of belt by running at fast throttle setting, then disengage clutch. Blade should stop in short period of time. Repeat adjustment as needed.

Models A-51, A-81, A-111, RR-532 and RR-832

ADJUSTMENT. On 32 inch mowers, idler pulley is mounted on slotted bracket and is used to adjust belt tension. Move pulley rearward to increase belt tension. When properly adjusted, belt should deflect one inch, under light finger pressure, at mid point between left spindle pulley and pto pulley.

On 36 inch mowers, adjust belt tension as follows: On early production mowers, mower hanger bracket has slotted mounting holes. Mower can be moved forward or rearward to increase or decrease tension. On late production mowers, belt tension is adjusted by tightening two adjusting bolts evenly against front mounting "J" pin, until belt deflects approximately one inch under finger pressure, at mid point between pulleys. Tighten jam nuts on adjusting

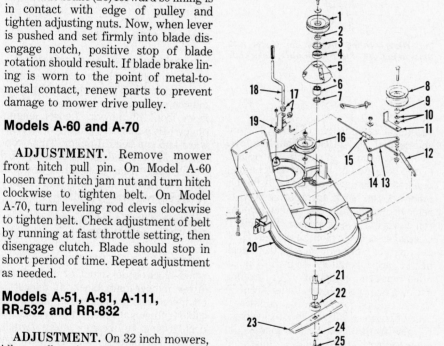

Fig. WH12—Exploded view of typical double spindle mower to show parts location and arrangement.

1. Blade pulley
2. Spacer
3. Snap ring
4. Ball bearing
5. Spindle housing
6. Needle bearing
7. Seal
8. Clutch idler pulley
9. Spacer
10. Washers
11. Belt retainer
12. Idler return spring
13. Idler arm
14. Pivot bushing
15. Clutch tension spring
16. Idler pulley
17. Bushings
18. Clutch lever
19. Pivot bracket
20. Mower deck
21. Spindle shaft
22. Spindle cup
23. Blade
24. Dome washer
25. Retainer bolt

bolts when adjustment is correct.

Models R-26, A-50, A-60 and A-70

REMOVE AND REINSTALL. Place mower in lowest position and disengage

blade clutch. Loosen belt guides as needed to remove belt from pulleys. Loosen idler pulley mounting bolt and slip belt between pulley and belt retainer and remove belt.

Renew belt by reversing removal procedure. Adjust belt guides to 1/16-inch clearance from belt. Position idler pulley belt retainer at right angle with idler arm.

Models A-51, A-81, A-111, RR-532 and RR-832

REMOVE AND REINSTALL. Remove mower unit as outlined in MOWER DECK paragraphs. Remove shields and loosen belt guides as necessary to remove belt from pulleys.

Wheel Horse

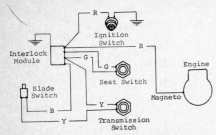

Fig. WH14—Typical wiring schematic for Models A-60, A-51 and A-81 with recoil start. Seat switch not used on all models.

1. BK-Black
2. O-Orange
3. R-Red
4. B-Brown
5. W-White
6. Y-Yellow
7. BL-Blue
8. G-Green

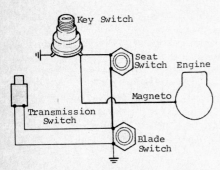

Fig. WH15—Wiring schematic for 1982 Models A-51 and RR-532.

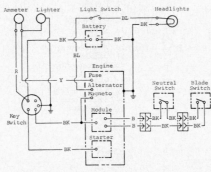

Fig. WH16—Wiring diagram for 1979 Models A-81 and A-111 with electric start. Refer to Fig. WH14 legend for wiring color code.

Fig. WH17—Wiring diagram for Model A-70. Refer to Fig. WH14 legend for wiring color code.

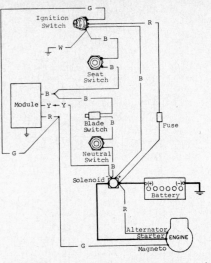

Fig. WH18—Wiring diagram for all other electric start models. Refer to Fig. WH14 legend for wiring color code.

Renew drive belt by reversing removal procedure. Adjust belt guides to 1/16-inch clearance from belt.

MOWER SPINDLE

All Models

R&R AND OVERHAUL. Refer to Fig. WH11, WH12, or WH13 for parts identification and arrangement. Remove mower unit as outlined in MOWER DECK paragraphs. Place wood block between blade and housing and unbolt and remove blade and drive pulley. Unbolt and remove spindle housing from deck. Push spindle out of housing. Remove retaining ring and press bearings out of housing.

Reassemble in reverse order of removal. When installing new bearings, press on outer race only. On mowers equipped with blade retainer bolt, torque to 22 ft.-lbs. If equipped with blade retainer nut, torque to 60-70 ft.-lbs. Lubricate with No. 2 multi-purpose lithium base grease.

MOWER DECK

Models A-51, A-60, A-70, A-81, A-111, RR-532 and RR-832

LEVEL ADJUSTMENT. With unit on level surface and tires properly in-

flated, measure height of blade at front and rear of deck. Blade should be level to ⅛-inch lower in front. To adjust, loosen bolts and move front deck hanger bracket up or down on Model A-60. On Model A-70, block front of mower to remove weight and disconnect leveling rod clevis at front of mower. Turn clevis clockwise to lower front of mower. Each 1½ turns of clevis will change front deck height approximately ⅛-inch. On all other models equipped with leveling cable, place wood block under front of mower and disconnect cable trunnion at rear of mower. Turn trunnion clockwise to raise front of mower. On models not equipped with leveling cable, adjustment is made with a jam bolt at front hanger bracket. With front of mower setting at specified height, turn jam bolt until head of bolt just contacts front hanger bracket. Tighten jam nut to lock adjustment.

Models R-26 and A-50

REMOVE AND REINSTALL. Disconnect wiring from mower safety switch. Disengage blade clutch and remove front and rear support rods from hanger brackets. Move mower to rear, remove belt from engine pulley and slip belt out of belt guard. Turn front wheels full left and pull mower deck out right side. Reinstall by reversing removal procedure.

Models A-60 and A-70

REMOVE AND REINSTALL. Disengage blade clutch and place mower in lowest position. Disconnect wiring from mower safety switch. Remove pull pin from front of mower. Loosen engine pulley belt guides, then move deck rearward and slip belt off pulley. Turn front wheels full left, then move deck forward to disengage rear deck guide and pull deck out right side of unit. Reinstall by reversing removal procedure.

Models A-51, A-81, A-111, RR-532 and RR-832

REMOVE AND REINSTALL. Place wood block under front of mower and lower mower onto block. If mower is equipped with leveling cable, disconnect cable trunnion at rear of mower. Disconnect wiring to mower safety switch. Loosen idler pulley to relieve belt tension and remove "J" bolts from front yoke, then slip drive belt out of mower pulleys. Remove lower nuts, washer and spacer from clutch-brake bracket stud (Fig. WH5). Hold pto pulley and remove clevis pin, then lower clutch-brake plate and remove belt. Turn front wheels full left and pull mower out from right side. Reinstall by reversing removal procedure.

ELECTRICAL

All Models

All units are equipped with safety interlock starting system to prevent starting with drives engaged. Refer to Fig. WH14 through WH18.

WHITE

WHITE OUTDOOR PRODUCTS
2625 Butterfield Road
Oak Brook, IL 60521

| Model | Make | Engine | | Cutting |
		Model	Horsepower	Width, In.
R50	B&S	130000	5	26
R80	B&S	190000	8	34
R82	B&S	190000	8	30

FRONT AXLE AND STEERING SYSTEM

Model R50

R&R AND OVERHAUL. To remove axle main member (15 – Fig. W1), first remove steering shaft (4) as follows: Remove steering wheel cap, retaining nut and washer, then lift off steering wheel (1) and cover (2). Unbolt and remove shaft extension (3), speed nut (5) and washer (6). Disconnect tie rods (11) from spindles (10 and 13), raise front of unit and withdraw steering shaft (4) from bottom. Block up under frame and remove front wheels. Remove cotter pins and washers, then lower spindles (10 and 13) from axle. Unbolt and remove axle main member (15) from frame (9). Flange bushings (7 and 12) and spindle bushings (14) can now be removed if necessary.

Clean and inspect all parts and renew any showing excessive wear or other damage. When reassembling, reverse disassembly procedure and lubricate bushings with SAE 30 oil. The rods (11) are non-adjustable.

Model R80

R&R AND OVERHAUL. To remove axle main member (15 – Fig. W2), support front of unit and remove front wheels. Disconnect tie rods (14) from spindles. Remove cotter pins and washers, then lower spindles (12 and 18) from axle main member. Remove pivot bolt (17) and unhook mower springs from axle support (16). Unbolt axle support from frame (9) and separate axle main member (15) from axle support. In-

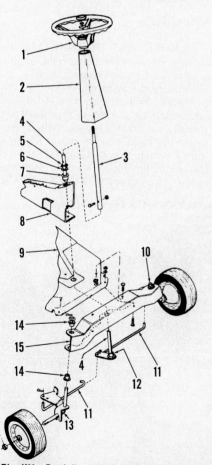

Fig. W1 — Partially exploded view of front axle and steering gear assembly used on Model R50.

1. Steering wheel
2. Cover
3. Shaft extension
4. Steering shaft
5. Speed nut
6. Washer
7. Flange bushing
8. Support
9. Frame
10. Spindle L.H.
11. Tie rods
12. Flange bushing
13. Spindle R.H.
14. Spindle bushings
15. Axle main member

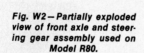

Fig. W2 — Partially exploded view of front axle and steering gear assembly used on Model R80.

1. Steering wheel
2. Spacer tube
3. Shaft extension
4. Steering shaft
5. Cover
6. Washer
7. Flange bushing
8. Console
9. Frame
10. Support
11. Rack & pinion assy.
12. Spindle R.H.
13. Spindle bushings
14. Tie rods (2 used)
15. Axle main member
16. Axle support
17. Pivot bolt
18. Spindle L.H.

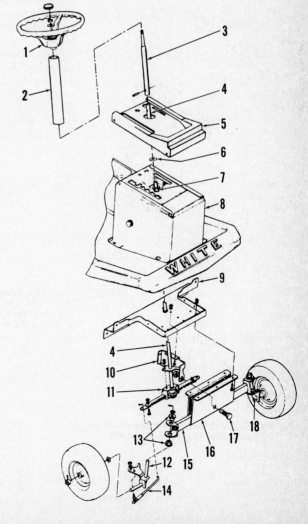

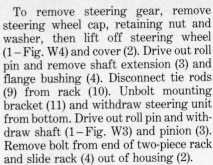

spect spindle bushings (13) for excessive wear and renew as necessary. Reassemble by reversing the disassembly procedure. Adjust tie rods as required for 1/8-inch toe-in.

To remove steering gear, remove steering wheel cap, retaining nut and

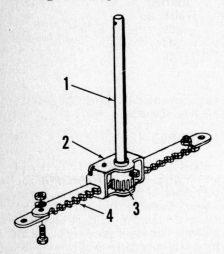

Fig. W3 — Steering rack and pinion assembly removed from Model R80.

1. Steering shaft
2. Steering housing
3. Pinion
4. Rack (2 piece)

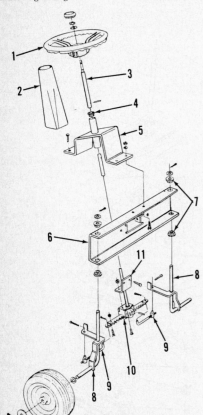

Fig. W4 — Exploded view of front axle and steering gear assembly used on Model R82.

1. Steering wheel
2. Cover
3. Shaft extension
4. Flange bushing
5. Steering support
6. Axle
7. Spindle bushings
8. Spindles
9. Tie rods
10. Steering rack
11. Mounting bracket

washer, then lift off steering wheel (1). Remove spacer tube (2), drive out roll pin and remove shaft extension (3) and washer (6). Raise front of unit and block securely. Disconnect tie rods from ends of rack. Unbolt steering housing (2 – Fig. W3) from support (10 – Fig. W2) and withdraw steering unit from bottom. Refer to Fig. W3 and drive out roll pin securing pinion (3) to steering shaft (1). Withdraw shaft and remove pinion. Remove bolt from one end of two-piece rack (4) and slide rack out of housing (2). Clean and inspect all parts and renew any showing excessive wear or other damage. Flange bushing (7 – Fig. W2) can be removed and a new bushing installed if necessary. Reassemble by reversing disassembly procedure. Lubricate rack and pinion with multi-purpose lithium grease. Lubricate flange bushing (7), spindle bushings (13), pivot bolt (17) and front wheels with SAE 30 oil.

Model R82

R&R AND OVERHAUL. To remove axle main member (6 – Fig. W4), support front of unit and remove front wheels. Disconnect tie rods (9) from spindles (8). Remove cotter pins and washers, then withdraw spindles and bushings (7) from axle ends. Unbolt and remove axle from frame.

Inspect for excessive wear and renew as needed. Reassemble by reversing removal procedure and lubricate spindle bushings and front wheels with SAE 30 oil. Adjust tie rods as required for 1/8-inch toe-in.

To remove steering gear, remove steering wheel cap, retaining nut and washer, then lift off steering wheel (1 – Fig. W4) and cover (2). Drive out roll pin and remove shaft extension (3) and flange bushing (4). Disconnect tie rods (9) from rack (10). Unbolt mounting bracket (11) and withdraw steering unit from bottom. Drive out roll pin and withdraw shaft (1 – Fig. W3) and pinion (3). Remove bolt from end of two-piece rack and slide rack (4) out of housing (2).

Clean and inspect parts and renew as needed. Reassemble by reversing removal procedure. Lubricate rack and pinion with multi-purpose lithium grease and flange bushings with SAE 30 oil.

ENGINE

All Models

For overhaul and repair procedures on engines listed in Specification Table, refer to Small Air Cooled Engines Service Manual.

Model R50

REMOVE AND REINSTALL. To remove engine assembly, raise engine hood and disconnect spark plug wire, ignition wire at magneto and throttle control cable. Pull starter rope out a few inches and tie a knot in rope between starter and starter safety switch. Remove rope handle and feed rope through safety switch. Knot will prevent rope from coiling inside starter. Remove mower as outlined in MOWER DECK

Fig. W5 — Bottom view of Model R50, with mower unit removed, showing removal of engine pulley belt guard.

paragraphs. Unbolt and remove engine pulley belt guard. See Fig. W5. Unhook variable drive springs, then remove retaining nut and washer from variable speed pulley. Lower variable speed pulley and remove primary traction drive belt from engine pulley. Unbolt and remove engine pulley. Remove engine mounting bolts and lift engine from frame.

Reinstall engine assembly by reversing removal procedure. Make certain belt guard and belt keepers are properly installed.

Model R80

REMOVE AND REINSTALL. To remove engine assembly, raise engine cover and disconnect spark plug wire. Disconnect battery cables, ignition wire from magneto, cable from starter and throttle control cable. Unbolt exhaust pipe from engine. Place lift and blade clutch lever in disengaged position. Remove belt keepers at engine pulley. Remove mower drive belt from engine pulley. Move lift and blade clutch lever to lowest notch. Unbolt and remove engine pulley belt guard. Refer to Fig. W6 and unhook main drive belt clutch idler springs. Remove shoulder bolt near transmission input pulley. Remove main drive belt from transmission pulley first, then from engine pulley. Unbolt and remove the engine pulley. Remove engine mounting bolts, then lift engine from frame.

Reinstall engine assembly by reversing removal procedure. Make certain belt guard and belt keepers are properly installed.

Fig. W6 – Bottom view of Model R80, with mower unit removed. Note the heavy and light clutch idler springs.

Model R82

REMOVE AND REINSTALL. To remove engine, disconnect spark plug wire, battery cables, ignition wire from magneto, cable from starter and throttle control cable at engine. Engage blade clutch lever. Loosen mower deck front mounting bolts and remove rear mounting bolts. Raise mower and slip mower drive belt off engine pulley. Remove engine pulley belt guard and shoulder bolts at transmission pulley. Slip traction drive belt off transmission pulley first, then off engine pulley. Remove engine mounting bolts and lift engine from frame.

Reinstall by reversing removal procedure.

TRACTION DRIVE CLUTCH AND DRIVE BELT

All Models

The traction drive clutch used on all models is belt idler type. On Model R50, variable speed pulley is also the clutch idler. There is no adjustment on clutch idler. If drive belt slips during normal operation due to excessive belt wear or stretching, belt must be renewed.

Model R50

REMOVE AND REINSTALL. To remove traction drive belts, raise engine hood and disconnect spark plug wire. Remove belt keepers at engine pulley. Remove mower as outlined in MOWER DECK paragraphs. Refer to Fig. W5 and remove engine pulley belt guard. Unhook variable speed and clutch springs (6 and 8 – Fig. W7). Remove retaining nuts and washers, then lower variable speed pulley (7) and transmission input pulley (12). Remove primary drive belt (9) and secondary drive belt (11).

Fig. W7 – View of traction drive belts, variable speed and clutch linkage and variable speed pulley used on Model R50.

1. Variable speed & clutch idler arms
2. Speed control bracket
3. Locking knob
4. Clutch pedal
5. Clutch rod
6. Spring
7. Variable speed pulley
8. Variable speed spring
9. Primary drive belt
10. Engine pulley
11. Secondary drive belt
12. Transmission pulley
13. Roll pin (2 used)
14. Upper sheave half
15. Movable sheave half
16. Ball bearing
17. Sleeve
18. Spacer
19. Ball bearing
20. Lower sheave half

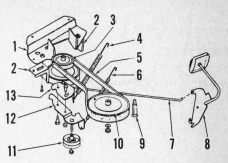

Fig. W8 – Exploded view of traction drive clutch assembly used on Model R80.

1. Engine pulley belt guard
2. Belt keepers
3. Engine pulley
4. Clutch spring (heavy)
5. Main drive belt
6. Clutch spring (light)
7. Clutch rod
8. Clutch pedal
9. Shoulder bolt
10. Transmission pulley
11. Idler pulley
12. Idler arm
13. Idler pivot plate

Install new belts by reversing removal procedure. Make certain belt guard and belt keepers are properly installed.

Model R80

REMOVE AND REINSTALL. To remove traction drive belt, raise engine cover and disconnect spark plug wire. Place lift and blade clutch lever in disengaged position. Remove belt keepers (2 – Fig. W8) at engine pulley and slip mower drive belt from engine pulley. Move lift and blade clutch lever to lowest notch. Unbolt and remove engine pulley belt guard (1), then unhook clutch idler springs (4 and 6). Remove shoulder bolt (9) near transmission input pulley. Remove traction drive belt (5) from transmission pulley first, then from engine pulley.

Install new belt by reversing removal procedure. Make certain belt guard and belt keepers are properly installed.

Model R82

REMOVE AND REINSTALL. To remove traction drive belt, disconnect

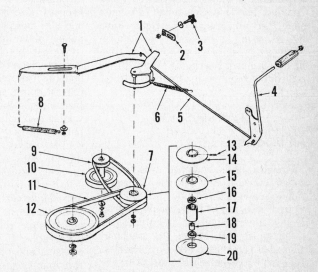

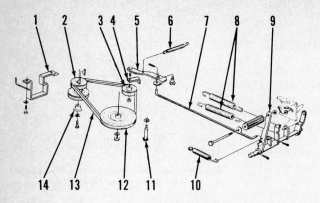

*Fig. W9 — Exploded view of
traction drive belt and clutch
linkage used on Model R82.*

1. Engine belt guide
2. Engine pulley
3. Clutch idler pulley
4. Belt keeper
5. Idler arm bracket
6. Idler spring
7. Clutch rod
8. Traction springs
9. Clutch-brake pedal assy.
10. Brake spring
11. Shoulder bolt
12. Transmission pulley
13. Traction drive belt
14. Step washer

spark plug wire and proceed as follows: Engage blade clutch lever. Loosen mower deck front mounting bolts and remove rear mounting bolts. Unbolt rear axle center bearing support bracket. Loosen engine mounting bolts, then raise engine for clearance and slide support bracket and bearing away from engine pulley. Raise mower deck and slip mower drive belt off engine pulley. Remove engine pulley belt guard (1 – Fig. W9) and transmission pulley shoulder bolts (11). Unbolt and remove idler pulley (3). Slip belt (13) off transmission pulley first, then off engine pulley.

Install new belt by reversing removal procedure. Make certain belt guard and keepers are properly installed.

VARIABLE SPEED PULLEY

Model R50

R&R AND OVERHAUL. To remove variable speed pulley (7 – Fig. W7), remove traction drive belts (9 and 11) as outlined previously. Then, with pulley assembly removed, drive out roll pins (13) and separate movable sheave half (15) and sheave halves (14 and 20) from sleeve (17). Press bearings (16 and 19) and spacer (18) from the sleeve.

Clean and inspect all parts and renew any showing excessive wear or other damage. Movable sheave half (15) must slide freely on sleeve (17). Reassemble by reversing disassembly procedure. Apply a dry lubricant to movable sheave half and sleeve. Bearings (16 and 19) are sealed and require no additional lubrication.

TRANSMISSION

All Models

Model R50 is equipped with single speed forward and reverse transmissions manufactured by J.B. Foote Foundry Co. Model R80 is equipped with Model 515 transmission and Model R82 is equipped with Model 701 transmission, both manufactured by Peerless Division of Tecumseh Products Co. For overhaul procedures, refer to Foote or Peerless paragraphs in TRANSMISSION REPAIR section of this manual.

Model R50

REMOVE AND REINSTALL. To remove transmission assembly, raise engine hood and disconnect spark plug wire. Remove belt keepers at engine pulley. Remove mower as outlined in MOWER DECK paragraphs. Unbolt and remove engine pulley belt guard. Unhook variable speed and clutch springs (6 and 8 – Fig. W7). Remove retaining nuts and washers, then lower variable speed pulley (7) and transmission input pulley (12) with main drive belts (9 and 11). Locate master link and disconnect drive chain. Disconnect transmission shift rod, then unbolt and remove transmission assembly.

Reinstall by reversing removal procedure. Make certain belt guard and belt keepers are properly installed. If drive chain requires adjustment, refer to DRIVE CHAIN paragraphs.

Model R80

REMOVE AND REINSTALL. To remove transmission assembly, raise engine cover and disconnect spark plug wire. Place lift and blade clutch lever in disengaged position. Remove belt keepers (2 – Fig. W8) at engine pulley and slip mower drive belt from engine pulley. Move lift and blade clutch lever to lowest notch. Unhook clutch idler springs (4 and 6) and remove shoulder bolt (9) near transmission input pulley (10). Remove traction drive belt from transmission pulley and remove nut, washer and pulley from transmission input shaft. Locate master link and disconnect drive chain. Disconnect transmission shift rod, then unbolt and remove transmission assembly.

Reinstall by reversing removal procedure. Make certain belt keepers are properly installed. If drive chain requires adjustment, refer to DRIVE CHAIN paragraphs. If transmission shift lever contacts end of shift quadrant slot before fourth or reverse gears are fully en-

gaged, adjust ferrule at rear of shift rod as required to correct condition.

Model R82

REMOVE AND REINSTALL. To remove transmission, disconnect spark plug wire and proceed as follows: Remove traction clutch idler pulley. Remove shoulder bolts at transmission pulley, then slip traction belt off pulley. Unbolt and remove transmission pulley. Disconnect drive chain at master link. Remove transmission mounting bolts and lift transmission from frame.

Reinstall by reversing removal procedure. If drive chain requires adjustment, refer to DRIVE CHAIN paragraphs.

DRIVE CHAIN

Model R50

ADJUSTMENT. The drive chain should deflect approximately ½-inch when depressed with thumb midway between sprockets. To adjust chain, loosen transmission mounting nuts slightly. Turn draw bolt, located under right side of main frame, clockwise to tighten or counter-clockwise to loosen drive chain. When chain tension is correct, retighten transmission mounting nuts.

Model R80-R82

ADJUSTMENT. Drive chain tension should be checked periodically. Chain should deflect approximately ½-inch when depressed with thumb midway between sprockets. To adjust drive chain, loosen chain idler mounting nuts and slide chain idler rearward to tighten or forward to loosen drive chain. When chain tension is correct, retighten idler mounting nuts.

All Models

REMOVE AND REINSTALL. To remove drive chain, rotate differential sprocket to locate master link. Disconnect master link and remove chain. Clean and inspect chain and renew if excessively worn.

Install drive chain and adjust chain tension as outlined in previous paragraphs. Lubricate chain with a light coat of SAE 30 oil.

DIFFERENTIAL

All Models

MTD Products differentials are used on all models. For overhaul procedures, refer to DIFFERENTIAL REPAIR section of this manual.

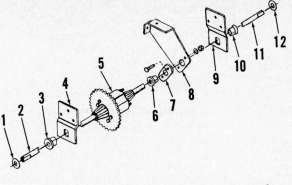

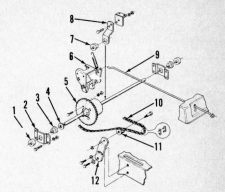

Fig. W10—Differential as-
sembly, rear axles and axle
bearings used on Model R50.

1. Washer
2. Axle shaft R.H.
3. Axle bearing (outer)
4. Bearing bracket
5. Differential assy.
6. Axle bearing (center)
7. Bearing plate
8. Bearing bracket
9. Bearing bracket
10. Axle bearing (outer)
11. Axle shaft L.H.
12. Washer

Fig. W12—Differential, rear axles and bearings
used on Model R82.

1. Spacer
2. Axle bracket
3. Axle bushing (outer)
4. Washer
5. Differential & axle assy.
6. Disc brake
7. Axle bushing (inner)
8. Support bracket
9. Brake rod
10. Drive chain
11. Idler
12. Idler arm

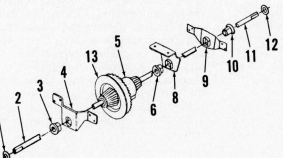

Fig. W11—Differential as-
sembly, rear axles and axle
bearings used on Model R80.

1. Washer
2. Axle shaft R.H.
3. Axle bearing (outer)
4. Bearing bracket
5. Differential assy.
6. Axle bearing (center)
7. Bearing bracket
8. Bearing bracket
9. Bearing bracket
10. Axle bearing (outer)
11. Axle shaft L.H.
12. Washer
13. Brake disc

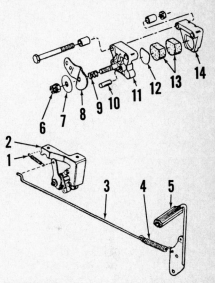

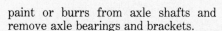

Model R50

REMOVE AND REINSTALL. To
remove differential and rear axle assembly, raise rear of unit and support
securely. Locate master link in drive
chain and disconnect chain. Remove
rear wheel assemblies. Refer to Fig.
W10 and unbolt three axle bearing
brackets from main frame. Remove differential and axle assembly, being careful not to damage brake caliper assembly. Clean all rust, paint or burrs from
axle shafts and remove axle bearings
and brackets.

Inspect axle bearings and renew as
necessary. When reinstalling, flanges on
outer axle bearings must face outward
(against wheel hubs). Flange on inner
axle bearing must be toward differential
housing. Lubricate axle bearing with
SAE 30 oil.

Model R80

REMOVE AND REINSTALL. To
remove differential and rear axle assembly, place mower lift and blade clutch
lever in disengaged position. Remove
belt keepers at engine pulley and slip
mower drive belt from engine pulley.
Raise rear of unit and support securely.
Locate master link in drive chain and
disconnect chain. Remove rear wheel assemblies and disconnect brake rod. Unbolt and remove left axle bearing and
bracket. See Fig. W11. Unbolt right and
center axle bearing brackets from main
frame, move differential assembly forward to clear the brake caliper unit, then
remove differential and rear axle assembly out toward right side. Clean all rust,

paint or burrs from axle shafts and
remove axle bearings and brackets.

Clean and inspect axle bearings and
renew as necessary. When reinstalling,
flanges on outer axle bearings must face
outward (against wheel hubs). Flange on
inner axle bearing must be toward differential housing. Balance of reassembly
is reverse of disassembly procedure.
Lubricate axle bearings with SAE 30 oil.

Model R82

REMOVE AND REINSTALL. To
remove differential and rear axle assembly (5 – Fig. W12), raise rear of unit and
block securely. Disconnect drive chain at
master link. Unbolt axle support bracket
(8). Unbolt axle bearing brackets (2) and
lower and remove differential assembly
from frame. Remove rear wheels and
spacers (1). Remove paint, rust or burrs
from axle and slide bearings off axle.

Reinstall in reverse order of removal.

GROUND DRIVE BRAKE

All Models

The brake used on all models is caliper
disc type. The differential chain
sprocket is the brake disc.

ADJUSTMENT. To adjust brake,
tighten adjusting nut (6 – Fig. W13) on
actuating lever (8) ½-turn. Recheck
brake and repeat adjustment if necessary.

R&R AND OVERHAUL. To remove
brake caliper assembly (2 – Fig. W13),
disconnect return spring (1) and brake

rod (3). Unbolt mounting bracket from
frame and remove caliper assembly. Remove adjusting nut (6), washer (7), cam
lever (8), spring (9) and actuating pins
(10). Remove through bolts, then remove mounting bracket and separate
brake pads (13) and back-up washer (12)
from carriers (11 and 14).

Clean and inspect all parts and renew
any showing excessive wear or other
damage. When reassembling, install
back-up washer (12) and thick brake pad
in carrier (11). Install actuating pins so
rounded end is toward cam lever (8).
Reinstall brake by reversing removal
procedure. Adjust brake as outlined in
previous paragraph.

Fig. W13—View showing caliper brake used on
Model R50, other models are similar. Chain
sprocket on differential is also the brake disc.

1. Return spring
2. Caliper assembly
3. Brake rod
4. Brake tension spring
5. Brake pedal
6. Adjusting nut
7. Washer
8. Cam lever
9. Spring
10. Actuating pin (2)
11. Carrier (outer)
12. Back-up washer
13. Brake pads
14. Carrier (inner)

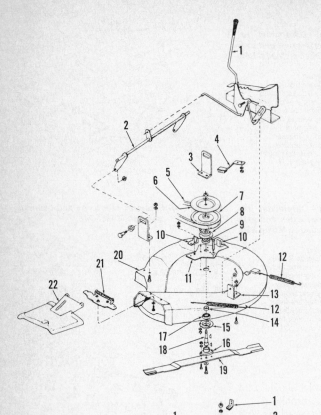

Fig. W14—Exploded view of the 26" single blade mower unit used on Model R50.

1. Lift & blade clutch lever
2. Lift arms & shaft assy.
3. Rear lift brackets
4. Blade brake pad
5. Mower belt
6. Blade brake disc
7. Mower pulley
8. Bearing cover
9. Ball bearing
10. Belt keepers
11. Mounting plate
12. Mower tension springs
13. Front lift bracket
14. Spacer
15. Bearing cover
16. Blade adapter
17. Ball bearing
18. Blade spindle
19. Blade 26"
20. Mower housing
21. Deflector bracket
22. Deflector

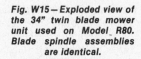

Fig. W15—Exploded view of the 34" twin blade mower unit used on Model R80. Blade spindle assemblies are identical.

1. Blade brake pad
2. Brake disc
3. Mower pulley
4. Mower belt
5. Washer
6. Plate
7. Belt keeper
8. Mower tension springs
9. Mower housing
10. Bearing retainers
11. Ball bearings
12. Blade spindle
13. Blade adapter
14. Blade
15. Deflector bracket
16. Deflector
17. Gage wheel
18. Pivot bar
19. Adjusting lever
20. Wheel bracket

MOWER DRIVE BELT

Model R50

REMOVE AND REINSTALL. To remove mower drive belt (5—Fig. W14), place lift and blade clutch lever in disengaged position. Remove belt keepers at engine pulley and slip mower drive belt from engine pulley. Move lift and blade clutch lever fully forward and unhook both tension springs (12) from front of mower housing. Disconnect lift rod from mower front lift bracket (13) and unbolt lift arms from rear lift brackets (3). Remove mower from under

right side. Unbolt and remove belt keepers (10) and shoulder bolt at mower pulley, then remove mower belt.

Install new belt by reversing removal procedure. Make certain belt keepers are properly installed.

Model R80

REMOVE AND REINSTALL. To remove mower drive belt (4—Fig. W15), place lift and blade clutch lever in disengaged position. Remove belt keepers at engine pulley and slip mower drive belt from engine pulley. Move lift and blade clutch lever to lowest notch. Un-

bolt and remove belt keeper (7) at left mower pulley, then remove mower belt.

Install new belt by reversing removal procedure. Make certain belt keepers are properly installed.

Model R82

REMOVE AND REINSTALL. To remove mower drive belt (5—Fig. W16), disconnect spark plug wire and proceed as follows: Place blade clutch lever in engaged position. Loosen mower deck front mounting bolts (8) and remove rear mounting bolts (7). Unbolt rear axle inner bearing support bracket. Loosen engine mounting bolts, then raise engine for clearance and slide support bracket and bearing away from engine pulley. Remove belt keeper (6) and idler pulley (3) from mower deck, then slip belt off pulleys.

Install new belt in reverse order of removal.

MOWER SPINDLE

Model R50

R&R AND OVERHAUL. To remove spindle, first remove mower assembly as outlined in MOWER DECK paragraphs. Unbolt and remove belt keepers (10—Fig. W14) and shoulder bolt at mower pulley, then remove mower belt. Remove nut, washer, blade brake disc (6) and mower pulley (7) from top of blade spindle. Remove blade center bolt, blade (19) and adapter (16) from bottom of spindle. Unbolt bearing retainers (8 and 15), then remove retainer (8) and bearing (9). Remove and separate spindle (18), bearing retainer (15), bearing (17) and spacer (14).

Clean and inspect all parts and renew any showing excessive wear or other damage. Reassemble by reversing disassembly procedure. Torque blade adapter mounting bolts to 20 ft.-lbs. and blade center bolt to 35 ft.-lbs.

Model R80

R&R AND OVERHAUL. To remove spindle, first remove mower assembly as outlined in MOWER DECK paragraphs. Unbolt and remove belt keeper (7—Fig. W15) and remove mower belt. Remove nut, washer, blade brake disc (2), mower pulley (3) and washer (5) from top of blade spindle. Remove blade center bolt, blade (14) and adapter (13) from bottom of spindle. Remove bolts securing bearing retainers (10) to mower housing, then remove and separate spindle (12), bearing retainers (10) and bearings (11).

Clean and inspect all parts and renew any showing excessive wear or other damage. Reassemble by reversing dis-

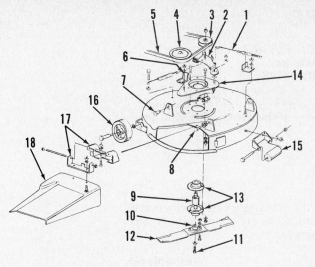

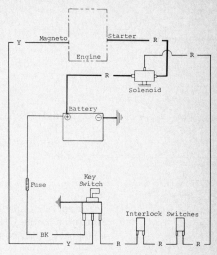

Fig. W16 – Exploded view of 30" single blade mower unit used on Model R82.

1. Clutch cable
2. Idler & brake arm
3. Idler pulley
4. Blade pulley
5. Mower belt
6. Belt keeper
7. Rear mounting bolt
8. Front mounting bolt
9. Spindle & bearing assy.
10. Blade adapter
11. Blade center bolt
12. Blade
13. Bearing retainers
14. Spindle mounting bracket
15. Roller
16. Gage wheel
17. Deflector hinge assy.
18. Deflector

Fig. W18 – Wiring diagram for Model R80.

1. R-Red
2. Y-Yellow
3. BK-Black

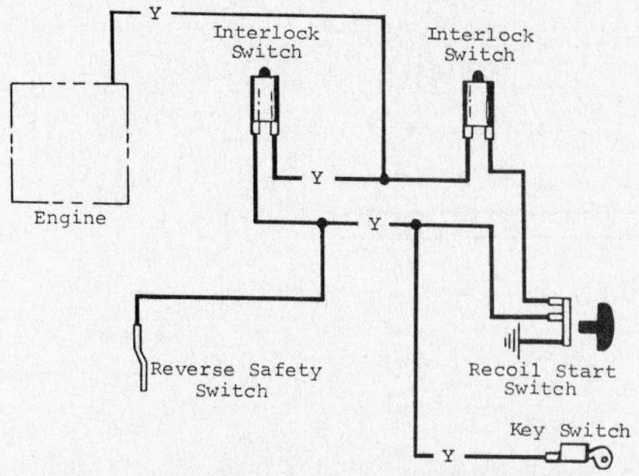

Fig. W17 – Wiring diagram for Model R50. Wires marked (Y) are yellow.

MOWER DECK

Model R50

REMOVE AND REINSTALL. Disconnect spark plug wire and place blade clutch lever in disengaged position. Remove belt keepers at engine pulley and slip mower drive belt off engine pulley. Move blade clutch lever to engaged position and unhook tension springs from front of mower deck. Disconnect lift rod from mower front lift bracket and unbolt lift arms from rear lift brackets. Remove mower from under right side.

Reinstall by reversing removal procedure.

Model R80

REMOVE AND REINSTALL. Disconnect spark plug wire and place blade clutch lever in disengaged position. Remove belt keepers at engine pulley and slip mower drive belt off pulley. Move blade clutch lever to engaged position and unhook tension springs at front of deck. Unpin lift linkage from mower lift brackets and remove mower unit from under right side.

Model R82

REMOVE AND REINSTALL. Disconnect spark plug wire and place blade clutch lever in engaged position. Loosen mower deck front mounting bolts and remove rear mounting bolts. Unbolt rear axle center bearing support

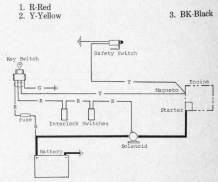

Fig. W19 – Wiring diagram for Model R82.

1. R-Red
2. Y-Yellow
3. G-Green

assembly procedure. Torque blade adapter mounting bolts to 20 ft.-lbs. and blade center bolt to 35 ft.-lbs.

Model R82

R&R AND OVERHAUL. To remove spindle, disconnect spark plug wire and remove mower as outlined in MOWER DECK paragraphs. Unbolt and remove mower pulley (4 – Fig. W16) from top of spindle. Remove blade center bolt (11), blade (12) and adapter (10) from bottom of spindle. Unbolt bearing retainers (13) from bottom of deck, then separate retainers and spindle assembly (9).

Clean and inspect all parts and renew any showing excessive wear or other damage. Reassemble by reversing removal procedure. Torque blade adapter mounting bolts to 20 ft.-lbs. and blade center bolt to 35 ft.-lbs.

bracket. Loosen engine mounting bolts and raise engine for clearance, then slide support bracket and bearing away from engine pulley. Raise deck and slip mower drive belt off engine pulley. Disconnect blade clutch cable and safety switch wires. Remove deck front mounting bolts, then remove mower from under right side.

Reinstall in reverse order of removal procedure.

ELECTRICAL

All Models

All models are equipped with safety interlock switches to prevent engine from starting with mower or traction drive engaged. Refer to Fig. W17, W18 or W19.

WIZARD

WESTERN AUTO SUPPLY CO.
2107 Grand Avenue
Kansas City, MO 64108

Model	Make	Engine Model	Horsepower	Cutting Width, In.
7110	Tecumseh	TVM-220	10	38
7115	B&S	250000	11	38
7380	B&S	250000	10	36

For service information and procedures, use the following model cross reference and refer to the GILSON section of this manual. Parts are not necessarily interchangeable and should be obtained from Western Auto Co.

Wizard Models	Gilson Models
7110	52064
7115	52073
7380	52051C

TRANSMISSION REPAIR

The transmissions covered in this section are used in one or more makes and models of riding mowers. This section outlines the transmission overhaul procedures. Refer to the individual riding mower section for transmission R&R procedures.

Eaton transmissions are manufactured by the Eaton Corporation. Foote transmissions are manufactured by the J.B. Foote Foundry Co. Indus Wheel transmissions are manufactured by the Indus Wheel Co., Division of Carlisle Corporation. Peerless transmissions are manufactured by the Peerless Division of Tecumseh Products Co. Simplicity transmissions are manufactured by Simplicity Manufacturing Co.

EATON HYDROSTATIC

Model 7

OVERHAUL. The Model 7 transmission uses a ball piston type pump and motor as shown in flow diagram (Fig. E1). Two directional valves are used to maintain hydraulic pressure by allowing oil from the reservoir to enter the system. Transmission operation is as follows:

The ball piston pump is reversible and direction of high pressure oil to the ball piston motor is determined by location of shift lever. The ball piston motor will rotate according to the high pressure oil from the pump and transfer power to the reduction gears and differential. When the shift lever is in the neutral position, oil pressure is equal on both sides of motor and it will not turn. Directional valves open and close to allow oil from the reservoir to enter the low pressure line to replace oil which is lost from the system due to oil seepage or excess pressure. If direction valve (V2) is forced to close due to high pressure in the adjoining oil line, then directional

valve (V1) can open if low pressure in the adjoining line is not sufficient to keep valve (V1) closed and oil from reservoir can enter the system. If pump rotation is reversed, directional valve (V1) will be the high pressure valve and directional valve (V2) will be the low pressure valve.

CAUTION: If oil expansion reservoir is removed, precautions should be taken to prevent the entrance of dirt or other foreign material into transmission.

Before disassembling the transmission thoroughly clean exterior of unit. Remove venting plug, invert assembly and drain fluid from unit. Remove transmission housing cap screws and place unit (output shaft downward) in a holding fixture similar to the one shown in Fig. E2. Remove aluminum housing (9 – Fig. E3) with control shaft and input shaft assemblies.

CAUTION: Do not allow pump and cam ring assemblies (24 through 27) to lift with housing. If pump rotor is raised with housing, ball pistons may fall out of rotor.

The ball pistons (25) are selective fitted to rotor bores to a clearance of 0.0002-0.0006 inch and are not interchangeable. Place a wide rubber band around pump rotor (24) to prevent balls (25) from falling out of rotor. Remove cam ring (27) and pump race (26). Hold motor rotor (31) down and remove pintle assembly (30). Remove free wheeling valve bracket (18) and springs (19). Place a wide rubber band around motor rotor (31) to prevent balls (33) and springs (32) from falling out of rotor. Remove motor assembly and motor race (39).

Remove snap ring (47), gear (46), spacer (45), retainer (44), snap ring (43) and key (38). Support body (40) and press output shaft (37) out of bearing (42) and oil seal (41). Ball bearing (42)

and oil seal (41) can now be removed from body (40).

Remove retainer (1) and withdraw ball bearing (3) and input shaft (4). Bearing can be pressed from input shaft after removal of snap ring (2). Oil seal (5) can be removed from outside of housing. To remove the control shaft (22), drill an 11/32-inch hole through aluminum housing (9) directly in line with center line of dowel pin. Press dowel pin from control shaft, remove snap ring (13) and washer (14) and withdraw control shaft. Remove oil seal (21). Thread the drilled hole with a ⅛-inch pipe tap.

To remove the directional check valves from pintle (30), drill through the pintle with a drill bit that will pass freely through roll pins. Redrill the holes from the opposite side with a ¼-inch drill bit. Press roll pins from pintle. Using a 5/16-18 tap, thread the inside of check valve bodies (36) and remove valve bodies using a draw bolt or a slide hammer puller. Remove check valve balls (35) and snap rings (34).

Number the piston bores (1 through 5) on pump rotor and on motor rotor. Use a plastic ice cube tray or equivalent and mark the cavities 1P-5P for the pump and 1M-5M, for the motor. Remove ball pistons (17) one at a time from pump rotor and place each ball in the correct cavity in the tray. Remove ball pistons (41) and springs (40) from motor rotor in the same manner.

Clean and inspect all parts and renew any showing excessive wear or any other damage. Ball pistons are selective fitted to 0.0002-0.0006 inch clearance and must be reinstalled in their original bores. If rotor bushings are scored or badly worn (0.002 inch or more clearance on pintle journals) renew pump rotor or motor rotor assemblies. Install ball pistons (25) in pump rotor (24) and

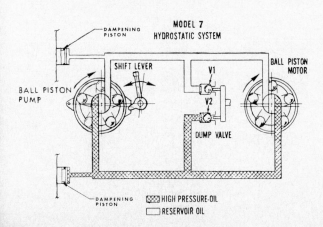

Fig. E — Flow diagram of Eaton Model 7 hydrostatic transmission.

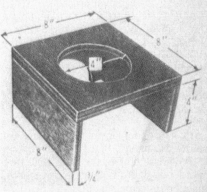

Fig. E2 — View showing dimensions of wooden stand which may be used to disassemble and reassemble Eaton Model 7 hydrostatic transmission.

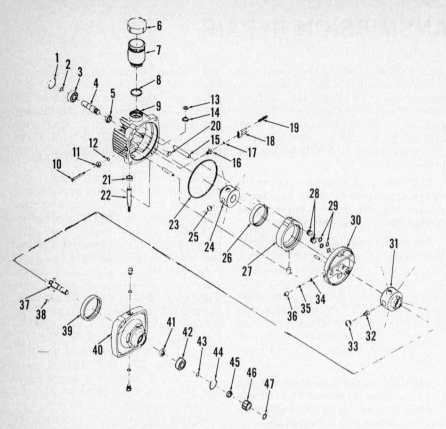

Fig. E3—Exploded view of Eaton Model 7 hydrostatic transmission.

1. Retainer	13. Retaining ring	24. Pump rotor	36. Check valve body
2. Snap ring	14. Control shaft	25. Pump ball pistons	37. Output shaft
3. Ball bearing	washer	26. Pump race	38. Key
4. Input shaft	15. Pivot pin	27. Pump cam ring	39. Motor race
5. Oil seal	16. Guide	28. Dampening pistons	40. Body
6. Reservoir cap	17. "O" ring	29. "O" ring	41. Oil seal
7. Reservoir	18. Bracket	30. Pintle	42. Ball bearing
8. Gasket	19. Spring	31. Motor rotor	43. Snap ring
9. Housing	20. Dowel pin	32. Spring	44. Retainer
10. Free wheeling valve	21. Oil seal	33. Motor ball pistons	45. Spacer
11. Nut	22. Control shaft	34. Retaining ring	46. Drive gear
12. "O" ring	23. Square cut seal ring	35. Ball, Grade 200	47. Snap ring

ball pistons (33) and springs (32) in motor rotor (31) and use wide rubber bands to hold pistons in their bores. Install snap rings (34), check valve balls (35) and valve bodies (36) in pintle (30) and secure with new roll pins. Renew oil seals (5 and 21) and reinstall control shaft and input shaft in housing (9) by reversing the removal procedure. When installing oil seals (5, 21 or 41), apply a thin coat of "Loctite" grade #271 to a ⅛-inch pipe plug and install plug in the drilled and tapped disassembly hole. Tighten plug until snug. Do not overtighten. Renew oil seal (41) and reinstall output shaft (37), bearing (42), snap rings, retainer, spacer and gear in body (40).

All components must be clean and dry before assembly. Place the aluminum housing assembly in the holding fixture with input shaft (4) pointing downward. Install pump cam ring (27) and race (26) on pivot pin (15) and dowel pin. Cam ring insert must be installed in cam ring with the hole to the outside. If insert is

installed upside down, it will contact housing and interfere with assembly. Cam ring must move freely from stop to stop. Install the pump rotor assembly and remove the rubber band used to retain pistons. Install free-wheeling valve

Fig. FT1—Exploded view of typical Models 35 and 3500 Foote single speed, forward-reverse transmissions.

1. Hi-pro key
2. Output shaft
3. Woodruff key
4. Output sprocket
5. Retaining ring
6. Flanged bushing
7. Case half
8. Bevel gear
9. Shift dog
10. Shift fork
11. Case half
12. Detent ball
13. Spring
14. Retaining ring
15. Washer
16. Flanged bushing
17. Bevel gear
18. Snap ring
19. Input bevel pinion
20. Flanged bushings
21. Input shaft
22. Washer
23. Snap ring

components (11, 12, 16 and 17). Install free-wheeling valve bracket (18) and springs (19). Install pintle assembly over cam pivot pin and into pump rotor. Place new "O" ring (23) in position on housing. Lay housing assembly on its side on a clean surface. Place the body assembly in the holding fixture with output shaft pointing downward. Install motor race (39) in body, then install motor rotor assembly aligning the rotor slot with drive pin on output shaft. Remove rubber band used to retain pistons in rotor. Place body and motor assembly on its side so that motor rotor is facing pintle in housing assembly. Slide the assemblies together and align the two assembly bolt holes. Install the two transmission housing cap screws and tighten them to a torque of 15 ft.-lbs. Rotate input shaft and output shaft. Both shafts should rotate freely. If not, disassemble the unit and correct as needed.

Screw free wheeling valve into transmission and install transmission in tractor. Fill reservoir to level indicated on reservoir. Check and adjust linkage as needed.

FOOTE

Models 35-3500

OVERHAUL. To disassemble the forward-reverse single speed transmissions, refer to Fig. FT1 and remove retaining ring (5), output sprocket (4) and key (3). Unbolt and remove case half (7). Remove bevel gear (8) and input shaft assembly (18 thru 23). Remove retaining ring (14) and washer (15), then withdraw output shaft (2) and key (1). Remove shift dog (9) and shift fork (10), being careful not to lose detent ball and spring (12 and 13). Bevel gear (17) can now be removed. Remove snap ring (18), input bevel pinion (19) and flanged bushings

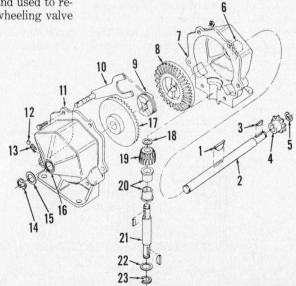

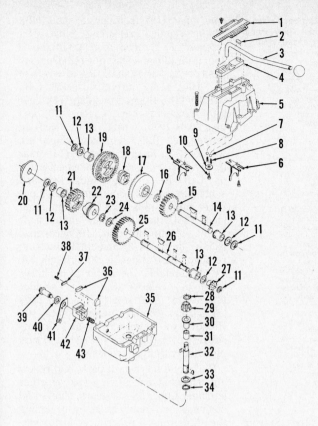

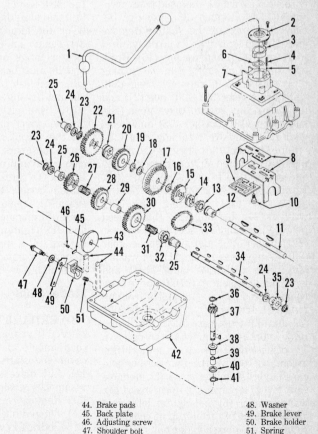

Fig. FT2—Exploded view of typical Models 2140 and 2240 Foote 2-speed transmissions.

1. Shifter cover
2. Nylon slide (2)
3. Shift lever
4. Cam plate
5. Transmission cover
6. Shift forks
7. Detent ball (2)
8. Spring (2)
9. Washer
10. Shoulder screw
11. Retaining ring
12. Washer
13. Flanged bushing
14. Drive shaft
15. Spur gear (22T)
16. Washer
17. Bevel gear
18. Shift dog
19. Bevel spur gear
20. Brake disc
21. Spur gear (22T)
22. Shift dog
23. Snap ring
24. Washer
25. Spur gear (28T)
26. Output shaft
27. Output sprocket
28. Snap ring
29. Input bevel pinion
30. Flanged bushing
31. Straight bushing
32. Input shaft
33. Washer
34. Snap ring
35. Transmission case
36. Brake pads
37. Back plate
38. Adjusting screw
39. Shoulder bolt
40. Washer
41. Brake lever
42. Brake caliper
43. Spring

ring (28) and bevel pinion (29) and withdraw input shaft (32) from bottom of case (35). Unbolt and remove brake assembly (36 through 43). To disassemble the cover and shifter assembly, remove the five shoulder screws (10), shift forks (6), detent balls, springs and washer (7, 8 and 9). Remove the four screws and lift off shifter cover (1), shift lever (3) and cam plate (4) with nylon slides (2).

Clean and inspect all parts and renew any showing excessive wear or other damage. Lubricate all gears, shafts and bushings and reassemble by reversing the disassembly procedure. Make certain that locator tangs on flanged bushings (13) are seated in notches in case (35). Backlash between input bevel pinion (29) and bevel gears (17 and 19) should be 0.001-0.015. End play should be 0.001-0.015 for drive shaft (14) and 0.001-0.012 for output shaft (26). Fill case and cover the gears with approximately 14 oz. of Shell Darina "O" grease or equivalent lithium grease. Install cover (5) and tighten cap screws to a torque of 80-90 in.-lbs. Renew brake pads (36) as necessary and reinstall brake assembly.

Models 2010-2210

OVERHAUL. To disassemble the 3-speed transmission, refer to Fig. FT3 and place shift lever in neutral position.

(20) from input shaft (21). If necessary, flanged bushings (6 and 16) can be removed from case halves (7 and 11).

Clean and inspect all parts and renew any showing excessive wear or other damage. Lubricate shafts, gears and bushings and reassemble by reversing the disassembly procedure. Fill case halves with approximately 5 oz. of Shell Darina "AX" (35) or "O" (3500) or equivalent grease. Install housing bolts and tighten securely.

Models 2140-2240

OVERHAUL. To disassemble the 2-speed transmissions, refer to Fig. FT2, then unbolt and remove the cover and shifter assembly (1 through 10). Lift drive shaft and output shaft assemblies straight upward out of case (35). Remove retaining ring (11), washer (12), flanged bushing (13), bevel spur gear (19) and shift dog (18) from drive shaft (14). Then, remove retaining ring (11), washer (12), flanged bushing (13), spur gear (15), Woodruff key, washer (16) and bevel gear (17) from opposite end of shaft. Remove brake disc (20), retaining ring (11), washer (12), flanged bushing (13), gear (21) and shift dog (22) from output shaft (26). Remove retaining ring (11), output sprocket (27), washer (12), flanged bushing (13), gear (25), Woodruff key and washer (24). Remove snap

Fig. FT3—Exploded view of typical Models 2010 and 2210 Foote 3-speed transmissions.

1. Shift lever
2. Shifter cover
3. Nylon insert
4. Spring (2)
5. Detent ball (2)
6. Wave washer
7. Transmission cover
8. Shift forks
9. Interlock plate
10. Shoulder bolt (4)
11. Drive shaft
12. Flanged bushing
13. Reverse drive sprocket
14. First & Rev. shift dog
15. First drive gear
16. Washer
17. Bevel gear
18. Washer
19. Snap ring
20. Second drive gear
21. Second & third shift dog
22. Third drive gear
23. Retaining ring
24. Washer
25. Flanged bushing
26. Third speed gear
27. Spring
28. Second speed gear
29. Spacer
30. First speed gear
31. Spring
32. Reverse driven sprocket
33. Chain
34. Output shaft
35. Output sprocket
36. Snap ring
37. Input shaft & bevel pinion
38. Flanged bushing
39. Straight bushing
40. Washer
41. Snap ring
42. Transmission case
43. Brake disc
44. Brake pads
45. Back plate
46. Adjusting screw
47. Shoulder bolt
48. Washer
49. Brake lever
50. Brake holder
51. Spring

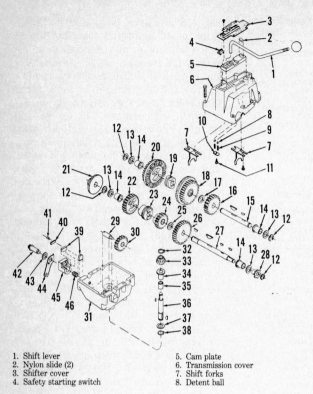

Fig. FT4—Exploded view of typical Model 2500 Foote 3-speed transmission.

9. Spring
10. Retaining plate
11. Shoulder screw
12. Retaining ring
13. Washer
14. Flanged bushing
15. Drive shaft
16. Spur gear
17. Washer
18. Bevel spur gear
19. Shift dog
20. Bevel & spur gear
21. Brake disc
22. Spur gear (25T)
23. Shift dog
24. Spur gear (17T)
25. Washer
26. Spur gear (28T)
27. Output shaft
28. Output sprocket
29. Idler shaft
30. Idler gear
31. Transmission case
32. Snap ring
33. Input bevel pinion
34. Flanged bushing
35. Straight bushing
36. Input shaft
37. Washer
38. Snap ring
39. Brake pads
40. Back plate
41. Adjusting screw
42. Shoulder bolt
43. Washer
44. Brake holder
45. Caliper
46. Spring

1. Shift lever
2. Nylon slide (2)
3. Shifter cover
4. Safety starting switch
5. Cam plate
6. Transmission cover
7. Shift forks
8. Detent ball

Unbolt and remove transmission cover and shifter assembly (1 through 10). Lift drive shaft and output shaft assemblies straight upward out of case (42). Move sprocket end of shafts together until flanged bushing (12) and reverse drive sprocket (13) can be removed from drive shaft (11) and chain (33). Remove the chain to separate the shafts. Remove shift dog (14) and its hi-pro key, first drive gear (15) and washer (16). From opposite end of shaft, remove bushing (25), washer (24), retaining ring (23), third drive gear (22), shift dog (21) and its hi-pro key, second drive gear (20), snap ring (19), washer (18) and bevel gear (17). Remove retaining ring (23), output sprocket (35), washer (24), bushing (25), reverse driven sprocket (32) and spring (31) from output shaft (34). From opposite end of shaft, remove brake disc (43), retaining ring (23), washer (24), bushing (25) third speed gear (26), spring (27), second speed gear (28), spacer (29) and first speed gear (30). Remove snap ring (41) and washer (40), then withdraw input shaft and bevel pinion assembly (37). Remove snap ring (36) and press input pinion from input shaft. If necessary, remove bushings (38 and 39) from case (42). Unbolt and remove brake assembly (44 through 51). To disassemble the cover and shifter assembly, remove the four screws and carefully raise shifter cover (2). Remove shift lever (1), cover (2), nylon insert (3), wave washer (6) and detent balls and springs (4 and 5) from top of cover (7). Remove the four shoulder bolts (10), in-

terlock plate (9) and shift forks (8) from bottom of cover.

Clean and inspect all parts and renew any showing excessive wear of other damage. Lubricate all gears, shafts and bushings and reassemble by reversing the disassembly procedure. Make certain that shift dogs (14 and 21) slide freely on the hi-pro keys. When installing drive shaft and output shaft assemblies, locator tangs on flanged bushings (12 and 25) must be seated in notches in case (42). Backlash between input bevel pinion and bevel gear (17) should be 0.001-0.015. End play should be 0.001-0.015 for the drive shaft (11) and 0.001-0.012 for the output shaft (34). Fill case and cover the gears, reverse sprockets and chain with approximately 14 oz. of Shell Darina "O" grease or equivalent No. 2 lithium grease. Install cover and shifter assembly, making certain that shift forks (8) engage slots in shift dogs (14 and 21). Tighten transmission cover cap screws to a torque of 80-90 in.-lbs. Renew brake pads (44) as necessary and reinstall brake assembly.

Model 2500

OVERHAUL. To disassemble this 3-speed transmission, refer to Fig. FT4, then unbolt and remove transmission cover and shifter assembly (1 through 11). Lift drive shaft and output shaft assemblies straight upward out of case (31). Remove retaining ring (12), washer (13), flanged bushing (14), gear (16) and its Woodruff key, washer (17) and bevel

spur gear (18) from drive shaft (15). Then, remove retaining ring (12), washer (13), flanged bushing (14), bevel spur gear (20) and shift dog (19) from opposite end of drive shaft. Remove retaining ring (12), output sprocket (28), washer (13), flanged bushing (14), gear (26), Woodruff key, washer (25) and gear (24) from output shaft. Remove brake disc (21), retaining ring (12), washer (13), flanged bushing (14), gear (22) and shift dog (23) from opposite end of shaft. Remove idler gear (30) and shaft (29). Remove snap ring (32), input bevel pinion (33) and key from input shaft (36), then withdraw input shaft from bottom of case. If necessary, remove bushings (34 and 35). Unbolt and remove brake assembly (39 through 46). To disassemble the cover and shifter assembly, remove the five shoulder screws (11), shift forks (7), detent balls (8), springs (9) and retaining plate (10). Remove the four screws and lift off shifter cover (3), shift lever (1) and cam plate (5) with nylon slides (2).

Clean and inspect all parts and renew any showing excessive wear or other damage. Lubricate all gears, shafts and bushings and reassemble by reversing the disassembly procedure. Make certain that shift dogs (19 and 23) slide freely on the hi-pro keys. When installing drive shaft and output shaft assemblies, locator tangs on flanged bushings (14) must be seated in notches in case (31). Backlash between input bevel pinion (33) and bevel spur gears (18 and 20) should be 0.001-0.015. End play should be 0.001-0.015 for the drive shaft (15) and 0.001-0.012 for the output shaft (27). Fill case and cover the gears with approximately 12 oz. of Shell Darina "O" grease or equivalent No. 2 lithium grease. Install cover and shifter assembly, making certain that shift forks (7) engage slots in shift dogs (19 and 23). Tighten transmission cover cap screws to a torque of 80-90 in.-lbs. Renew brake pads (39) as necessary and reinstall brake assembly.

Model 4-Speed W/No Reverse

OVERHAUL. To disassemble the 4-speed transmission, refer to Fig. FT5 and place shift lever in neutral position. Unbolt and remove transmission cover and shifter assembly (1 through 10). Lift drive shaft and output shaft assemblies straight upward out of case (36). Remove snap ring (12), washer (13), flanged bushing (14), thrust washer (15), second drive gear (16), shift dog (17) and its hi-pro key, first drive gear (18) and washer (19) from drive shaft (11). Then, remove snap ring (12), washer (13), flanged bushing (14), fourth drive gear (25), shift dog (24) and its hi-pro key, third drive gear (23), snap ring (22),

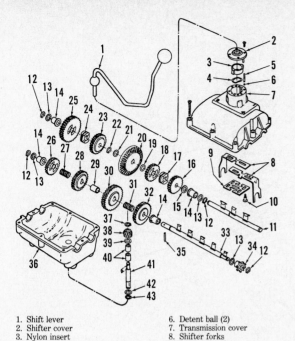

11. Drive shaft
12. Snap ring
13. Washer
14. Flanged bearing
15. Thrust washer
16. Second drive gear
17. Shift dog
18. First drive gear
19. Washer
20. Bevel gear
21. Washer
22. Snap ring
23. Third drive gear
24. Shift dog
25. Fourth drive gear
26. Fourth speed gear
27. Spring
28. Third speed gear
29. Spacer
30. First speed gear
31. Spring
32. Second speed gear
33. Output shaft
34. Output sprocket
35. Roll pin
36. Transmission case
37. Snap ring
38. Input bevel pinion
39. Thrust washer
40. Needle bearings
41. Input shaft
42. Washer
43. Snap ring

1. Shift lever
2. Shifter cover
3. Nylon insert
4. Wave washer
5. Spring (2)
6. Detent ball (2)
7. Transmission cover
8. Shifter forks
9. Interlock plate
10. Shoulder bolt (4)

Model 2600

OVERHAUL. To disassemble the 5-speed transmission, refer to Fig. FT6 and unbolt and remove transmission cover and shifter assembly (1 through 16). Raise output shaft (30) slightly and remove shift fork, fork pivot and pivot bushings (52 through 55). Lift drive shaft and output shaft assemblies straight upward out of case (56). Move sprocket end of shafts together until flanged bushing (17) and reverse drive sprocket (26) can be removed from drive shaft (27) and chain (31). Remove the chain to separate the shafts. Remove first and second drive gear (25), spacer (24) and bevel gear (23) from drive shaft (27). Then, remove flanged bushing (17), fifth drive gear (18), fourth drive gear (19), third drive gear (20), washer (21) and spacer (22). Remove snap ring (28), output sprocket (29), flanged bushing (17), reverse driven sprocket (32), first speed gear (33) and second speed gear (34). From opposite end of output shaft, remove brake disc (43), flanged bushing (17), washer (42), fifth speed gear (41), fourth speed gear (40) and third speed gear (39). Carefully slide driving hubs (36) from output shaft. Identify and remove driving key assemblies (35), retaining washers (37) and collar (38). Right

washer (21) and bevel gear (20). Remove snap ring (12), output sprocket (34), washer (13), flanged bushing (14), second speed gear (32), spring (31) and first speed gear (30) from output shaft (33). From opposite end of shaft, remove snap ring (12), washer (13), flanged bushing (14), fourth speed gear (26), spring (27), third speed gear (28) and spacer (29). Remove snap ring (37), input bevel pinion (38) and thrust washer (39), then withdraw input shaft (41) from bottom of case (36). If necessary, press needle bearings (40) from case. To disassemble the cover and shifter assembly, remove the four screws and carefully raise shifter cover (2). Remove shift lever (1), cover (2), nylon insert (3), wave washer (4) and detent balls and springs (5 and 6) from top of cover (7). Remove the four shoulder bolts (10), interlock plate (9) and shift forks (8) from bottom of cover.

Clean and inspect all parts and renew any showing excessive wear or other damage. Lubricate all gears, shafts, needle bearings and bushings and reassemble by reversing the disassembly procedure. Keep the following points in mind: Shift dogs (17 and 24) must slide freely on the hi-pro keys. Locator tangs on flanged bushings (14) must be seated in notches in case (36). Backlash between input bevel pinion (38) and bevel gear (20) should be 0.001-0.015. End play should be 0.001-0.015 for the drive shaft (11) and 0.001-0.012 for the output shaft (33).

Fill case and cover the gears with approximately 10 oz. of Shell Darina "O" grease or equivalent No. 2 lithium

grease. Install cover and shifter assembly, making certain that shift forks (8) engage slots in shift dogs (17 and 24). Tighten transmission cover cap screws to a torque of 80-90 in.-lbs.

1. Pivot shaft
2. Washer
3. Snap ring
4. Shifter cover
5. Actuator cover
6. Retaining clips
7. Interlock plate
8. Interlock pawl & carrier
9. Safety starting switch
10. Springs
11. Detent pin
12. Detent spring
13. Torsion spring
14. Washer
15. Sleeve bushing
16. Transmission cover
17. Flanged bushing
18. Fifth drive gear
19. Fourth drive gear
20. Third drive gear
21. Washer
22. Spacer (short)
23. Bevel gear
24. Spacer (long)
25. First & second drive gear
26. Reverse drive sprocket
27. Drive shaft
28. Snap ring
29. Output sprocket
30. Output shaft
31. Chain
32. Reverse driven sprocket
33. First speed gear
34. Second speed gear
35. Driving key assemblies
36. Driving hubs
37. Key retaining washers
38. Collar
39. Third speed gear
40. Fourth speed gear
41. Fifth speed gear
42. Washer
43. Brake disc
44. Brake pads
45. Back plate
46. Adjusting screw
47. Shoulder bolt
48. Washer
49. Brake lever
50. Brake caliper
51. Spring
52. Pivot bushings
53. Shift fork
54. Fork pivot
55. Pivot bushings
56. Transmission case
57. Snap ring
58. Input shaft & pinion
59. Flanged bushing
60. "O" ring
61. Straight bushing
62. Washer
63. Snap ring

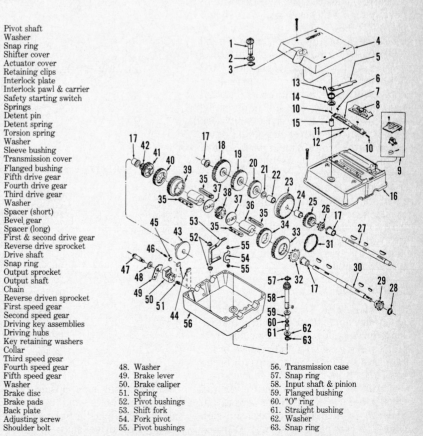

Fig. FT6 — Exploded view of typical Model 2600 Foote 5-speed transmission.

side driving key assemblies (1st, 2nd and reverse) have shortest distance between the key drive lugs. Do not interchange with left side key assemblies (3rd, 4th and 5th). Remove snap ring (63) and washer (62), then withdraw input shaft assembly from case. Remove snap ring (57) and bevel pinion from input shaft (58). Bushings (59 and 61) and "O" ring (60) can now be removed from case (56). Unbolt and remove brake assembly (44 through 51). To disassemble the cover and shifter assembly, remove shift lever, then unbolt and remove shifter cover (4). Remove snap ring (3) and washer (2) from bottom of cover (16) and remove pivot shaft (1), actuator lever (5), torsion spring (13) and washer (14) from top of cover. Remove clips (6) and lift out interlock plate (7) and springs (10). Remove pawl and carrier (8), detent pin (11) and spring (12). Press sleeve bushing (15) from cover. If necessary, remove safety starting switch (9).

Clean and inspect all parts and renew any showing excessive wear or other damage. Lubricate all gears, shafts and bushings and reassemble by reversing the disassembly procedure. Apply a light coat of switch lubricant to all moving parts when reassembling the shifter mechanism in cover. Keep the following points in mind: Locator tangs on flanged bushings (17) must be seated in notches in case (56). Input shaft end play should be 0.010-0.020. Backlash between input bevel pinion and bevel gear (23) should be 0.005-0.010. End play should be 0.001-0.015 for the drive shaft (27) and 0.001-0.012 for the output shaft (30).

Fill case to centerline of shafts (27 and 30) with Shell Darina "O" grease or equivalent No. 2 lithium grease. Install transmission cover and shifter assembly as follows: Place lever end of shift fork (53) directly over bevel gear (23). Center the actuator lever (5) and interlock pawl (8) in cover (16). Carefully set cover assembly on the case. Make certain that top pivot bushing (52) is seated properly in cover and that pin on interlock pawl (8) is engaged in slot in shift fork (53). Tighten transmission cover cap screws to a torque of 80-90 in.-lbs. Renew brake pads (44) as necessary and reinstall brake assembly.

Model 4000

OVERHAUL. To disassemble transaxle, place shift lever in neutral and remove drive pulley. Remove shoulder bolt from brake assembly and remove brake caliper (69 – Fig. FT7), spring (70), brake pads (71), brake disc (72) and Woodruff key. Unbolt and remove shift lever and cover. Remove the two set screws (10 and 12) from case, turn transmission over and catch detent springs (13) and balls.

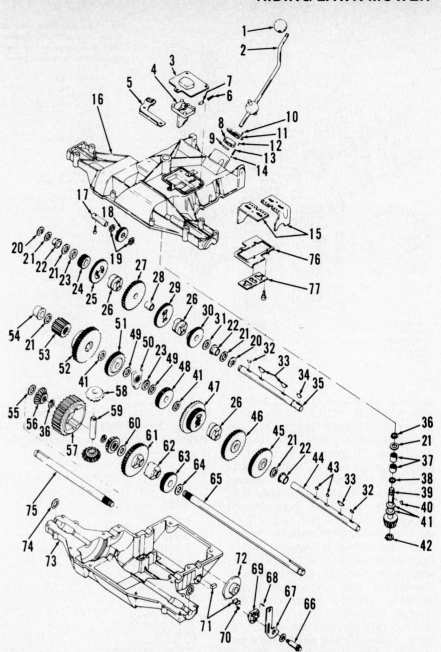

Fig. FT7 – Exploded view of Foote Model 4000 sliding gear transmission.

1. Knob	21. Shim washer	41. Shim washer
2. Shift lever	22. Bearing	42. Snap ring
3. Cover plate	23. Spacer	43. Woodruff key #3
4. Shift fork	24. Spur gear (13T)	special
5. Hi-Lo shift lever	25. Spur gear (25T)	44. Drive shaft
6. Detent spring	26. Clutch collar	45. Gear (20T)
7. Pin	27. Spur gear (30T)	46. Gear (33T)
8. Nylon insert	28. Spacer	47. Bevel gear assy.
9. Wave washer	29. Spur gear (25T)	48. Gear (20T)
10. Set screw	30. Spur gear (20T)	49. Shim washer
11. Nylon cover	31. Shim washer	50. Shaft support assy.
12. Set screw	32. Woodruff key #3	51. Gear (25T)
13. Spring	33. Hi-pro key	52. Gear (37T)
14. Detent ball	34. Woodruff key #61	53. Gear assy. (12T)
15. Shift fork	35. Intermediate shaft	54. Bearing
16. Case half	36. Snap ring	55. Shim
17. Idler shaft	37. Needle bearings	56. Bevel gear (15T)
18. Reverse idler gear	38. "O" ring	splined
19. Washer	39. Input shaft	57. Spur gear (32T)
20. "E" ring	40. Key	

58. Bevel gears
59. Cross shaft
60. Shim
61. Spur gear (35T)
62. Gear lock
63. Spur gear (22T)
64. Shim
65. Axle shaft R.H.
66. Shoulder bolt
67. Brake lever
68. Set screw
69. Brake caliper
70. Spring
71. Brake pads
72. Brake disc
73. Case half
74. Felt seal
75. Axle shaft L.H.
76. Support plate
77. Lock-out plate

With transmission upside down, remove case bolts and separate case halves with a plastic hammer or rubber mallet. Lift drive shaft assembly (44) out. All parts on drive shaft are a slip fit.

Lift intermediate shaft assembly out and remove "E" ring (20) from one end. Slide parts off shaft, being careful to keep parts in order. Push axles together as differential assembly is removed. Ax-

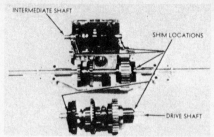

Fig. FT8 — View showing shim location in Foote Model 4000 transmission.

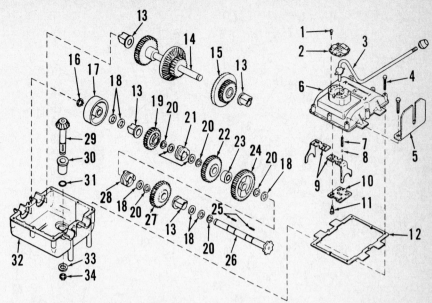

Fig. IWT1 — Exploded view of typical 3-speed Model 11385 Indus Wheel transmission.

1. Screw (4)	11. Shoulder bolt (4)	19. Third speed gear	27. Reverse gear
2. Plate	12. Gasket	20. Snap ring	28. Shift dog (1st &
3. Shift lever	13. Bearing	21. Shift dog (2nd & | Rev.)
4. Cap screw (8)	14. Drive shaft & gears	3rd)	29. Input shaft & pinion
5. Shift bracket	assy.	22. Second speed gear	30. Flanged bushing
6. Transmission cover	15. Bevel spur gear	23. Spacer	31. Seal ring
7. Spring	(rev. drive)	24. First speed gear	32. Transmission case
8. Detent ball	16. Snap ring	25. Flat keys	33. Thrust washer
9. Shift forks	17. Brake drum	26. Output shaft	34. Snap ring
10. Backing plate	18. Thrust washers		

le bevel gears are a press fit; all other parts are slip fit. Further disassembly is evident after examination.

Clean and inspect all parts and renew any showing excessive wear or damage. Before installing input shaft, pack needle bearings with grease. Install Hi-Lo shift mechanism and check for free action. Apply a light coating of grease to reverse idler shaft and gear and torque mounting bolt to 80-90 in.-lbs. Assemble intermediate shaft with light coating of grease and new "E" rings (20). Intermediate and drive shaft end play is 0.020 to 0.030 inch. End play is adjusted by changing shim thickness as shown in Fig. FT8. Input shaft end play is 0.010 to 0.020 inch. End play is adjusted by changing shim washer (21 – Fig. FT7) under snap ring (36). When installing detent balls, springs and set screws, tighten set screws until heads are flush with top of case.

Pack axle cavities in both case halves with grease. Use 24 oz. of Shell Darina "O" or equivalent grease spread equally in main part of gearcase. Tighten 14 case bolts to 80-90 in.-lbs. Tighten center bolt to 100-110 in.-lbs. Adjust brake assembly as previously described.

INDUS WHEEL

Model 1138

OVERHAUL. To disassemble the transmission, refer to Fig. IWT1 and remove the eight cap screws (4). Remove cover and shifter assembly (1 through 11) and gasket (12). Lift drive shaft assembly from case (32) and separate bearings (13) and reverse drive gear (15) from drive shaft and gear assembly (14). Remove the output shaft assembly and remove snap ring (16), brake drum (17), thrust washers (18), bearing (13) and third speed gear (19). Remove snap ring (20), thrust washer (18), second and third shift dog (21), flat key (25) and thrust washer (18). Remove snap ring (20), second speed gear (22), spacer (23) and first speed gear (24). Remove snap ring (20), thrust washer (18), first and reverse shift dog (28), flat key (25) and thrust washer (18). Remove snap ring

(20), reverse gear (27), bearing (13), thrust washers (18) and last snap ring (20) from output shaft (26). Remove snap ring (34) and thrust washer (33), then withdraw input shaft and pinion assembly (29). Flanged bushing (30) and seal ring (31) can now be removed from bore in case.

To disassemble the cover and shifter assembly, hold downward on plate (2) and remove the four retaining screws (1). Carefully raise plate (2) and shift lever (3) so that springs (7) do not fly out. Invert the cover (6) and remove springs and detent balls (8). Remove the four shoulder bolts (11), backing plate (10) and shift forks (9).

Clean and inspect all parts and renew any showing excessive wear or other damage. When reassembling, install new seal ring (31) and install bushing (30). Lubricate bushing and input shaft (29), then install input shaft, thrust washer (33) and snap ring (34). Place approximately 7 oz. of Shell EPRO #71030 grease or equivalent in the case (32). Lubricate all gears, bearings and shafts, then using Fig. IWT1 as a guide, reassemble both shaft assemblies. Place shaft assemblies in case and apply approximately 3.5 oz. of the recommended grease over all gears. Make certain the flats on bearings (13) are level with case surface and install new gasket (12). Reassemble cover and shifter assembly by reversing the disassembly procedure. Reinstall cover and shifter assembly,

making certain that shift forks (9) engage slots in shift dogs (21 and 28). Install shift bracket (5) and tighten cover retaining cap screws (4) securely.

PEERLESS

Series 600 Transaxles

OVERHAUL. To disassemble the transaxle, remove drain plug and drain lubricant. Remove brake assembly, input pulley and rear wheel assemblies. Place shift lever in neutral position, then unbolt and remove shift lever and housing assembly. Unbolt and remove axle housings (16 and 52 – Fig. PT1). Place unit in a vise so that heads of socket head capscrews are pointing upward. Drive dowel pins out of case and cover. Unscrew socket head cap screws and lift off cover (55). Install two or three socket head screws into case to hold center plate (76) down while removing the differential assembly. Pull differential assembly straight up out of case. It may be necessary to gently bump lower axle shaft to loosen differential assembly. Remove center plate (76). Hold shifter rods (19) together and lift out shifter rods, forks (20), shifter stop (23), shaft (27), sliding gears (25 and 26) and spur gear (24). On early model transaxle, remove idler shaft (29) and gear (30) as individual parts. On late model transaxle, remove idler shaft and gear as a one-piece assembly as shown in 29A – Fig.

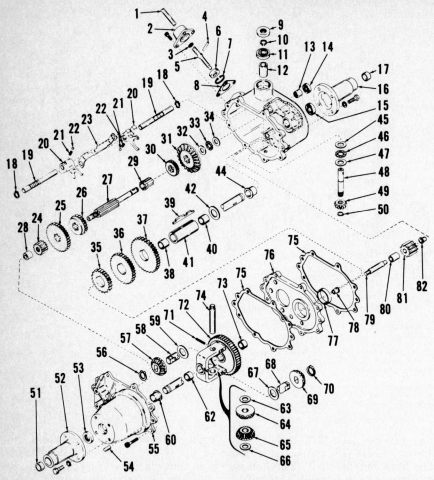

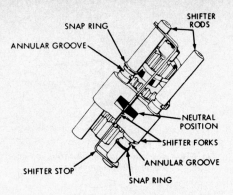

Fig. PT2 — To position shifter assembly in neutral for reassembly, align notches in shifter forks with notch in shifter stop.

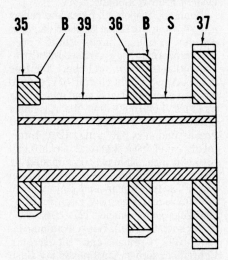

Fig. PT3 — Note position of bevels (B) on gears (35 & 36) and short section (S) of key (39) between gears (36 & 37) used on early 600 series transaxle.

Fig. PT1 — Exploded view of an early Peerless Series 600 3-speed transaxle. Late model is similar, refer to Fig. PT3A for internal parts difference.

1. Shift lever	22. Detent ball	41. Sleeve
2. Lever housing	23. Shifter stop	42. Thrust washer
3. Quad ring	24. Spur gear	44. Bushing
4. Roll pin	25. Sliding gear (1st &	45. Washer
5. Shift lever	reverse)	46. Thrust bearing
6. Retainer	26. Sliding gear (2nd &	47. Washer
7. Snap ring	3rd)	48. Input shaft
8. Gasket	27. Shift & brake shaft	49. Pinion gear
9. Oil seal	28. Needle bearing	50. Snap ring
10. Snap ring	29. Idler shaft	51. Bushing
11. Ball bearing	30. Gear	52. Axle housing
12. Bushing	31. Bevel gear	53. Oil seal
13. Needle bearing	32. Washer	54. Dowel pin
14. Oil seal	33. Thrust bearing	55. Cover
15. Oil seal	34. Washer	56. Snap ring
16. Axle housing	35. Gear (25 teeth)	57. Side gear
17. Bushing	36. Gear (34 teeth)	58. Axle shaft
18. Snap ring	37. Gear (39 teeth)	59. Thrust washer
19. Shift rod	38. Bushing	60. Bushing
20. Shift fork	39. Key	62. Bushing
21. Spring	40. Bushing	63. Thrust washer

64. & 65. Differential	
pinions	
66. Thrust washer	
67. Thrust washer	
68. Axle shaft	
69. Side gear	
70. Snap ring	
71. Roll pin	
72. Differential carrier	
& gear	
73. Bushing	
74. Drive pin	
75. Gasket	
76. Center plate	
77. Bushing	
78. Bushing	
79. Reverse idler shaft	
80. Spacer	
81. Reverse idler gear	
82. Bushing	

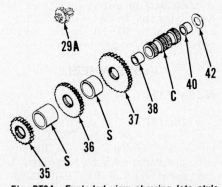

Fig. PT3A — Exploded view showing late style countershaft assembly and one-piece idler shaft and gear assembly used on 600 series transaxles.

C. Countershaft	36. Gear (34 teeth)
S. Spacer	37. Gear (39 teeth)
29A. Idler shaft & gear	38. Bushing
assy.	40. Bushing
35. Gear (25 teeth)	42. Thrust washer

PT3A. On all models remove reverse idler shaft (79 – Fig. PT1), spacer (80) and gear (81). On early model transaxle, with reference to Fig. PT1 remove cluster gears (35, 36 and 37) on sleeve (41) and thrust washer (42). On late model transaxle, with reference to Fig. PT3A remove cluster gears (35, 36 and 37), spacers (S) on countershaft (C) and thrust washer (42). On all models remove bevel gear (31 – Fig. PT1), washers (32 and 34) and thrust bearing (33). Remove input shaft oil seal (9), snap ring (10), input shaft (48) and gear (49). Washers (45 and 47) and thrust bearing (46) are removed with input

shaft. Remove bearing (11) and bushing (12).

To disassemble cluster gear assembly, press gears and key from sleeve (41). Bushings (38 and 40) are renewable in sleeve (41).

To disassemble the differential, drive roll pin (71) out of drive pin (74). Remove drive pin, thrust washers (63 and 66) and differential pinions (64 and 65). Remove snap rings (56 and 70) and withdraw axle shafts from side gears (57 and 69). Remove side gears.

Clean and inspect components for excessive wear or other damage. Renew all seals and gaskets. Check for binding of

shift forks on shift rods. When reassembling, position shift forks in neutral position by aligning notches on shift rods with notch in shifter stop. See Fig. PT2. Install input shaft assembly by reversing the removal procedure. Position case so

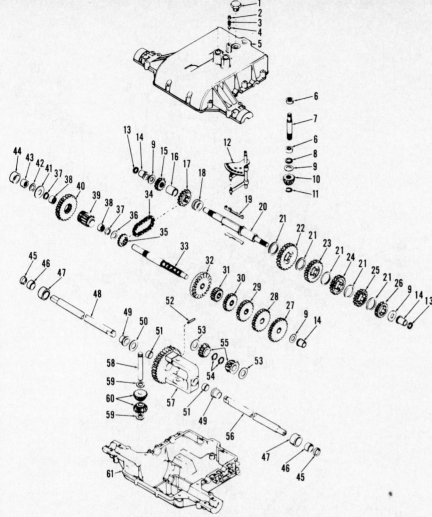

Fig. PT4 — Exploded view of Series 800 Transaxle.

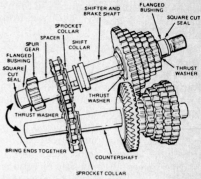

Fig. PT5 — Mark position of chain on sprocket collars, angle shafts together and remove chain.

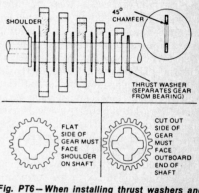

Fig. PT6 — When installing thrust washers and gears on brake shaft, 45 degree chamfer on inside diameter of thrust washers must face shoulder on brake shaft.

1. Plug
2. Set screw
3. Spring
4. Ball
5. Cover
6. Needle bearing
7. Input shaft
8. Square cut ring
9. Thrust washer
10. Input pinion
11. Snap ring
12. Shift fork assy.
13. Square cut ring
14. Bushing
15. Spur gear (12 or 15 teeth)
16. Spacer
17. Sprocket (18 teeth)
18. Shift collar
19. Key
20. Brake shaft
21. Thrust washer
22. Spur gear (35 teeth)
23. Spur gear (30 teeth)
24. Spur gear (25 teeth)
25. Spur gear (22 teeth)
26. Spur gear (20 teeth)
27. Gear (30 teeth)
28. Gear (28 teeth)
29. Gear (25 teeth)
30. Gear (20 teeth)
31. Spur gear (12 or 15 teeth)
32. Bevel gear (42 teeth)
33. Countershaft
34. Roller chain
35. Sprocket (9 teeth)
36. Flat washer
37. Square cut ring
38. Needle bearing
39. Output pinion
40. Output gear
41. Flat washer
42. Square cut seal
43. Needle bearing
44. Spacer
45. Oil seal
46. Needle bearing
47. Spacer
48. Axle shaft (13¾ inches)
49. Bushing
50. Washer
51. Bushing
52. Pin
53. Thrust washer
54. Snap ring
55. Bevel gear
56. Axle shaft (14-7/8 inches)
57. Differential gear assy.
58. Drive pin
59. Thrust washer
60. Bevel pinion
61. Case

that open side is up. Install needle bearing (13–Fig. PT1) and oil seal (14), then install idler shaft (29) and gear (30) on early model transaxle and on late model transaxle install one-piece idler shaft and gear (29A–Fig. PT3A). On all models, install bevel gear (31–Fig. PT1), washers (32 and 34) and thrust bearing (33). Be sure thrust bearing is positioned between washers. Reverse idler shaft (79) may be used to temporarily hold idler gear assembly in position. On early model transaxle, place cluster gear (35, 36 and 37) on key (39) so that bevel on gears (35 and 36) is

toward large gear (37) and short section of key (39) is between middle gear (36) and large gear (37) as shown in Fig. PT3. Press gears and key on sleeve (41–Fig. PT1). On late model transaxle, install thrust washer (42–Fig. PT3A), countershaft (C), cluster gears (35, 36 and 37) and spacers (S). On all models, install shifter assembly (18 through 27–Fig. PT1) in case, making certain that shifter rods are properly seated. Install reverse idler shaft (79), gear (81) and spacer (80). Beveled edge of gear should be up. Install gasket (75), center plate (76), then second gasket (75) on

case. Assemble differential by reversing disassembly procedure. Install differential assembly in case with longer axle pointing downward. Install locating dowel pins and secure cover (55) to case. Install oil seals (15 and 53), axle housings (16 and 52) and shift lever assembly (1 through 8).

Fill transaxle housing after unit is installed to level plug opening with SAE 90EP gear oil. Capacity is approximately 1½ pints.

Series 800 Transaxle

OVERHAUL. To disassemble the transaxle, first remove drain plug and drain lubricant. Place shift lever in neutral and remove shift lever. Remove setscrew (2–Fig. PT4), spring (3) and index ball (4). Unbolt cover (5) and push shift fork assembly (12) in while removing cover. Before removing gear shaft assemblies, shift fork (12) may be removed. It will be difficult to keep parts from falling off. Note position of parts before removal. Remove gear and shaft assemblies from case taking care not to disturb drive chain (34). Remove needle bearing (43), flat washer (41), square cut seals (42), output gear (40) and output pinion (39) from the countershaft. Angle the two shafts together (Fig. PT5). Mark the position of chain on sprocket collars and remove chain. Remove sprocket (35–Fig. PT4), bevel gear (32), spur gears (27, 28, 29, 30 and 31), thrust washer (9) and flange bushing (14). All gears are splined to the countershaft. Disasembly of brake shaft is self-evident from observation. Remove snap ring (11), input bevel gear (10) and pull input shaft (7) through cover.

To disassemble the differential, drive roll pin out of drive pin (58) and remove drive pin. Remove pinion gears (60) by rotating gears in opposite directions. Remove snap rings (54), side gears (55), thrust washers (53) and slide axles out.

Clean and inspect all parts and renew any showing excessive wear or other damage. When installing new inner input shaft needle bearings, press bearing in to a depth of 0.135-0.150 inches below flush. When installing thrust washers and shifting gears on brake shaft, the 45° chamfer on inside diameter of thrust washers must face shoulder on brake shaft. The flat side of gears must face shoulder on shaft. Complete assembly and torque case to cover cap screws to 80-100 in.-lbs.

Reinstall transaxle by reversing removal procedure and pack transaxle with 24 oz. of E.P. lithium grease. Adjust brake and drive belt tension as required.

Series 2300 Transaxle

OVERHAUL. To disassemble the transaxle, first remove drain plug and drain lubricant. Place shift lever in neutral position, then unbolt and remove shift lever assembly. Remove axle housings (14 and 64–Fig. PT6A). Remove seal retainers (11) with oil seals (12) and "O" rings (13) by pulling each axle shaft out of case and cover as far as possible. Place transaxle unit on the edge of a bench with left axle pointing downward. Remove cap screws securing case (16) to cover (66) and drive aligning dowel pins out of case. Lift case (16) up 1½ to 2 inches, tilt case about 45 degrees, rotate case clockwise and remove it from the assembly. Input shaft (32) and input gear (33) will be removed with the case. Withdraw differential and axle shaft assembly and lay aside for later disassembly. Remove the 3-cluster gear (44) with its thrust washer (46) and spacer (42). Lift out reverse idler gear (25), spacer (24) and shaft (23). Hold upper ends of shifter rods together and lift out shifter rods, forks, shifter stop (21), sliding gears (30 and 31) and shaft (28) as an assembly. Remove low reduction gear (57), reduction shaft (56) and thrust washer (55), then remove 2-cluster gear (40) from brake shaft. Lift out the output gear (50), shaft (51) and thrust washers (49 and 52). To remove brake

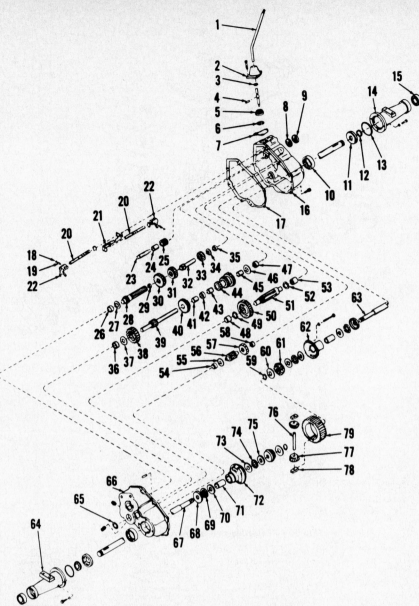

Fig. PT6A – Exploded view of Peerless Model 2300 4-speed transaxle.

1. Shift lever	22. Shifter fork	40. 2-cluster gear	60. Thrust washer
2. Shift lever housing	23. Reverse idler shaft	41. Bushing	61. Axle gear
3. Seal ring	24. Spacer	42. Spacer	62. Differential carrier
4. Roll pin	25. Reverse idler gear	43. Bushing	63. Axle shaft R.H.
5. Retainer	26. Needle bearing	44. 3-cluster gear	64. Axle housing L.H.
6. Snap ring	27. Thrust washer	45. Bushing	65. Oil seal
7. Gasket	28. Shifter shaft	46. Thrust washer	66. Transaxle cover
8. Ball bearing	29. Needle bearing	47. Needle bearing	67. Axle shaft L.H.
9. Oil seal	30. 1st, 2nd & reverse gear	48. Needle bearing	68. Thrust washer
10. Carrier bearing	31. 3rd & 4th gear	49. Thrust washer	69. Thrust bearing
11. Seal retainer	32. Input shaft	50. Output gear	70. Thrust washer
12. Oil seal	33. Input gear	51. Output shaft	71. Bushing
13. "O" ring	34. Thrust washer	52. Thrust washer	72. Differential carrier
14. Axle housing R.H.	35. Needle bearing	53. Needle bearing	73. Thrust washer
15. Axle outer bearing	36. Needle bearing	54. Needle bearing	74. Thrust bearing
16. Transaxle case	37. Thrust washer	55. Thrust washer	75. Thrust washer
17. Gasket	38. Idler gear	56. Low reduction shaft	76. Drive pin
18. Detent ball	39. Brake & cluster shaft	57. Low reduction gear	77. Bevel pinion gear
19. Spring		58. Needle bearing	78. Drive block
20. Shifter rod		59. Snap ring	79. Ring gear
21. Shifter stop			

shaft (39) and gear (38) from cover (66), block up under gear (38) and press shaft out of gear.

CAUTION: Do not allow cover or low reduction gear bearing boss to support any part of the pressure required to press brake shaft from gear.

Remove input shaft (32) with input gear (33) and thrust washer (34) from case (16).

To disassemble the differential, remove the four cap screws and separate axle shaft and carrier assemblies from ring gear (79). Drive blocks (78), bevel

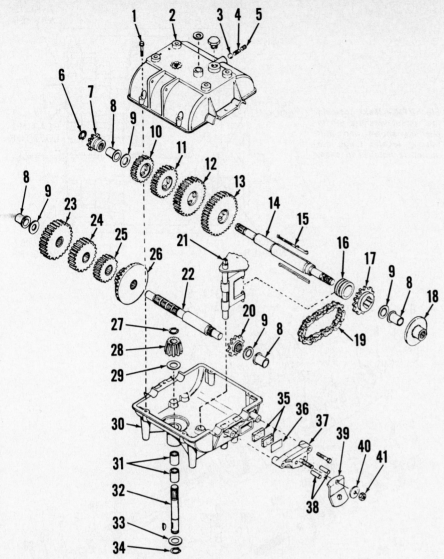

Fig. PT7 — Exploded view of typical 4-speed Series 500 Peerless transmission. Series 500 3-speed transmission is similar. Refer to text.

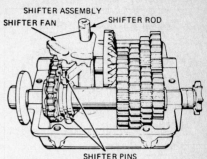

Fig. PT8 — Series 500 transmission with cover removed. Shifter rod, fan, fork and pins are removed as an assembly.

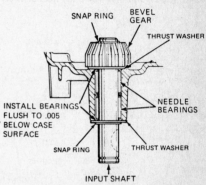

Fig. PT9 — Input shaft needle bearings must be installed flush to 0.005 below case surfaces.

1. Cap screw (6)	13. First speed gear	21. Shifter assy.	31. Needle bearings
2. Cover	14. Output & brake shaft	22. Countershaft	32. Input shaft
3. Detent ball	15. Shifter (drive) keys	23. Fourth drive gear	33. Thrust washer
4. Spring	16. Shifter collar	24. Third drive gear	34. Snap ring
5. Set screw	17. Reverse driven sprocket	25. Second drive gear	35. Brake pads
6. Snap ring	18. Brake disc	26. Bevel & spur (first drive) gear	36. Back-up plate
7. Output sprocket	19. Chain	27. Snap ring	37. Brake caliper
8. Flanged bushing	20. Reverse drive sprocket	28. Input bevel gear	38. Actuating pins
9. Thrust washer		29. Thrust washer	39. Brake lever
10. Fourth speed gear		30. Transmission case	40. Washer
11. Third speed gear			41. Adjusting nut
12. Second speed gear			

pinion gears (77) and drive pin (76) can now be removed from ring gear. Remove snap rings (59) and withdraw axle shafts (63 and 67) from axle gears (61) and carriers (62 and 72).

Clean and inspect all parts and renew any showing excessive wear or other damage. When installing new needle bearings, press bearing (29) in spline shaft (28) to a depth of 0.010 inch below end of shaft and low reduction shaft bearings (54 and 58) 0.010 inch below thrust surfaces of bearing bosses. Carrier bearings (10) should be pressed in from inside of case and cover until bear-

ings are 0.290 inch below face of axle housing mounting surface. All other needle bearings are to be pressed in from inside of case and cover to a depth of 0.015-0.020 inch below the thrust surfaces.

Renew all seals and gaskets and reassemble by reversing the disassembly procedure, keeping the following points in mind: When installing brake shaft (39) and idler gear (38), beveled edge of gear teeth must be up away from cover. Install reverse idler shaft (23), spacer (24) and reverse idler gear (25) with rounded end of teeth facing spacer. Install input

gear (33) and shaft (32) so that chamfered side of input gear is facing case (16).

Tighten transaxle cap screws to the following torque:

Differential cap screws . . . . . 7-10 ft.-lbs.
Case to cover cap screws . . . 8-10 ft.-lbs.
Axle housing cap screws . . . 15-18 ft.-lbs.
Shift lever housing cap
screws . . . . 8-10 ft.-lbs.

Reinstall transaxle by reversing the removal procedure and fill transaxle unit to the level plug opening with SAE 90 EP gear oil. Capacity is approximately 4 pints. Adjust brake and drive belt tension as required.

Series 500 Transmission

OVERHAUL. The Series 500 transmission may have three or four forward speeds and one reverse. Models 503, 505 and 509 are 3-speed and Models 501, 508 and 515 are 4-speed units. The transmissions are very similar except that on 3-speed units, gear (13 – Fig. PT7) is not used. Output sprocket (7) and brake assembly may be located on either side of transmission. Service procedures are similar for all models. The following procedures are for the 4-speed unit.

To disassemble the transmission, place shift lever in neutral position, then remove shift lever and safety starting switch if so equipped. Refer to Fig. PT7

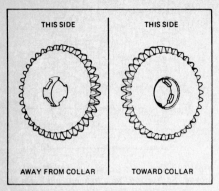

Fig. PT10 — Install speed gears (10 through 13 — Fig. PT7) with flat side away from shift collar.

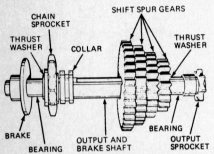

Fig. PT11 — Output and brake shaft properly assembled for installation.

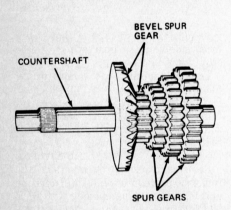

Fig. PT12 — View showing correct installation of bevel spur gear and spur gears on countershaft.

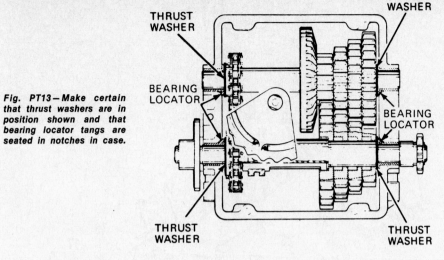

Fig. PT13 — Make certain that thrust washers are in position shown and that bearing locator tangs are seated in notches in case.

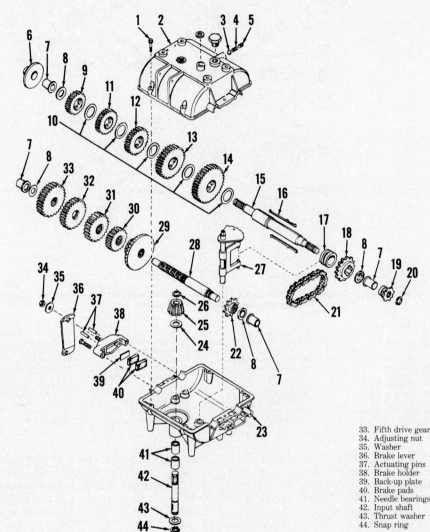

Fig. PT14 — Exploded view of typical 5-speed Series 700 Peerless transmission. Series 700 4-speed is similar. Refer to text.

and remove set screw (5), spring (4) and detent ball (3). Remove six cap screws (1) and lift off cover (2). Pull shifter assembly (21) upward and remove from case. See Fig. PT8. Lift both gear and shaft assemblies straight upward out of case (30 — Fig. PT7). Move reverse sprockets (17 and 20) together until bushing (8), thrust washer (9) and reverse drive sprocket (20) can be removed from countershaft (22) and chain (19). Remove chain and separate the shaft assemblies. Remove bushing, thrust washer, spur gears (23, 24 and 25) and bevel spur gear (26) from countershaft (22). Remove snap ring (6), output sprocket (7), bushing (8), thrust washer (9) and spur gears (10, 11, 12 and

1. Cap screw (6)	10. Thrust washers	18. Reverse driven
2. Cover	11. Fourth speed gear	sprocket
3. Detent ball	12. Third speed gear	19. Output sprocket
4. Spring	13. Second speed gear	20. Snap ring
5. Set screw	14. First speed gear	21. Chain
6. Brake disc	15. Output & brake	22. Reverse drive
7. Flanged bushing	shaft	sprocket
8. Thrust washer	16. Shifter (drive) keys	23. Transmission case
9. Fifth speed gear	17. Shifter collar	24. Thrust washer

25. Input bevel gear	33. Fifth drive gear
26. Snap ring	34. Adjusting nut
27. Shifter assy.	35. Washer
28. Countershaft	36. Brake lever
29. Bevel spur (first	37. Actuating pins
drive) gear	38. Brake holder
30. Second drive gear	39. Back-up plate
31. Third drive gear	40. Brake pads
32. Fourth drive gear	41. Needle bearings
	42. Input shaft
	43. Thrust washer
	44. Snap ring

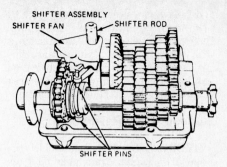

Fig. PT15 — Series 700 transmission with cover removed. Shifter rod, fan, fork and pins are removed as an assembly.

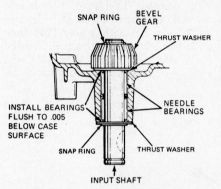

Fig. PT16 — Input shaft needle bearings must be installed flush to 0.005 below case surfaces.

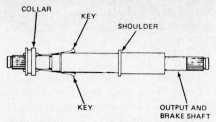

Fig. PT17 — Install shifter collar and shifter (drive) keys on output shaft as shown. Thick side of collar must face shoulder on shaft.

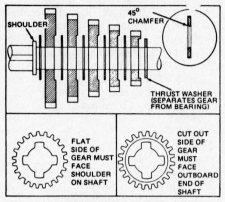

Fig. PT18 — View showing correct installation of thrust washers and gears on output shaft. The 45° inside chamfer on thrust washers must face the shoulder on the shaft.

13) from output and brake shaft (14). Then, remove brake disc (18), bushing (8), thrust washer (9) and reverse driven sprocket (17) from opposite end of the shaft. Slide shifter collar (16) and shifter (drive) keys (15) from shaft. Remove snap ring (27), input bevel gear (28) and thrust washer (29), then withdraw input shaft (32) from case. Unbolt and remove brake assembly (35 through 41).

Clean and inspect all parts and renew any showing excessive wear or other damage. If needle bearings (31) are being renewed, press bearings in until they are flush to 0.005 below case surfaces. See Fig. PT9. Apply a light coat of E.P. lithium grease to bearings, shafts and gears, then reassemble by reversing the disassembly procedure. When installing gears (10 through 13 – Fig. PT7), refer to Fig. PT10 and install gears with flat side away from shifting collar. Reverse drive sprocket (20 – Fig. PT7) must be installed with large hub side of sprocket facing away from bevel spur gear (26). Make certain that thrust washers are in positions shown in Fig. PT13 and that bearing locator tangs are seated in notches in case. Install shifter assembly, then cover gears, shafts, reverse sprockets and chain with 12 oz. of E.P. lithium grease. Install cover (2 – Fig. PT7) and tighten cap screws (1) to a torque of 90-100 in.-lbs. Install detent ball (3), spring (4) and set screw (5) and tighten set screw two full turns below

flush. Renew brake pads (35) as necessary and reinstall brake assembly.

Series 700 Transmission

OVERHAUL. The Series 700 transmission may be equipped with four or five speeds forward and one reverse. The transmissions are very similar except that on 4-speed units, gears (11 and 32 – Fig. PT14) are not used. Output sprockets (19) and brake assembly may be located on either side of transmission. Service procedures are similar for all models. The following procedures are for the 5-speed unit.

To disassemble the transmission, place shift lever in neutral position, then remove shift lever and safety starting

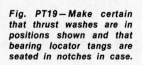

Fig. PT19 — Make certain that thrust washes are in positions shown and that bearing locator tangs are seated in notches in case.

switch if so equipped. Refer to Fig. PT14 and remove set screw (5), spring (4) and detent ball (3). Remove the six cap screws (1) and lift off cover (2). Pull shifter assembly (27) upward and remove from case. See Fig. PT15. Lift both gear and shaft assemblies straight upward out of case (23 – Fig. PT14). Move reverse sprockets (18 and 22) together until bushing (7), thrust washer (8) and reverse drive sprocket (22) can be removed from countershaft (28) and chain (21). Remove chain and separate shaft assemblies. Remove bushing, thrust washer, spur gears (33, 32, 31 and 30) and bevel spur gear (29) from countershaft (28). Remove brake disc (6), bushing (7), thrust washer (8), spur gears (9, 11, 12, 13 and 14) and thrust washers (10) from output and brake shaft (15). Then, remove snap ring (20), output sprocket (19), bushing (7), thrust washer (8) and reverse driven sprocket (18) from opposite end of shaft. Slide shifter collar (17) and shifter (drive) keys (16) from shaft. Remove snap ring (26), input bevel gear (25) and thrust washer (24), then withdraw input shaft (42) from bottom of case. Unbolt and remove brake assembly (34 through 40).

Clean and inspect all parts and renew any showing excessive wear or other damage. If needle bearings (41) are being renewed, press bearings in until they are flush to 0.005 below case surfaces. See Fig. PT16. Apply a light coat of E.P. lithium grease to bearings, shafts and gears, then reassemble by reversing the disassembly procedure. Refer to Fig. PT17 and install shifter collar and shifter keys on output and brake shaft. Thick side of collar must face shoulder on shaft. When installing gears and thrust washers (9 through 14 – Fig. PT14), flat side of gears and the 45° inside chamfer on thrust washers must face the shoulder on shaft. See Fig. PT18. Reverse drive sprocket (22 – Fig. PT14) must be installed with large hub side of sprocket facing towards bevel spur gear (29).

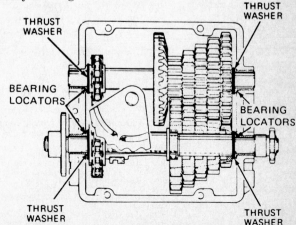

Make certain that thrust washers are installed in positions shown in Fig. PT19 and that bearing locator tangs are seated in notches in case. Install shifter assembly, then cover gears, shafts, reverse sprocket and chain with 12 oz. of E.P. lithium grease. Install cover (2-Fig. PT14) and tighten cap screws (1) to a torque of 90-100 in.-lbs. Install detent ball (3), spring (4) and set screw (5) and tighten set screw two full turns below flush. Renew brake pads (40) as necessary and reinstall brake assembly.

SIMPLICITY

2-Speed W/Reverse

OVERHAUL. With transmission case (14-Fig. ST1), shift links (10, 17 and 20), retaining rings (7) and springs (8) removed as outlined in the riding mower section, proceed as follows: Remove shaft and gear assemblies (W, X, Y and V), then remove set collar (23) and withdraw differential and axle assembly (Z). Remove brake shaft and gear assembly (U). Drive roll pins (11) from shafts (9, 16 and 21) and slide off thrust collars (13) and gears (12, 18 and 22). Remove the four through-bolts (36) and carefully separate differential pinions (32), spacers (33) and pinion spindles (31) from drive gear (27) and differential plate (35). Note location of washer (26), bushing (28), thrust cup (29) and bushing (34) and remove right axle (30) and drive gear (27) from left axle (25).

Clean and inspect all parts and renew any showing excessive wear or other damage. Inspect bushings in rear frame and renew as necessary. Reassemble by reversing the disassembly procedure. Lubricate all gears, shafts and bushings with general purpose lithium grease during assembly. Additional lubrication should be pumped through grease zerk in case (14) each 10 hours of operation.

2-Speed W/No Reverse

OVERHAUL. To disassemble the 2-speed transaxle, drain gear oil from unit and remove input pulley and rear wheel assemblies. Place the transaxle in a vise with left axle housing (37-Fig. ST2) pointing downward. Unbolt and remove transaxle cover (3). Remove right axle (31) with spacer (34), then withdraw left axle (32) from differential right side. Tilt drive gear (33) and remove drive gear and differential as-

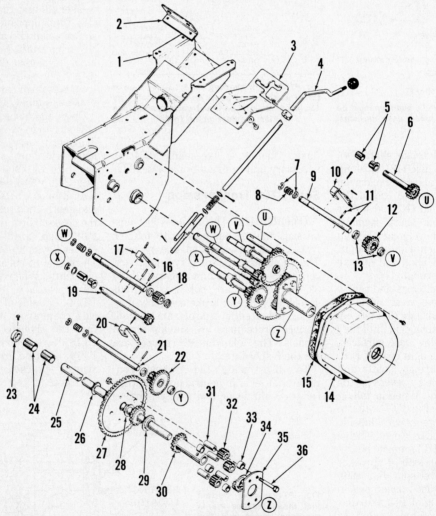

Fig. ST1—Exploded view of typical transmission and differential assembly used on Simplicity Models 305, 315, 355, 3005 and 3008-2.

1. Rear frame	11. Roll pin	19. Input shaft & gear
2. Shift rod guide	12. Low speed gear	20. Shift link
3. Shift quadrant	13. Thrust collars	21. Shifter shaft (Hi)
4. Shift rod	14. Gear case	22. High speed gear
5. Bronze bushings	15. Gasket	23. Set collar
6. Brake shaft	16. Shifter shaft	24. Bushings
7. Retaining rings	(Reverse)	25. Axle shaft L.H.
8. Spring	17. Shift link	26. Washer
9. Shifter shaft (low)	18. Reverse gear	27. Drive gear
10. Shift link		

28. Bushing
29. Thrust cup
30. Axle shaft R.H.
31. Pinion spindle (4)
32. Differential pinion (4)
33. Spacer (4)
34. Bushing
35. Differential plate
36. Through-bolt (4)

sembly and second spacer (34). Remove washer (21), withdraw brake shaft (18) and remove cluster gear (22) and washer (23). Remove shift lever assembly (2). Remove nut and lockwasher from shift rail (8), then remove rail, shift fork (5), input shaft (15) and sliding gear (13) as an assembly. Remove either snap ring (12) and slide gear (13) from input shaft. If necessary, remove shift fork from rail after first removing cap screw, detent ball (6) and spring (7). Remove the four through-bolts and separate differential plate (26), spindles (27), pinions (28), spacers (29) and springs (30) from drive gear (33). Unbolt and remove axle housing (37) from transaxle case (1).

Clean and inspect all parts and renew any showing excessive wear or other damage. Do not remove needle bearing (10) or bushings (16, 20, 24, 36 and 38) unless need for renewal is indicated. Renew gasket (4), seal ring (35) and all oil seals and reassemble by reversing the disassembly procedure. Fill transaxle housing, after unit is installed, to level plug opening with SAE 90 gear oil. Capacity is approximately 2½ pints.

3-Speed

OVERHAUL. To disassemble either 3-speed transaxle (Fig. ST3 or ST4), first remove rear wheel and hub assem-

blies, brake drum (1) and input pulley (20). Unbolt and remove side plates on models so equipped. Remove plug and drain lubricant. Remove cap screws securing transaxle cover (2) to case (26). Drive out the alignment roll pins at edge of cover and using a screwdriver, pry the cover off the case. Remove left axle shaft (34). Align differential pinion teeth and remove right axle shaft (33) out left side of differential. Lift out drive gear (28) and differential assembly. Remove brake shaft and cluster gear (36 and 37) and low reduction gear and shaft (18). Remove the nuts from shift rails (11 and 39) and reverse shaft (14). Remove shift rod and lever assembly. Withdraw shift

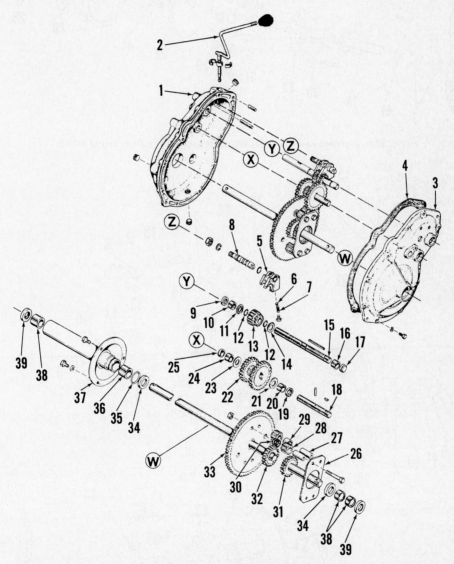

Fig. ST2—Exploded view of transmission and differential used on Simplicity Model 808 riding mower.

1. Transaxle case	11. Washer	21. Washer	30. Spring (2)
2. Shift lever	12. Snap ring	22. Cluster gear	31. Axle shaft R.H.
3. Transaxle cover	13. Hi-Lo sliding gear	23. Washer	32. Axle shaft L.H.
4. Gasket	14. Washer	24. Bushing	33. Drive gear
5. Shift fork	15. Input shaft	25. Expansion plug	34. Spacer
6. Detent ball	16. Bushing	26. Differential plate	35. Seal ring
7. Spring	17. Expansion plug	27. Pinion spindle (4)	36. Bushing
8. Shift rail	18. Brake shaft	28. Differential pinion	37. Axle housing
9. Oil seal	19. Oil seal	(4)	38. Bushings
10. Needle bearing	20. Bushing	29. Spacer (2)	39. Oil seal

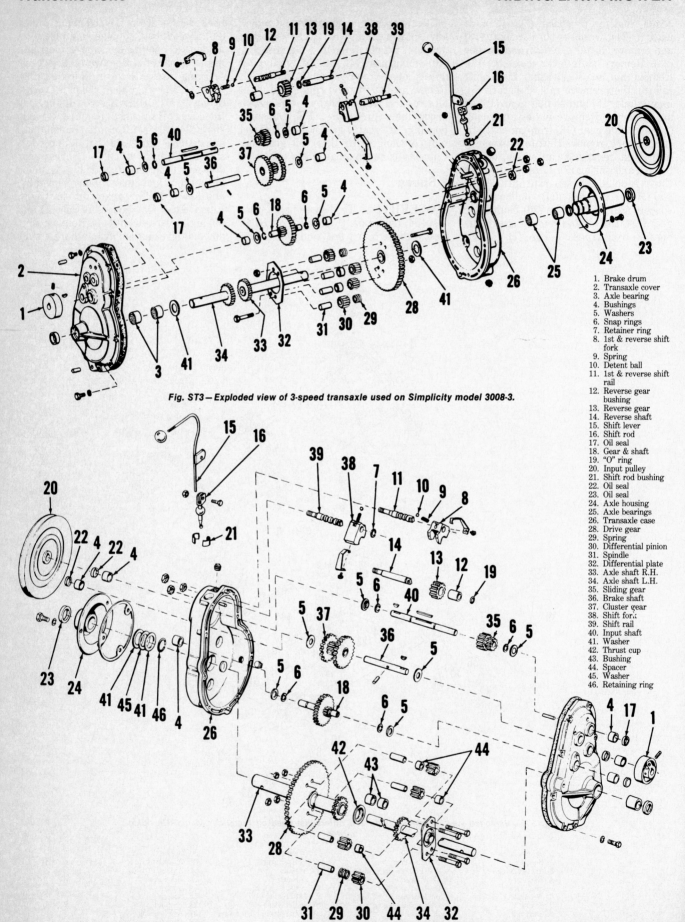

Fig. ST3 — Exploded view of 3-speed transaxle used on Simplicity model 3008-3.

1. Brake drum
2. Transaxle cover
3. Axle bearing
4. Bushings
5. Washers
6. Snap rings
7. Retainer ring
8. 1st & reverse shift fork
9. Spring
10. Detent ball
11. 1st & reverse shift rail
12. Reverse gear bushing
13. Reverse gear
14. Reverse shaft
15. Shift lever
16. Shift rod
17. Oil seal
18. Gear & shaft
19. "O" ring
20. Input pulley
21. Shift rod bushing
22. Oil seal
23. Oil seal
24. Axle housing
25. Axle bearings
26. Transaxle case
28. Drive gear
29. Spring
30. Differential pinion
31. Spindle
32. Differential plate
33. Axle shaft R.H.
34. Axle shaft L.H.
35. Sliding gear
36. Brake shaft
37. Cluster gear
38. Shift fork
39. Shift rail
40. Input shaft
41. Washer
42. Thrust cup
43. Bushing
44. Spacer
45. Washer
46. Retaining ring

Fig. ST4 — Exploded view of 3-speed transaxle used on Simplicity model 3008-FE3. Refer to Fig. ST3 for identification of parts.

fork (38), shift rail (39), sliding gear (35) and input shaft (40). Shift fork (8), shift rail (11), reverse gear (13) and shaft (14) can now be removed. Note location of springs (29) and spacers (44) and remove the four through-bolts. Separate differential pinions, spacers, springs and spindles from differential plate (32) and drive gear (28).

Clean and inspect all parts and renew any showing excessive wear or other damage. Renew oil seals and gaskets and reassemble by reversing disassembly procedure. Tighten differential through-bolts to a torque of 20 ft.-lbs. and nuts securing reverse shaft and shift rails to case to a torque of 50 ft.-lbs. Install shims (45 – Fig. ST4) as required to obtain 0.010-0.040 end play on axle shaft (33).

Fill transaxle housing, after unit is installed, to level plug opening with SAE 90 gear oil. Capacity is approximately 2½ pints.

DIFFERENTIAL REPAIR

Differentials used in most riding mowers are manufactured by other than OEM plant. Exceptions are noted as they occur. The differential and axle shaft suppliers normally provide axle ends of custom design and in lengths to match hub and wheel designs specified by the base manufacturer.

Differentials are identified by make and models, according to latest available information, in each riding mower section of this manual.

Removal and reinstallation procedures are covered in each unit manufacturer's section. Also see manufacturer's section for such external adjustments as axle alignment and correct setting of drive chain tension.

DEERE AND COMPANY

OVERHAUL. NOTE: Complete differential assembly is not identified by a model number; therefore, if parts are needed, be sure to specify mower model number.

When rear axle-differential assembly has been removed, clean exterior thoroughly, then remove cotter pin from retainer collar (9 – Fig. JD1) and back off four nuts (7) from housing cap screws (1). This will allow separation of housing halves (4) and removal of all internal gears. Clean housings, gears and shafts and renew as necessary.

If axle support shaft (10) is damaged or badly worn, drive out groove pin (11) to separate support shaft (10) from left side axle shaft (12).

During reassembly, lubricate parts with a generous coating of JOHN DEERE No. AT30408 High Temperature grease or an equivalent quality EP lithium grease. Add a little excess to each case half.

Tighten assembly bolts to 13-16 ft.-lbs. torque to complete reassembly.

FOOTE

Model 2260

OVERHAUL. Disassemble removed axle-differential after exterior clean-up by removing four locknuts (5 – Fig. FD1). Separate case halves (7) and remove all internal parts for cleaning and inspection. Renew any parts which show signs of extensive wear or damage. Sleeve bearing (12) may have to be pressed from its bore in center of idler shaft (11) if renewal is necessary. If bushed axle bosses of housing halves (7) are worn excessively, housing will require renewal as bushings are not serviced separately. Do not overlook condition of four flat thrust washers (8). If worn badly (original thickness – 0.062), these should be renewed to prevent excessive looseness when differential is reassembled.

Coat all gear teeth and wear surfaces of shafts liberally with a good grade of EP lithium grease during reassembly and torque assembly bolts and nuts (4 and 5) to 12-15 ft.-lbs.

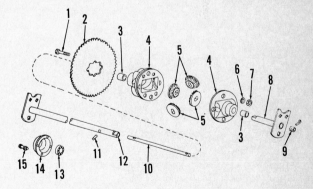

Fig. JD1 – Exploded view of differential assembly manufactured by John Deere for riding mower Models 66 and 68.

1. Cap screw 5/16X1¼ (4)
2. Sprocket
3. Bronze bearing (2)
4. Housing half (2)
5. Bevel gear (4)
6. Washer, 0.048 (4)
7. Elastic locknut (4)
8. Axle, RH
9. Retainer collar
10. Axle support shaft
11. Groove pin, 3/16
12. Axle, LH
13. Ball bearing
14. Axle bearing retainer
15. Cap screw ¼X½ (4)

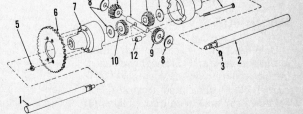

Fig. FD1 – Exploded view of Foote Model 2260 differential.

1. Short axle shaft
2. Long axle shaft
3. Snap ring (2)
4. Bolt, 5/16-18X3¾ (4)
5. Locknut, 5/16-18 (NC)
6. Sprocket
7. Housing half (2)
8. Flat washer 0.062 (4)
9. Pinion gear (2)
10. Miter (side) gear (2)
11. Idler shaft
12. Sleeve bearing

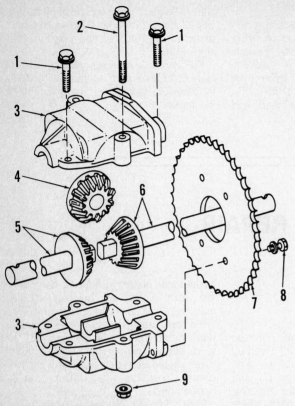

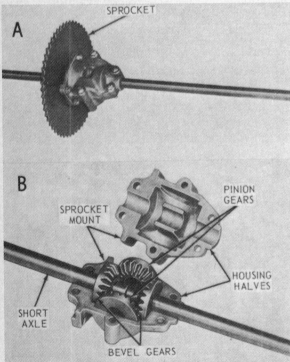

Fig. IW1 — Exploded view of Indus Wheel Company's Model 73DP differential. This model is also referred to as a Mast-Foos differential.

1. Bolt, 5/16 NF X 1¼ (4)
2. Bolt, 5/16 NF X 2-5/8 (2)
3. Housing half (2)
4. Pinion gear (2)
5. Long axle and bevel gear
6. Short axle and bevel gear
7. Sprocket
8. Cap screw, 5/16 X 1 (4)
9. Flange nut, 5/16 NF (6)

Fig. IW2 — Assembled view of Indus Wheel Model 73DP differential with sprocket installed (View A) and partially disassembled (View B). See text.

INDUS WHEEL (Mast-Foos)

Model 73DP

OVERHAUL. With rear axle-differential removed and thoroughly cleaned, unbolt and remove axle sprocket (7 – Fig. IW1). Unbolt and separate cast housings (3) then withdraw pinions (4)

Fig. IW3 — Exploded view of Indus Wheel Model 63DP differential.

1. Flanged axle bearing (2)
2. Wheel hub key (2)
3. Short axle shaft
4. Locknut (4)
5. Housing halves (2)
6. Bevel gear (2)
7. Snap ring (2)
8. Bevel pinion (2)
9. Cross
10. Pinion shaft
11. Long axle shaft
12. Assembly bolt (4)

and axle gears (5 and 6).

Use solvent to clean all parts, then inspect carefully for undue wear or damage.

Torque assembly bolts to 115-140 in.-lbs.

LUBRICATION. Use ¾ to 1¼ ounces of grease applied to gear teeth and to wear surfaces of axle shafts and castings. Shell EPRO 71030 (Alrania) is recommended by Indus Wheel Company. John Deere recommends use of their number AT30408 High Temperature or equivalent EP lithium grease. Lawn Boy (OMC) specifies use of Lubriplate number 630-AA. Roper recommends Shell Darina type AX. Do not substitute inexpensive lead soap greases or other lubricants of unknown quality or performance characteristics.

Model 63DP

OVERHAUL. With axle-differential assembly removed, thoroughly clean exterior of axle, differential housing and sprocket with a suitable solvent. Pay particular attention to axle shaft ends, threads and keyways. Remove burrs from axle.

Remove locknuts (4 – Fig. IW3) from assembly bolts (12) and pull bolts out as sprocket is removed. Separate housing halves and withdraw internal differential parts. To remove gears (6) from axles (3 and 11), first remove snap rings (7).

Renew all defective parts and reassemble in reverse of disassembly order.

During reassembly, lubricate all wear surfaces with a minimum of three ounces of Shell EPRO 71030 grease or equivalent.

Torque locknuts (4) in an even cross-pattern to 13-15 ft.-lbs. Heads of bolts (12) are on sprocket side.

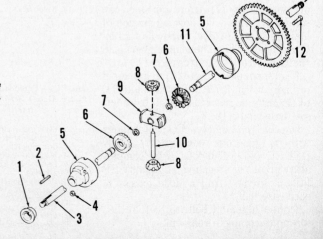

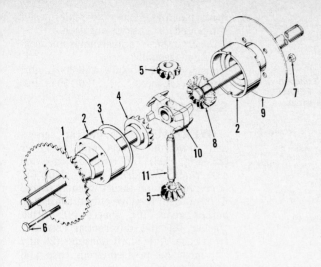

Fig. IW4—Exploded view of Indus Wheel differential known by Murray assembly number 21402. It is typical of Models 21000, 21179 and 20685.

1. Sprocket
2. Housing half (2)
3. Gasket
4. Gear & short axle
5. Bevel pinion (2)
6. Bolt, 5/16-24X3¼
7. Locknut, 5/16-24 (NF)
8. Gear & long axle
9. Brake disc*
10. Differential cross
11. Pinion shaft
   *Selected models only

Murray Assembly 21402

Overhaul procedure for differential unit shown in Fig. IW4 is very similar to that for Indus Wheel Model 63DP, which is described in the preceding section.

MTD PRODUCTS

All Models

OVERHAUL. Differential models shown in Fig. MTD1 and MTD2 are very similar in design and appearance and many internal parts can be interchanged from one model to the other. Side and pinion gears, drive pins, assembly dowels, spacer washers and assembly bolts and nuts for each model bear identical part numbers. Obviously, housings used are different and axle lengths and sprockets will vary among different riding mower models. The outstanding difference to be noted is that in Fig. MTD1, bevel gears (2) are attached to axles (3 and 11) by spring pins (1); whereas, as shown in Fig. MTD2, snap rings (8) are used. Housings (2—Fig. MTD2) also have renewable bushings (3) while flanged bearings are used in housing half assemblies (5—Fig. MTD1). On both models, the overhaul procedure will be obvious after removing the assembly bolts.

Approximately two ounces of high temperature (450°F.) EP lithium grease applied to friction surfaces during reassembly is required for lubrication.

Torque all four assembly bolts evenly to 13-15 ft.-lbs. in a cross pattern with locknut fitted on sprocket side of differential.

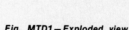

Fig. MTD1—Exploded view of MTD Products differential.

1. Spring pin
2. Bevel gear
3. Long axle
4. Washer, 0.76X1.49
5. Housing half (2)
6. Screw, 5/16-24X4 (4)
7. Dowel, 3/16
8. Flat washer 0.64 ID
9. Pinion gear (2)
10. Drive pin
11. Short axle
12. Sprocket
13. Lockwasher, 5/16 (8)
14. Locknut, 5/16-24 (NF)

PEERLESS

Series 100

OVERHAUL. The 100 Series PEERLESS differentials, when TECUMSEH-PEERLESS assembly numbers are used, may be identified by numbers 101 through 199, sometimes with a letter suffix. Number variations specify particular differences in axle shaft lengths and configuration of shaft ends for matching to various hub styles.

Before disassembly of removed axle-differential, thoroughly clean outside. Clear keyways and/or pin bores and carefully remove burrs from axle surfaces. Hardened shaft metals call for use of a stone to smooth finish.

Remove four locknuts which hold sprocket (1—Fig. PD1) to differential housing (2), then back out through-bolts (6) and separate housing halves. Lift out drive pin (12) together with pinion gears

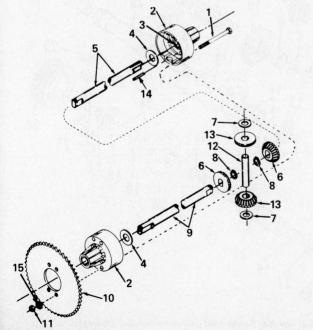

Fig. MTD2—Exploded view of MTD Products differential No. 10483, identified also by Jacobsen No. 501062.

1. Bolt 5/16-24X4 (4)
2. Housing half (2)
3. Sleeve bearing
4. Flat washer (2)
5. Long axle
6. Bevel gear
7. Flat washer (2)
8. Snap ring, ¾-in.
9. Short axle
10. Sprocket
11. Locknut, 5/16-24
12. Drive pin
13. Pinion gear (2)
14. Dowel pin, 3/16-in.
15. Lockwasher (4)

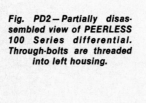

1. Sprocket
2. Differential housing (2)
3. Bushing (2)
4. Thrust washer (2)
5. Axle shaft (long)
6. Through-bolt (4) 5/16-inch
7. Flat washer (4)
8. Bevel gear (2)
9. Thrust washer (2)
10. Pinion gear (2)
12. Drive pin
13. Snap ring (2)
14. Axle shaft (short)

drive pin and gears. No seals, sealer compounds or gaskets are used.

Observe these torque values in checking final assembly:

Housing through-bolts . . 250-300 in.-lbs.
Sprocket locknuts . . . . . . 120-150 in.-lbs.

Series 1300

OVERHAUL. Unscrew cap screws and drive out dowel pins in cover (29 – Fig. PD3). Lift cover off case and axle shaft. Withdraw brake shaft (5), idler gear (4) and thrust washers (3 and 6) from case. Remove output shaft (11), output gear (10), spacer (9), thrust washer (8) and differential assembly from case. Axle shaft housings (20 and 22) must be pressed from case and cover.

To disassemble differential, unscrew four cap screws (17) and separate axle shaft and carrier assemblies from ring gear (28). Drive blocks (25), bevel pinion

Fig. PD2—Partially disassembled view of PEERLESS 100 Series differential. Through-bolts are threaded into left housing.

(10) and thrust washers (9). Remove snap ring (13), then bevel gears (8) and thrust washers (4). Examine all parts for wear, cracks, chips or galling after thorough clean-up with solvent.

Bushings (3) can be pressed from housing halves (2). PEERLESS bushing tool No. 670204 is available for reinstallation of bushings.

Renew all defective parts and reassemble differential in reverse of disassembly order. It will be noted that sprocket (1), threaded side of differential housing (2) and assembly nuts are all assembled on short axle side.

During assembly, lubricate differential using one ounce of EP (Extreme Pressure) lithium grease on wear points and gear teeth, apply grease to bushings (3) before inserting axles, and grease

1. Case
2. Gasket
3. Washer
4. Idler gear
5. Brake shaft
6. Washer
7. Bearing
8. Washer
9. Spacer
10. Output gear
11. Output shaft
12. Snap ring
13. Side gears
14. Thrust washers
15. Thrust bearing
16. Differential carrier
17. Bolt
18. Axle shaft R.H.
19. Bushing
20. Axle housing
21. Oil seal
22. Axle housing
23. Axle shaft L.H.
24. Differential carrier
25. Drive block
26. Drive pinion
27. Drive pin
28. Ring gear
29. Cover

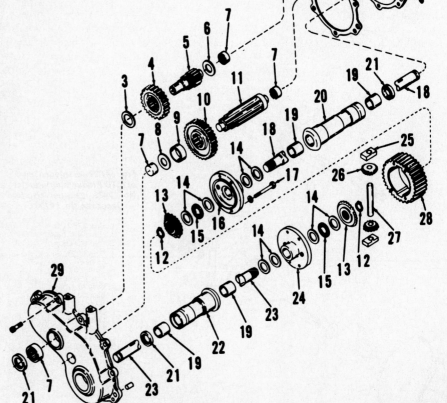

Fig. PD3—Exploded view of Peerless Series 1300 gear reduction and differential unit.

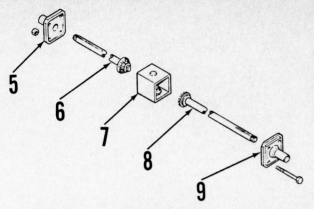

5. End cap, L.H.
6. Axle shaft, L.H.
7. Housing and spider gears
8. Axle shaft, R.H.
9. End cap, R.H.

gears (26) and drive pin (27) can now be removed from ring gear. Remove snap rings (12) and slide axle shafts (18 and 23) from axle gears (13) and carriers (16 and 24).

Clean and inspect all parts and renew any parts damaged or excessively worn.

When installing needle bearings, press bearings in from inside of case or cover until bearings are 0.015-0.020 inch below thrust surfaces. Be sure heads of differential cap screws (17) and right axle shaft (18) are installed in right carrier housing (16). Right axle shaft is installed

through case (1). Tighten differential cap screws to 7 ft.-lbs. and cover cap screws to 10 ft.-lbs. Differential assembly and output shaft (11) must be installed in case at same time. Remainder of assembly is reverse of disassembly procedure.

STEWART

Model 9500

OVERHAUL. To disassemble differential, remove bolts and nuts securing end caps (5 and 7 – Fig. ST1) to housing assembly (7). Separate end caps from housing, then clean and inspect all parts and renew any showing excessive wear or any other damage. Reassemble by reversing the disassembly procedure. Fill differential housing with 3/4-1 1/4 ounces of medium EP grease.

NOTES